# Fluids and Chemical Engineering Conference (FluidsChE 2017)



MATEC Web of Conferences Volume 111 (2017)

Kota Kinabalu, Malaysia
4 - 6 April 2017

**Editors:**

**H.A. Abdulbari**

ISBN: 978-1-5108-4327-1

**Printed from e-media with permission by:**

Curran Associates, Inc.
57 Morehouse Lane
Red Hook, NY 12571

**Some format issues inherent in the e-media version may also appear in this print version.**



Printed by Curran Associates, Inc. (2017)

For additional information, please contact EDP Sciences – Web of Conferences
at the address below.

EDP Sciences – Web of Conferences
17, Avenue du Hoggar
Parc d'Activité de Courtabœuf
BP 112
F-91944 Les Ulis Cedex A
France

Phone: +33 (0) 1 69 18 75 75
Fax: +33 (0) 1 69 28 84 91

contact-edps@webofconferences.org

**Additional copies of this publication are available from:**

Curran Associates, Inc.
57 Morehouse Lane
Red Hook, NY 12571 USA
Phone: 845-758-0400
Fax: 845-758-2633
Email: curran@proceedings.com
Web: www.proceedings.com

# TABLE OF CONTENTS

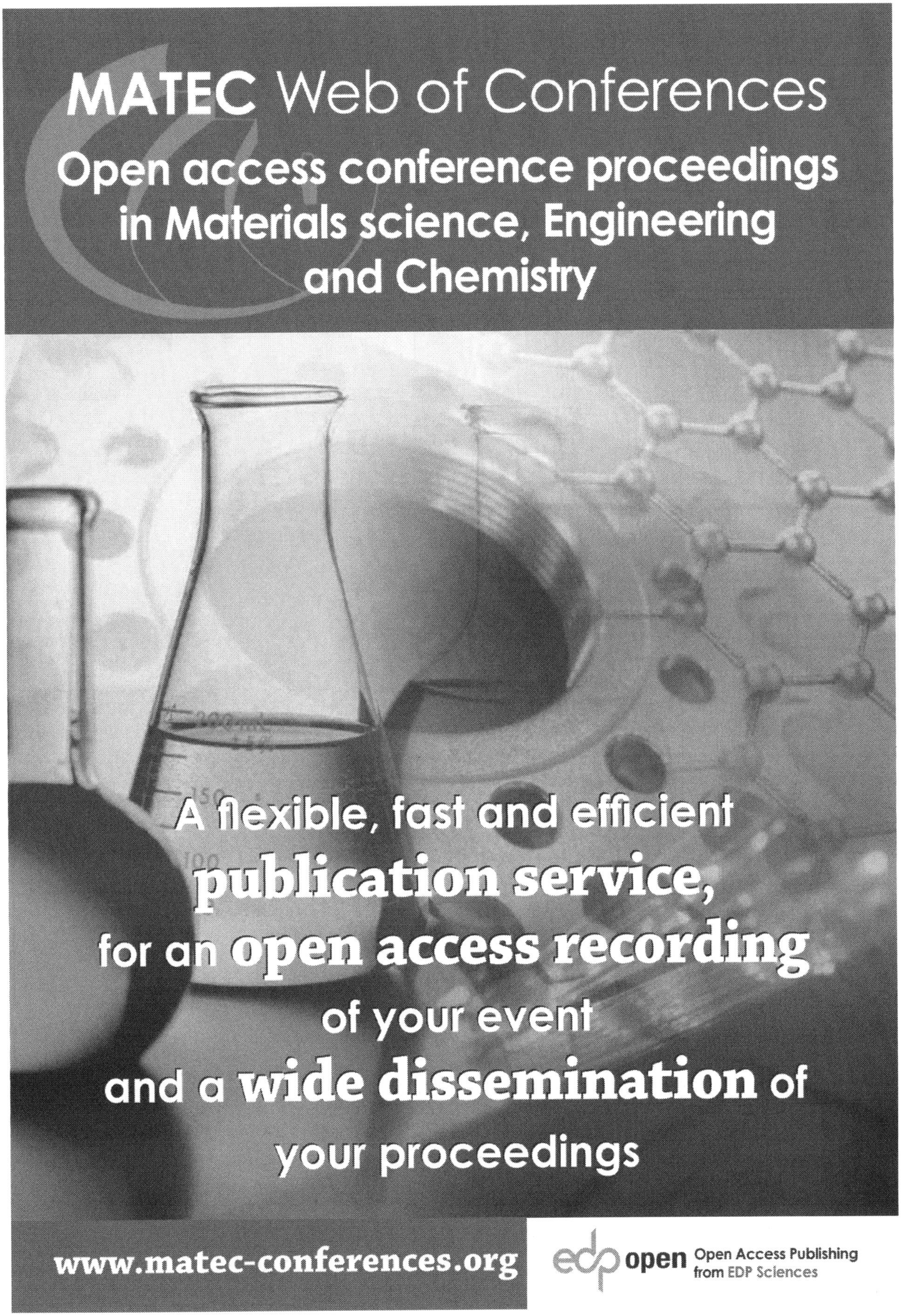
MATEC Web of Conferences
Open access conference proceedings in Materials science, Engineering and Chemistry
A flexible, fast and efficient publication service, for an open access recording of your event and a wide dissemination of your proceedings
www.matec-conferences.org
edp open Open Access Publishing from EDP Sciences

ISSN (electronic edition): 2261-236X

**INTERNATIONAL CONFERENCE ON FLUIDS & CHEMICAL ENGINEERING 2017 (FLUIDSCHE2017)**

4th to 6th of April 2017, TH Hotel Kota Kinabalu Sabah, Malaysia

The International Conference on Fluids and Chemical Engineering (FluidsChE 2017) is organised by CARIFF Universiti Malaysia Pahang, Malaysia. FluidsChE 2017 aims to provide a platform to discuss ideas and latest research findings. With the theme "Advancing Fluids Engineering for A Sustainable Future", we hope that FluidsChE 2017 will offer a scientific platform for researchers from all around the globe to present their research. The conference covers a wide scope of all branches of engineering and sciences. With the theme "Advancing Fluid Engineering for a Sustainable Future", FluidsChE 2017 is the right avenue to discuss current issues relating to fluids engineering as a sub discipline of chemical engineering and its significance beyond its traditional sphere. Not only that fluids engineering is widely applied in the oil and gas, oleo- chemical, biotechnology and pharmaceutical industries, its applicability in emerging areas such as microfluidic and nano-machining also provide new spheres for constructive intellectual discourse in this conference. Besides serving as the platform for intellectual exchange, the conference will also serve as a constructive meeting point for the delegates to create or enhance collaboration. This 2nd edition of FluidsChE 2017 has succeeded in attracting over 200 submissions of participants from different countries like Malaysia, Thailand, Pakistan, Iraq, Iran, Indonesia, India, and Kazakhstan; to exchange and transfer knowledge and expertise.

**Central organizing committee**

| | |
|---|---|
| Patron | Professor Dato' Dr. Daing Nasir Ibrahim |
| Advisor | Professor Dr. Mashitah M Yusoff |
| Conference chair | Prof. Dr. Hayder A Abdulbari |
| Co-chair | Assoc. Prof. Dr. Jolius Gimbun |
| Secretary I | Wan Farid Wan Rusli |
| Secretary II | Nurul Azra Bakaruddin |
| Treasury I | Assoc. Prof. Dr. Cheng Chin Kui |
| Treasury II | Noraini Mohd Zamri, CARIFF |
| Website | Assoc. Prof. Dr. Jolius Gimbun |
| | Wan Farid Wan Rusli, CARIFF |
| Technical unit | Dr. Siew Choo Chin |
| | Dr. Viet Vo |
| | Assoc. Prof. Ir. Dr. Chin Sim Yee |
| Protocol & Logistic | Dr. Sumaiya Zainal Abidin |
| | Nur Syuhada Asmar |
| Conference booklet & material | Suliha Akma Kamaruddin |
| IJST manager | Assoc. Prof. Dr. Jolius Gimbun |
| MATEC manager | Prof. Dr. Hayder A Abdulbari |
| BCREC manager | Assoc. Prof. Dr. Cheng Chin Kui |

**CONFERENCE SCOPE**

Fluids Engineering

- Flow assurance in pipeline
- Theoretical and experimental fluid flow
- Computational fluid dynamics
- Microfluidics
- Water science and technology

- Mixing and formulation
- Heat and mass transfer
- Oil and gas

Engineering & Science

- Engineering technology
- Manufacturing and mechanical engineering
- Biosensor & electrical technology
- Reaction engineering and catalysis
- Renewable energy and biofuels
- Food science and technology
- Advanced material & civil engineering
- Pharmaceutical and drug delivery
- Separation technology
- Natural products & industrial crops
- Safety and health management
- Process control and optimisation
- Environmental engineering
- Corrosion science
- Fuel cell
- Waste water
- Finite Element Analysis
- Modelling and simulation

edp sciences

MATEC Web of Conferences

# Statement of Peer review

In submitting conference proceedings to MATEC Web of Conferences, I certify to the Publisher that I adhere to the **Policy on Publishing Integrity** of the journal in order to safeguard good scientific practice in publishing.

1. All articles have been subjected to peer review administered by the proceedings editors.
2. Reviews have been conducted by expert referees, who have been requested to provide unbiased and constructive comments aimed, whenever possible, at improving the work.
3. Proceedings editors have taken all reasonable steps to ensure the quality of the materials they publish and their decision to accept or reject a paper for publication has been based only on the merits of the work and the relevance to the journal.

Title, date and place of the conference

FLUIDS AND CHEMICAL ENGINEERING CONFERENCE 2017, KOTA KINABALU, SABAH, MALAYSIA, 4TH to 6TH OF APRIL 2017

Proceedings editor(s):

PROF. DR. HAYDER A ABDULBARI

Date and editor's signature

01-06-2017

MATEC Web of Conferences **111**, 01001 (2017) DOI: 10.1051/matecconf/201711101001
*FluidsChE 2017*

# Drag reduction by natural polymeric additives in PMDS microchannel: Effect of types of additives

*Fiona* W.M. Ling[1]*, and *Hayder* A. Abdulbari [1]

[1] Centre of Excellence for Advanced Research in Fluid Flow, University Malaysia Pahang, Lebuhraya Tun Razak 26300 Kuantan, Pahang, Malaysia

**Abstract.** Drag reduction technology was used in medical applications to enhance the blood flow in semi-clogged blood streams which can be an alternative treatment for atherosclerosis. In this present study, natural polymeric drag reducing additives (DRA) was introduced to replace synthetic polymer which has the possibility of bringing side effects to human health. Three different sources, namely okra, aloe vera and hibiscus were utilized to extract the natural polymeric additives which were then tested in custom made microchannel simulating human heart blood vessels. The performance of different types of additives was evaluated using pressure measurements. The maximum drag reduction up to 63.48% is achieved using 300 ppm of hibiscus mucilage at operating pressure of 50 mbar. In this present work, hibiscus showed the best drag reduction performance, giving the highest %FI in most of the cases. This experimental results proved that these natural polymeric additives could be utilized as DRA in enhancing the blood flow in semi-clogged blood streams.

## 1 Introduction

Atherosclerosis, hardening and narrowing of the arteries, lead to cardiovascular diseases such as heart attacks, strokes, and peripheral vascular disease [1]. Such narrowed arteries can cause turbulence when the blood flows through it which results in increased pressure in the blood vessels and back pressure due to the creation of eddies [2]. The growth of plaques also can significantly reduce the blood flow through vessels. Rupture of plaques can cause blood clots and block the blood flow thus preventing the blood to travel to other parts of the body [3].

With the technologies nowadays, the standard treatment for atherosclerosis is through medication and surgery. However, the drugs [4] do not solve the problem where it only will slow down the progression of atherosclerosis and always come together with side effects. Besides, the degree of success with surgery is limited due to the cardiovascular complication [5] and may also have some side effects after surgery [6]. To solve these problems, several authors reported that a minute amount of long chain polymer can enhance the blood flow which could be an alternative treatment method for chronic coronary artery diseases [7].

This idea was developed from the pioneering work of Toms [8] who had observed the enhancement of flow after adding a minute amount of polymers in pipe [9], which known as drag reduction (DR). Since then, researchers started to gain attention in using natural polymeric drag reducing additives (DRA) to enhance the blood flow. It had been claimed that natural polymeric additives were more resistance to degradation compare to synthetic polymer in the presence of red blood cell [10] where mechanical degradation is a serious drawback in medical applications that will cause the polymer to lose the properties of DRA [11].

Aloe vera as natural polymeric DRA was widely used in researches due to the simplicity in extraction process [12] with minimal toxicity [13]. Polymers extracted from aloe vera was proven to have high performance on DR by improving the survival rate of the animal models of lethal hemorrhagic shock [14, 15]. Besides, it also reduced the number of animal mortality from myocardial ischemia [16]. Although there are some other natural polymeric additives [17-19] are well known for their performance as DRA, there is still limited studies investigating these polymers in enhancing the blood flow.

With the development of microfluidic technology nowadays, researchers started to utilize this technique in as an economical and reliable method for hypothesis testing of the related theoretical phenomena such as DR [20, 21]. *In vitro* microfluidic studies are crucial in investigating the mechanism behind DR in blood flow. Recently, several researchers proposed that, polymeric DRA substantially reduces the near-wall cell-free layer which facilitates the gas transport besides showing excellent DR properties [22, 23].

In this present work, two new natural polymeric additives, namely okra and hibiscus were introduced and investigated their performance on DR comparing with

* Corresponding author: fionalingwm@gmail.com

aloe vera. Five different concentration (100 to 500 ppm) of these natural polymers were extracted and experimented in two custom-made microchannels that having the real size of human heart vessels to simulate the flow in real human vessels. Note that this work is to test the feasibility of okra and hibiscus as flow enhancer in enhancing the low turbulence intensity flow which analogous to the blood flow using the microfluidic technique.

## 2 Methodology

### 2.1 Material

Fresh okra, aloe vera and hibiscus were purchased and cleaned before extraction. Okra and aloe vera were cut into small pieces while hibiscus leaves were picked for the extraction process. These plants were then soaked in 1 liters of ultrapure water separately for 5 hours at room temperature and kept in 4 °C freezer overnight. The extraction process was completed when the plants produced mucilage and turned the liquid into thick, viscous liquid. The solution was then filtered using a sieve to prevent any suspended solids from getting through. Water was evaporated at 40 °C and 90 rpm using vacuum rotary evaporator (model: IKA HB 10) to produce thick mucilage. The mucilage was then ready for dilution to obtain the required concentration.

### 2.2 Experimental procedure

Fig 1 shows the experimental setup for obtaining experimental data for all mucilage concentrations. In this work, two polydimethylsiloxane (PDMS) microchannels were used where these two microchannels have a real size of human heart vessel with the different clogging part.

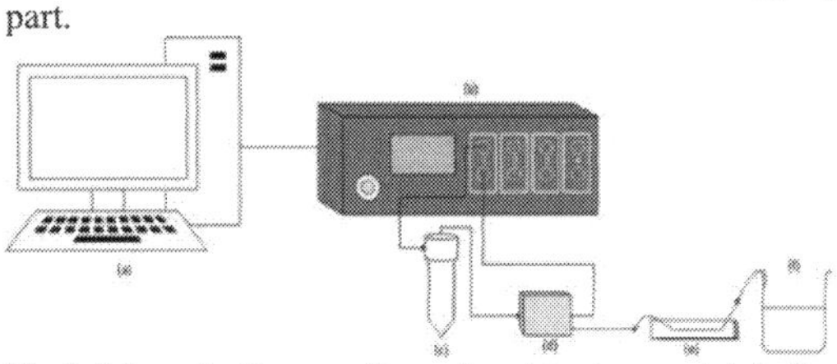

**Fig 1.** Schematic diagram of experimental setup consisted of (a) computer (b) pressure and vacuum controller (c) reservoir containing solution (d) flow sensor (e) custom made microchannel (f) beaker as storage tank

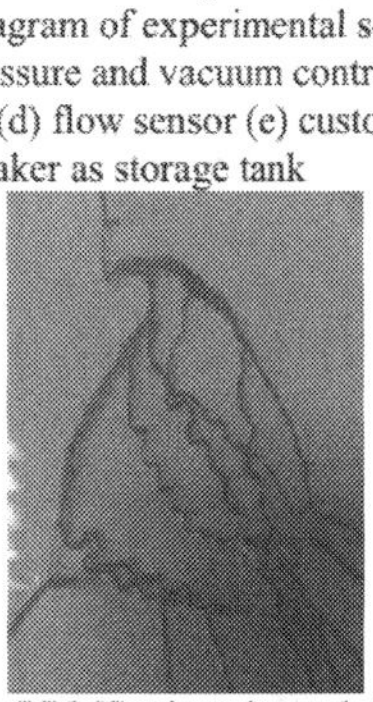

**Fig 2.** Custom-made PDMS microchannel that having similar size of human real heart vessel with a semi-clogged part

Mucilage was weighed accordingly and added to deionized water which acts as transported liquid and then being stirred manually to produce five different concentration solution ranging from 100 to 500 ppm.

In this present study, the effect of extracted mucilage on flow enhancement was experimented using an open loop microfluidic system. The operation started by executing commands to pressure and vacuum controller using Elveflow Smart Interface that allowed the operating pressure (50 to 500 mbar) to push the solution out from the reservoir. The flow rate corresponding to the operating pressure was recorded. The steps were repeated for all the five concentration of the three mucilage.

The percentage of flow rate increment (%FI) was then calculated using the formula given as:

$$\%FI = \frac{F_a - F_b}{F_a} \times 100 \qquad \text{(Eq. 1)}$$

where

$F_a$ = Flow rate after the addition of DRA (μL/min)
$F_b$ = Flow rate before the addition of DRA (μL/min)

## 3 RESULTS AND DISCUSSIONS

### 3.1. Characterization of mucilage

The extracted mucilage were brought for characterization analysis using Fourier transform-infrared (FTIR). The FTIR spectrum of the sample was recorded using Thermo Scientific Nicolet iS50 FT-IR Spectrometer in the range of 400 – 4000 cm$^{-1}$, in attenuated reflection mode (ATR).

The composition of components of extracted mucilage were analyzed and tabulated in Table 1. In DR operation, polysaccharides extracted from the natural plant's source are expected to be the main elements to reduce the frictional forces [14, 15]. However, there is no literature reported on the type of sugar that is responsible in DR. Thus, the type of the sugars present in extracted mucilage is the main interest in this study. As shown in Table 1, 19.07% and 20.94% of D-xylulose were detected in okra and aloe vera, respectively. D-xylulose was not detected in hibiscus, yet 27.25% of D-ribulose was detected.

**Table 1.** Components in extracted mucilage

| Components | Composite (%) | | |
|---|---|---|---|
| | Okra | Aloe vera | Hibiscus |
| Water | 62.19 | 65.34 | 57.79 |
| D-Xylulose | 19.07 | 20.94 | - |
| D-Ribulose | - | - | 27.25 |
| D-Pantethine anhydrous | 9.63 | - | - |

### 3.2 Experimental results

Fig 3 demonstrated the %FI for all three extracted mucilage varying the operating pressure for a different type of additives. It can be clearly seen that %FI was the highest for hibiscus among the three additives for all the cases where the maximum %FI was up to 60%. In this

DOI: 10.1051/matecconf/201711101001

work, okra has the lowest performance in DR giving %FI less than 18%. Aloe vera achieved the medium %FI where the maximum %FI for all the cases did not exceed 55%.

The performance of DR is expected to be affected by the type of additives where the polysaccharide composition in the mucilage plays an important role [24]. In this case, it is expected the composition of the polysaccharides present in the extracted mucilage would directly influence the DR performance. Recall back in Section 3.1, the polysaccharides content in the mucilage were analyzed using FTIR.

In this work, the higher composition of polysaccharides presents in the extracted mucilage resulting in the higher %FI. It can be seen that 27.25% of D-ribulose gives the highest %FI whereas 19.07% of D-xylulose that present in okra mucilage gives the lowest %FI indicating least DR performance. The small difference of the polysaccharides composition, which is less than 10%, in the mucilage gives the significant difference in %FI may be due to the difference type of the polysaccharides existing in mucilage and also the complex interaction with other elements available in the extracted mucilage.

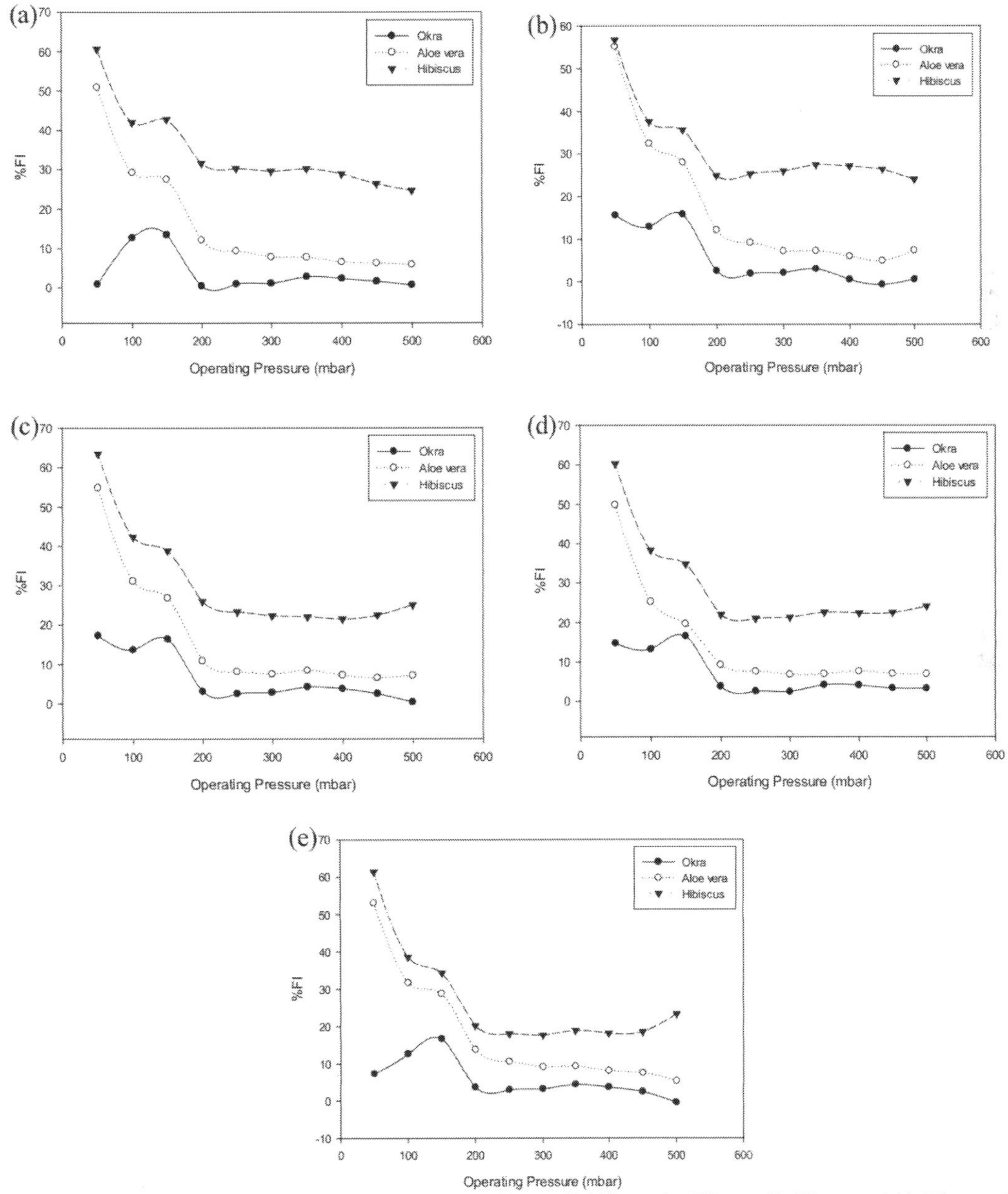


**Fig 3.** Variation of %FI at various operating pressure for (a) 100 ppm (b) 200 ppm (c) 300 ppm (d) 400 ppm and (e) 500 ppm of three different mucilage in Model 1

DOI: 10.1051/matecconf/201711101001

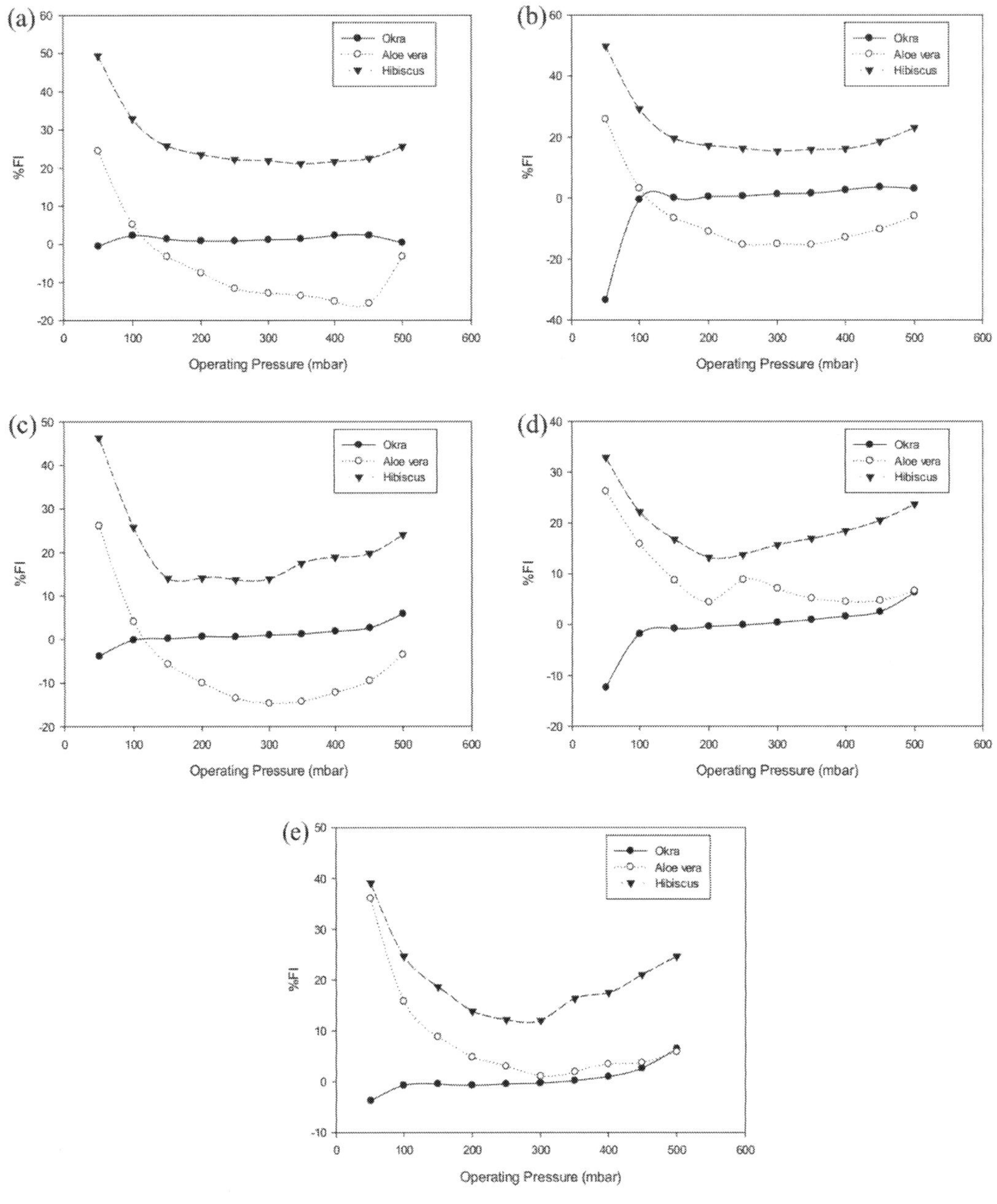


**Fig 4.** Variation of %FI at various operating pressure for (a) 100 ppm (b) 200 ppm (c) 300 ppm (d) 400 ppm and (e) 500 ppm of three different mucilage in Model 2

Similar phenomena also can be seen in Model 2 as illustrated in Fig 4. Although at a lower concentration, aloe vera did not show much DR effect, the %FI is higher when a higher concentration of aloe vera mucilage was utilized. In this case, okra still showed relatively low %FI with %FI that less than 10%. Hibiscus mucilage showed stable DRA performance where in all the cases, hibiscus still achieved the highest %FI.

The difference in the performance of the extracted mucilage in both Model 1 and Model 2 is expected due to the different clogging location in the design. However, the exact relationship of the location of the clogging area with DR is not reported in this work and other literature.

## 4 Conclusions

In this present work, the effect of type of natural polymeric additives extracted from okra, aloe vera and

 DOI: 10.1051/matecconf/201711101001

hibiscus were investigated using microfluidics technique. The experimental results proved that these natural polymeric additives could be utilized as DRA to enhance the blood flow in semi-clogged blood streams. The composition of the sugar (polysaccharides) group is expected to directly affecting the performance of these mucilage on DR performance. From the results, hibiscus mucilage is the best DRA by achieving the maximum %FI of 63.48% at the concentration of 300 ppm.

## References


1. Abdulbari, H.A., et al., *Investigating the effect of solid particle addition on the turbulent multiphase flow in pipelines.* International Journal of the Physical Sciences, 2011. **6**(15): p. 3672-3679.
2. Insull Jr, W., *The pathology of atherosclerosis: Plaque development and plaque responses to medical treatment.* The American Journal of Medicine, 2009. **122**(1, Supplement): p. S3-S14.
3. Brostow, W., *Drag reduction in flow: Review of applications, mechanism and prediction.* Journal of Industrial and Engineering Chemistry, 2008. **14**(4): p. 409-416.
4. Lewis, S.J., *Prevention and treatment of atherosclerosis: a practitioner's guide for 2008.* American Journal of Medicine, 2009. **122**(1 Suppl): p. S38-50.
5. García-González, V., et al., *Reality of a vaccine in the prevention and treatment of atherosclerosis.* Archives of Medical Research, 2015. **46**(5): p. 427-437.
6. Cirillo, F., et al., *Carotid atherosclerosis is associated with in-hospital mortality after cabg surgery.* International Journal of Angiology, 2002. **11**(4): p. 210-215.
7. Pacella, J.J., et al., *A novel hydrodynamic approach to the treatment of coronary artery disease.* European Heart Journal, 2006. **27**: p. 2362–2369.
8. Toms, B.A. *Proceedings of the International Congress on Rheology* 1949. Amsterdam: North-Holland.
9. Truong, V.-T., *Drag Reduction Technologies.* 2001, Australia: DSTO Aeronautical and Maritime Research Laboratory
10. Marhefka, J., et al., *Mechanical Degradation of Drag Reducing Polymers in Suspensions of Blood Cells and Rigid Particles.* Biorheology, 2008. **45**(5): p. 599-609.
11. Daly, A.R., et al., *Application of drag-reducing polymer solutions as test fluids for in vitro evaluation of potential blood damage in blood pumps.* American Society of Artificial Internal Organs Journal, 2010. **56**(1): p. 6-11.
12. Channe Gowda, D., B. Neelisiddaiah, and Y.V. Anjaneyalu, *Structural studies of polysaccharides from aloe vera.* Carbohydrate Research, 1979. **72**: p. 201-205.
13. Talmadge, J., et al., *Fractionation of Aloe vera L. inner gel, purification and molecular profiling of activity.* International Immunopharmacology, 2004. **4**(14): p. 1757-1773.
14. Kameneva, M.V., et al., *Blood soluble drag-reducing polymers prevent lethality from hemorrhagic shock in acute animal experiments.* Biorheology, 2004. **41**: p. 53–64.
15. Macias, C.A., et al., *Survival in a rat model of lethal hemorrhagic shock is prolonged following resuscitation with a small volume of a solution containing a drag-reducing polymer derived from aloe vera.* Shock, 2004. **22**(2): p. 151-6.
16. Sakai, T., et al., *Infusion of a drag-reducing polymer extracted from aloe vera prolonged survival time in a rat model of acute myocardial ischaemia.* British Journal of Anaesthesia 2007. **98**(1): p. 23-8.
17. Abdul Bari, H.A., K. Letchmanan, and R.M. Yunus, *Drag Reduction Characteristics Using Aloe Vera Natural Mucilage: An Experimental Study.* Journal of Applied Sciences, 2011. **11**: p. 1039-1043.
18. Hayder A., A., K. Siti Nuraffini , and N. A.H., *Grafted natural polymer as new drag reducing agent: An experimental approach.* Chemical Industry & Chemical Engineering Quarterly, 2012. **13**(3): p. 361-371.
19. Kamarulizam, H.A.A.B.S.N. and R.C. Man, *Investigating drag reduction characteristic using Okra Mucilage as new drag reduction agent.* Journal of Applied Sciences, 2011. **11**: p. 2554-2561.
20. Ming, F.L.W., et al., *Insoluble nano-powders additives enhancing the flow of liquid in microchannel: Effect of particle size.* ARPN Journal of Engineering and Applied Sciences, 2016. **11**(4): p. 2146-2152.
21. Abdulbari, H.A. and F.L.W. Ming, *Drag reduction properties of nanofluids in microchannels* The Journal of Engineering Research (TJER) 2015. **12**(2): p. 60-67.
22. Zhao, R., et al., *Drag-reducing polymers diminish near-wall concentration of platelets in microchannel blood flow.* Biorheology, 2010. **47**(3-4): p. 193-203.
23. Kameneva, M.V., *Microrheological effects of drag-reducing polymers in vitro and in vivo.* International Journal of Engineering Science, 2012. **59**: p. 168-183.
24. Landahl, M.T., *Drag reduction by polymer addition,* in *Theoretical and Applied Mechanics: Proceedings of the 13th International Congress of Theoretical and Applied Mechanics, Moskow University, August 21–16, 1972,* E. Becker and G.K. Mikhailov, Editors. 1973, Springer Berlin Heidelberg: Berlin, Heidelberg. p. 177-199.

# Experimental measurement and CFD simulation on the hydrodynamics of an internal-loop airlift reactor

*Shi Yan* Liew[1], *Zhi Yee* Lam[1], and *Jolius* Gimbun[1,2*]

[1]Faculty of Chemical and Natural Resources Engineering, University Malaysia Pahang, 26300 Gambang, Pahang, Malaysia
[2]Centre of Excellence for Advanced Research in Fluid Flow (CARIFF), University Malaysia Pahang, 26300 Gambang, Pahang, Malaysia

**Abstract.** This paper concerns with the experimental measurement and computational fluid dynamics simulation on local hydrodynamics of a gas-liquid internal-loop airlift reactor. The aim of this work is to study the sensitivity of the drag models and the significance of considering the lift force on the predictive accuracy of the simulation. The experimental analysis was carried out using laser Doppler anemometry at three different heights (i.e. $Y$ = 0.20 m, 0.30 m and 0.38 m) across the riser and downcomer at volumetric flow rate of 0.30 m$^3$/h to provide validation for the simulation results. A transient three-dimensional gas-liquid internal-loop airlift reactor was carried out using FLUENT 16.2 by implementing the two-fluid model approach. The Eulerian-Eulerian multiphase and standard $k$-$\varepsilon$ dispersed turbulence model were employed in this study. Results suggest that the spherical drag model performed poorly and that the drag model governed by Rayleigh-Taylor shows promising accuracy in the prediction of overall mean axial liquid velocity. On the other hand, the consideration of lift model shows slightly improvement in accuracy. These findings may serve as a guidance for future scale-up and design of airlift reactor studies

## 1 Introduction

The airlift reactor is a class of pneumatic reactor that is designed to address the shortcomings found in the bubble column [1]. This device is attractive for its lower likelihood for mechanical failures, relatively lower shear rate in the absence of mechanical agitation and minimal loss of sterility [2]. Scaling up an airlift reactor is still currently challenging as the complexity of the hydrodynamics is influenced by the geometry of the reactor.

Many empirical correlations have been proposed in the past relating to riser-to-downcomer ratio, gas holdup and superficial gas velocity which were used to aid in current industrial scale-up procedures [3-5]. These correlations were however derived based on global hydrodynamics measurements which may not reflect the local intrinsic information such as Kolmogorov effect and flow patterns of the fluid flow. Recent breakthrough in optical imaging techniques has made local hydrodynamics studies easier in turbulent gas-liquid dispersion systems. The laser Doppler anemometry (LDA) is one of the non-intrusive methods available to measure local velocities of the flow field. The LDA is capable of studying transient flow relatively close to the wall, providing single point time-averaged measurements [6-7]. This technique requires seeding particles which are assumed to track the flow dynamics following the principles of Stokes number [8]. LDA has been successfully employed in stirred tanks and bubble columns to measure mean axial liquid velocity profiles despite the challenges of capturing accurate two-phase flow field [6, 9-13]. Though experimental using LDA as a measurement method in airlift reactors are still limited [14-15]. In addition, both global and local hydrodynamics measurements used to derive empirical correlations for scale-up purposes is accurate only for similar geometrical designs, operating conditions and liquid properties.

Thus numerical approach becomes an attractive alternative method that allows flexible design solutions in a virtual experimentation environment where it screens potential design and configurations. This would reduce fabrication cost and time for re-optimization due to scale-up failures. Computational fluid dynamics (CFD) has garnered interested in the field of numerical studies for its capabilities to provide an in-depth understanding into the hydrodynamics and mass transfer of airlift reactors with reasonable accuracy. The modelling approach of an airlift reactor depends on its flow regime. In bubbly flow, the two-fluid model is widely employed across literature whereby the bubble sizes are assume to be mono-dispersed [16-18]. The two-fluid model implements the Eulerian framework whereby the conservation equations for both the liquid and gas phases are ensemble-averaged reducing the computational cost to predict the gas-liquid flow fields in an airlift reactor. However, this procedure inevitably eliminates microscopic information describing the interacting forces between the gas-liquid phases. Detailed information that describes the contact between the bubbles with liquid phase for effective mass transfer

* Corresponding author: jolius@ump.edu.my

are important for a successful scale-up and design of an airlift reactor. Hence, additional closure models are introduced to account the two-phase momentum exchanges.

As of current, no universal set of closure models have been concluded for bubbly flow in an airlift reactor although a diverse set of closure models have been proposed across literature [19-22]. These closure models which composed of interfacial momentum forces can be divided down further as drag and non-drag forces.

The drag force is known to be the pre-dominant contributing force whereby it determines the velocity of the gas phase and overall flow pattern in the gas-liquid system [23]. Past airlift reactor studies have favoured Schiller and Naumann (1935) model over other proposed drag models to predict the airlift reactor hydrodynamics [16, 24]. This model is however meant for spherical bubbles which may not be suitable for more turbulent flows where bubble may deform due to turbulent forces. Hence, comparison studies have been conducted over the decade to evaluate the sensitivity of these drag models on local hydrodynamics. A study carried out in an internal-loop airlift reactor have shown superior performance through the drag model Karamanev and Nikolov (1992) as a function of Reynolds number when compared with other models governed by Eötvös number [19]. Another study showed that the combination of both Reynolds and Eötvös number (i.e. Dijkhuizen et al., 2010) gave improvements in radial profile of the axial liquid velocity prediction [20]. However, both the these studies were conducted in a two-dimensional spatial simulation whereby its accuracy can be limited to lower superficial gas velocity and sensitive to gas inlet geometry [22, 25]. Comparison on drag model performance in a three-dimensional airlift reactor showed better performance in predicting gas velocity when the drag coefficient as a function of superficial phase velocity was used [21]. Presently, there is absent in literature on the evaluation of the performance of the Universal drag model which is governed by different flow regimes (i.e. function of Reynolds number, Rayleigh-Taylor instability and gas holdup) in a three-dimensional airlift reactor.

On the other hand, the lift force is a non-drag force that is not obvious in comparison of the drag force but has a role in radial gas holdup. A lack of clarity on the effects of considering the lift model has result some studies to neglect it [21, 26]. A past study found that the lift force assist in correcting the radial profiles of the gas holdup, axial liquid velocity and turbulent kinetic energy [27]. In addition, another study pointed that lift should be considered in order to capture the bubble plume oscillations [28]. Plume oscillations are responsible in predicting the instantaneous bubble velocity. However, another study conducted in a pipe showed that the effect of lift and turbulent dispersion force cancels out each other near walls in the gas holdup and liquid velocity [29]. Due to the discrepancy, much clarity I still needed on the significance of considering the lift model in airlift reactors, hence is studied in our work as well.

The outline of this paper consists of the experimental measurement on local hydrodynamics of an internal-loop airlift reactor as the first part. The measurements are obtained using LDA taken across the riser and downcomer at three different heights and will be used to validate with the simulation sensitivity study of the interfacial forces. The second part concerns with a transient simulation of the internal-loop airlift reactor to evaluate the performance of different drag models and the effect of considering the lift force for a closure model of a two-phase internal-loop airlift reactor. The drag models considered in this study consists of a drag coefficient governed by Reynolds number, Eötvös number and bubble swarm which accommodates the bubble swarm correction model. In addition, drift velocity was considered in this simulation to take into account the effects of fluctuating liquids on the gas phase [30].

## 2 Experimental procedure

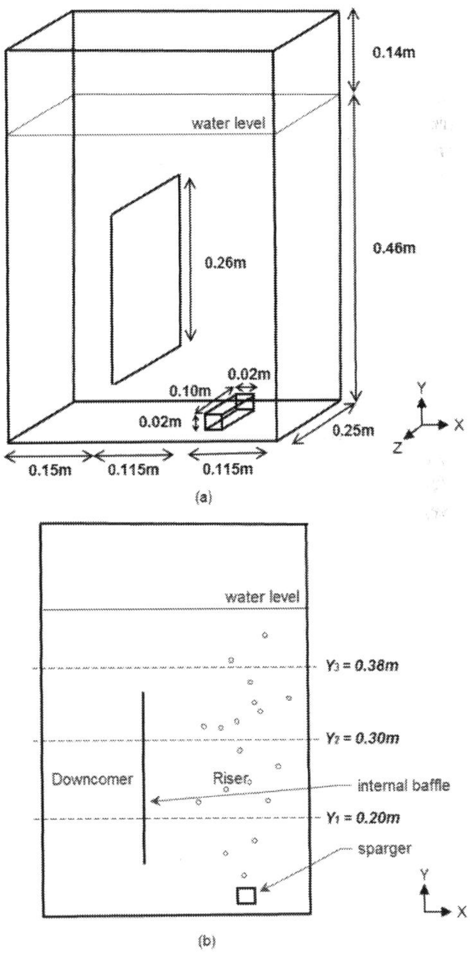


**Fig. 1.** Laboratory scale setup of an internal-loop airlift reactor, (a) Dimensions of the airlift reactor and (b) LDA measurements taken at three positions.

Figure 1(a) shows a laboratory scale internal-loop airlift reactor made of transparent Perspex materials with measurements of 0.40 m (width) x 0.60 m (height) x 0.25 m (depth). An internal baffle with a height of 0.26 m is positioned 0.10 m from the bottom of the reactor. The top of the reactor was left open. Air is introduced into the airlift reactor through a 4" Classica air stone with the dimensions of 0.10 m in length and 0.20 m in height located at the base of the reactor. The reactor was filled with water at a height of 0.46 m and aerated with a volumetric flow rate of $Q_g$ = 0.30 m$^3$/h using a vacuum pump. In order to obtain a non-coalescence system, 4 kg/m$^3$ of salt was added into the reactor [31].

Measurements of the axial liquid velocity component is taken using two component FlowExplorer. LDA system from Dantec Dynamics whereby the system is operating in back-scattered mode. The gas-liquid system was seeded with polyamide seeding particles of diameter size 20 μm and density of 1030 kg/m$^3$ from Dantec Dynamics. The measurements were taken at heights 0.20 m, 0.30 m and 0.38 m of the column as shown in Figure 1(b). These measurements were taken radially at an interval of 10 mm between each points with a 2.5 mm gap away from the wall. Both 15 mW He-Ne and 5 W Ar-ion lasers manufactured by Spectra-Physics were used as light sources emitting wavelengths of 532 nm and 561 nm. The laser intensity is set at 40 mW for both lasers. A 500 mm focal length lens with a probe volume of 2.1 mm$^3$ was used as a receiving lens in this study. The laser and focal lens were mounted on a bi-directional traverse controlled using a workstation.

Data validation and signal processing were carried out using BSA3 Processor v6.10 software collected from the site using burst mode data collection method. It is worth noting that the data rate of acquisition drops as the laser approaches the bubble region. This is especially observed when the laser moves from the wall to the bubble plume region in the riser. When light hits the bubble, the intensity of the light reflected will be higher [32]. However, these lights are reflected and scattered, lowering the probability of the lights hitting the detector. This results in the reduction in sample count obtained.

Several measurements were performed at these locations repetitively by adjusting the gain, sensitivity, and anode to improve the accuracy of the time-averaged values. The duration of time specified to collect the samples is set at 60 s which is sufficient enough to obtain time-averaged value with an accuracy of ±0.01 m/s. Within that timeframe and setup as describe in section 2, 10$^2$ to 10$^3$ amount of samples were obtained. Although longer sampling would be encouraged, it was found large data quantities would not result to significant improvements in accuracy [26].

## 3 CFD simulation

### 3.1. Geometry and computational grid

A transient simulation of a three-dimensional internal-loop airlift reactor was carried out using ANSYS FLUENT 16.2. The reactor was prepared in GAMBIT 2.4.6 according to the dimensions as described in section 2 as illustrated in Figure 2 using hexahedral cells. The headspace was extended with an additional 0.30 m to prevent recirculation and complex flow features from affecting flow field results. The governing laws and interfacial forces adopted in this simulation were further elaborated in section 3.2.

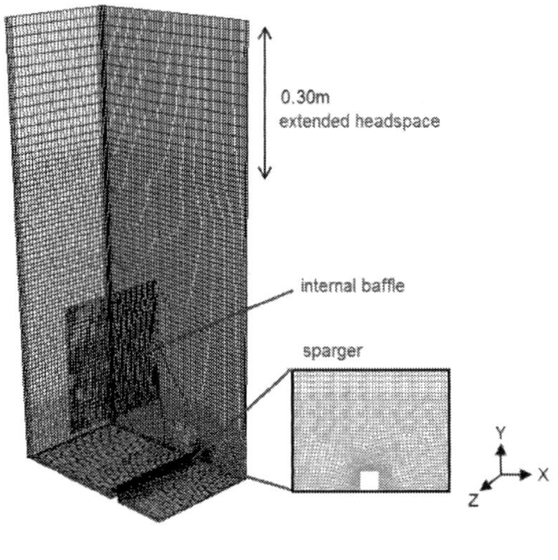


**Fig. 2.** Geometry dimensions of airlift reactor as prepared in GAMBIT 2.4.6.

No-slip boundary conditions were applied on the walls and the top of the headspace region was set as pressure outlet. A gas with a volumetric flow rate, $Q_g$ = 0.30 m$^3$/h was injected and treated as a continuous source of gas at the sparger. The pressure and temperature were set to ambient. The CFD simulation was performed using mean bubble size correlation [33]. All residuals were set below 1x10$^{-8}$ for each time step to achieve good convergence. The axial liquid velocity at height 0.30 m was monitored and the iterations were halted once a constant value was observed. The data was averaged once a pseudo-steady condition was achieved. The time step was set at 0.0001 s. The CFD simulation was performed using six units of HP Compaq Pro 6300 MT with a quad core processor (3.2 GHz i7-3770, 4 GB of RAM).

### 3.2. Multiphase modelling

The Eulerian-Eulerian framework was adopted in this simulation work, whereby the mass and momentum equations for the continuous and dispersed phases were weighted by their corresponding volume fractions. Each of the phases are ensemble-averaged and that the total sum of the volume fractions should satisfy to a unity as described in the following governing equations:

$$\frac{\partial(\alpha_k \rho_k)}{\partial t} + \nabla \cdot (\alpha_k \rho_k \mathbf{u}_k) = 0 \quad (k = g, l) \qquad (1)$$

$$\frac{\partial(\alpha_k\rho_k\mathbf{u}_k)}{\partial t}+\nabla\cdot(\alpha_k\rho_k\mathbf{u}_k\mathbf{u}_k)=-\alpha_k\nabla p+\nabla\bar{\bar{\tau}}+\nabla\alpha_k\rho_k g+F_k \quad (2)$$

where $\alpha$ is the volume fraction, $\rho$ is the density, $t$ is the time and $\mathbf{u}$ is the velocity vector, respectively. The interphase mass transfer is not considered in this study, hence the mass source term in the right hand side of (1) is neglected. Meanwhile in the first part of the right hand-side of expression (2) describes the pressure gradient, followed by the tensor stress between the phases, gravity and lastly the interfacial momentum transfer between the two-phases, respectively.

$$F_k=\sum F_{l,g}=F_D+F_L+F_{TD} \quad (3)$$

The interfacial momentum forces as shown in (3) compose of drag, $F_D$ and non-drag forces whereby, the non-drag forces that are considered in this study consists of lift, $F_L$ and turbulent dispersion force, $F_{TD}$ . Virtual mass force is neglected in this simulation study as it is computationally costly to simulate with no significant improvements [22].

### 3.3. Turbulence modelling

This work employs a standard $k$-$\varepsilon$ dispersed turbulence model based on Tchen-theory for systems where the concentration of the dispersed phase is dilute. The transport equation for $k$ and $\varepsilon$ in the dispersed model are given as:

$$\frac{\partial(\alpha_l\rho_l k_l)}{\partial t}+\nabla\cdot(\alpha_l\rho_l\mathbf{u}_l k_l)= \nabla\cdot\left(\alpha_l\left(\mu_l+\frac{\mu_{t,l}}{\sigma_k}\right)\nabla k_l\right)+\alpha_l G_{k,l}-\alpha_l\rho_l\varepsilon_l+\alpha_l\rho_l\Pi_{k,l} \quad (4)$$

$$\frac{\partial(\alpha_l\rho_l\varepsilon_l)}{\partial t}+\nabla\cdot(\alpha_l\rho_l\mathbf{u}_l\varepsilon_l)=\nabla\cdot\left(\alpha_l\left(\mu_l+\frac{\mu_{t,l}}{\sigma_\varepsilon}\right)\nabla\varepsilon_l\right) +\alpha_l\frac{\varepsilon_l}{k_l}\left(C_{1,\varepsilon}G_{k,l}-C_{2,\varepsilon}\rho_l\varepsilon_l\right)+\alpha_l\rho_l\Pi_{\varepsilon,l} \quad (5)$$

$G_{k,l}$ is the production rate of the turbulent kinetic energy and its form is similar to the one applied for single flow. The values of the constant in (4) and (5) are given by $\sigma_k=1$, $\sigma_\varepsilon=1.3$, $C_{1,\varepsilon}=1.44$ and $C_{2,\varepsilon}=1.92$.

### 3.4. Interfacial momentum forces

#### *3.4.1 Drag force*

Drag force is the resistance force acting on the body (i.e. bubble) moving in a surrounding fluid (i.e. liquid phase). The drag force is governed by inertia whereby the force becomes increasing significant at larger Reynolds number. A layer of viscous stress and pressure distribution forms around the moving body, affecting the bubble size and shape resulting to the drag force embodying the bubble. The drag force can be describe as a function of local velocity as described in (6):

$$F_D=\frac{3}{4}\frac{\alpha_g\alpha_l\rho_l C_D}{d_b}|v_r|v_r \quad (6)$$

where $d_b$ represents the Sauter mean bubble diameter, $C_D$ is the drag coefficient which can be retrieved from the drag model and the relative velocity, $v_r=(\mathbf{u}_g-\mathbf{u}_l)/v_d$ whereby $v_d$ represents the drift velocity accounted by Simonin and Viollet (1990) model as shown in (17). N this study, the drag models Schiller and Naumann (1935), Tomiyama et al. (1999) and Universal drag were considered for comparison.

Schiller and Naumann (1935) drag model is derived from a single spherical bubble moving in a stagnant liquid. It has been a favourable drag model applied widely in airlift reactor studies even in those which consist of polydisperse bubbles in their study [23-25, 35]. Hence, it is chosen as one of the drag models for evaluation:

$$C_D=\begin{cases}\frac{24}{\mathrm{Re}_b}\left(1+0.15\mathrm{Re}_b^{0.687}\right) & if\ \mathrm{Re}_b<1000\\ 0.44 & if\ \mathrm{Re}_b\ge 1000\end{cases} \quad (7)$$

where $\mathrm{Re}_b$ is the bubble Reynolds number defined as $\mathrm{Re}_b=(\rho_l d_b|v_r|)/\mu_l$ and $\mu_l$ is the dynamic liquid viscosity.

The Tomiyama et al. (1999) drag model is derived from a single bubble in an infinite stagnant liquid studied under a wide range of bubble sizes, gravity acceleration, different degrees of contamination and different fluid properties. The expression of the model is given by :

$$C_D=\max\left(\min\left(\frac{24}{\mathrm{Re}_b}\left(1+0.15\mathrm{Re}_b^{0.687}\right),\frac{72}{\mathrm{Re}_b}\right),\frac{8E\ddot{o}}{3(E\ddot{o}+4)}\right) \quad (8)$$

$E\ddot{o}$ is the Eötvös number defined as $E\ddot{o}=\left(gd_b^2(\rho_l-\rho_g)\right)/\sigma$ where $\sigma$ is the surface tension. The Eötvös number measures the distortion of the bubbles based on the gravitational force to surface tension ratio. When the Eötvös number is $E\ddot{o}<1$, the surface tension governs which retain the sphere of the bubble which is likely to be true in disperse homogeneous regime. As the Eötvös number increases, the gravitational force is more prominent resulting to bubble deformation.

On the other hand, the Universal drag considers a wide range of flow regimes. The model divides the flow regimes into viscous, distorted and capped bubble regimes. In addition, the model also incorporates the bubble swarm correction as described below [36]:

$$C_{D,vis}=\frac{24}{\mathrm{Re}_b}\left(1+\mathrm{Re}_b^{0.75}\right) \quad (9)$$

$$C_{D,dis} = \frac{2d_b}{3\lambda_{RT}}\left(\frac{1+17.67\beta^{6/7}}{18.67\beta}\right) \quad \text{where, } \beta = (1-\alpha_l)^{1.5} \quad (10)$$

$$C_{D,cap} = \frac{8}{3}(1-\alpha_g)^2 \quad (11)$$

where, $\lambda_{RT} = \left(\sigma / \left(g\left|\rho_l - \rho_g\right|\right)\right)^{0.5}$ is the Rayleigh-Taylor instability wavelength. The drag coefficient is calculated based on conditions that define each respective regimes. When $C_{D,dis} < C_{D,vis}$ the system is considered a viscous regime, then $C_D = C_{D,vis}$. If $C_{D,vis} < C_{D,dis} < C_{D,cap}$ the drag coefficient for a distorted bubble regime, $C_D = C_{D,dis}$ is applied. When $C_{D,dis} > C_{D,cap}$, the system is considered a capped bubble regime, $C_D = C_{D,cap}$.

### *3.4.2 Lift force*

Lift force is the interfacial force acting on the lateral direction, perpendicular to the direction of the flow. This results to the migration of the bubbles in a shear field through the traverse lift force. The migration of the bubbles can be translated through the shear-induced lift force model, (12) as given by [37-39]:

$$F_L = \alpha_g \rho_l C_L \left(\mathbf{u}_g - \mathbf{u}_l\right) \times \nabla \mathbf{u}_l \quad (12)$$

where, $C_L$ represents the lift coefficient. The lift coefficient can be integrated either as a non-dimensional constant or through a lift model as a function of the bubble Eötvös and bubble Reynolds number. As lift coefficient is affected by bubble size, this study opt to incorporate lift coefficient using lift model instead of assigning a fixed constant value as it tends to be influenced by bubble size and the proposed constant values varies across literature [40-41]. Tomiyama et al. (2002) lift model is implemented to calculate the lift coefficient in this study as shown in (13). The principle behind this correlation was derived from a single bubble moving in a glycerol-water solution. This model has also shown success in air-water system and thus is being implemented in this study [42].

$$C_D = \begin{cases} \min\left[0.288\tanh(0.121\mathrm{Re}_b), f(E\ddot{o}_m)\right] & \text{if } E\ddot{o}_m < 4 \\ f(E\ddot{o}_m) & \text{if } 4 < E\ddot{o}_m < 10 \\ -0.27 & \text{if } 10 < E\ddot{o}_m \end{cases} \quad (13)$$

$$f(E\ddot{o}_m) = 0.00105E\ddot{o}_m^3 - 0.0159E\ddot{o}_m^2 - 0.0204E\ddot{o}_m + 0.474 \quad (14)$$

The modified Eötvös number is given by $E\ddot{o}_m = \left(gd_a^2\left(\rho_l - \rho_g\right)\right)/\sigma$. $d_a$ is the major dimension of the bubble which can be calculated through an empirical correlation of the aspect ratio as given [43]:

$$d_a = d_b\sqrt[3]{1+0.163E\ddot{o}^{0.757}} \quad (15)$$

### *3.4.3 Turbulent dispersion force*

The turbulent dispersion force describes the effect of the turbulent fluctuations of liquid velocity on the bubbles. In this study, the Simonin and Viollet (1990) model on drift velocity is being implemented in both the sensitivity study in the drag models and consideration of the lift force. The correlation assumes the bubbles are to be caught up in turbulent eddies of the liquid phase and carried from high concentration regions to lower concentration regions as described in the following:

$$v_d = -D'_{g,l}\left(\frac{\nabla\alpha_g}{\alpha_g} - \frac{\nabla\alpha_l}{\alpha_l}\right) \quad (16)$$

$$F_{TD} = \frac{C_{TD}k_{l,g}D'_{g,l}}{\sigma_{TD}}\left(\frac{\nabla\alpha_g}{\alpha_g} - \frac{\nabla\alpha_l}{\alpha_l}\right) \quad (17)$$

where $v_d$ is the drift velocity, $D'_{g,l}$ is the gas-liquid turbulent dispersion term, coefficient $C_{TD} = 1$ and $\sigma_{TD}$ is the Prandtl number set as 0.75. Details of the derivation of $D'_{g,l}$ is elaborated in [19] and will not be repeated here.

## 4 Results and discussion

### 4.1 Measurement of mean axial liquid velocity from airlift reactor

The mean axial liquid velocity was measured using LDA in an internal-loop airlift reactor and the measurements were illustrated in Figure 3. 0.00 m < $X_R$ < 0.25 m represents the horizontal position of the riser meanwhile -0.15 m < $X_D$ < 0.00 m represents the horizontal position of the downcomer.

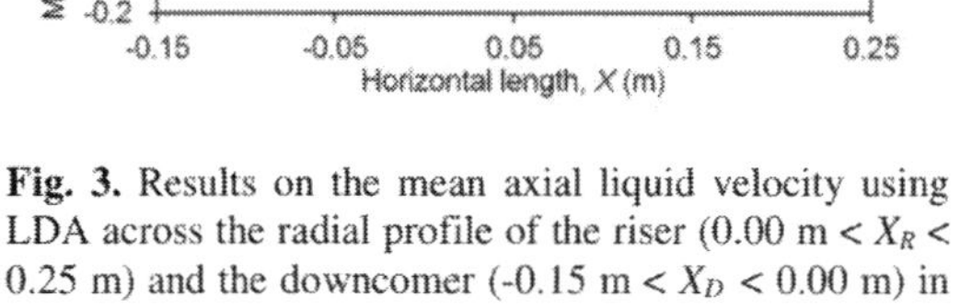


**Fig. 3.** Results on the mean axial liquid velocity using LDA across the radial profile of the riser (0.00 m < $X_R$ < 0.25 m) and the downcomer (-0.15 m < $X_D$ < 0.00 m) in the airlift reactor at height, $Y_1$ = 0.20 m, $Y_2$ = 0.30 m and $Y_3$ = 0.38 m.

At height 0.20 m and 0.30 m the axial liquid velocity shows negative in the downcomer capturing the downward flow direction. Meanwhile, the riser at all heights showed an upward flow as shown through the positive values of the mean axial liquid velocity. This indicates there is a liquid circulation occurring in the airlift reactor which is vital for mixing. The axial velocity in the riser region shows greater fluctuation in the riser than those in the downcomer which is similar in trend with [14] study. Nevertheless, results obtained from the LDA is satisfactory within a ±0.01 m/s range of discrepancy between the repetitive results as recommended [15].

## 4.2 Comparison between different drag models

Prior to the CFD simulation, a grid sensitivity study was conducted by preparing a coarse grid (236k cells), intermediate grid (373k cells) and fine grid (717k cells). Figure 4 shows the results from the grid independence study, whereby the Universal drag model, Tomiyama lift model and drift velocity were used for the evaluation. At height $Y_1$ = 0.20 m, all grids were in close agreement with the experimental data especially in the downcomer. At higher heights where the liquid velocity of more driven by the buoyancy of the bubble, both the intermediate and fine grid achieved good agreement with experimental data obtained from LDA. Meanwhile, large discrepancy occurs with experimental data using the coarse grid at higher heights. The computational time spent by each grids respectively are; coarse grid (0.12 s/iteration), intermediate grid (0.26 s/iteration) and fine grid (0.35 s/iteration), respectively. From the overall result, the intermediate grid (373k cells) yielded a grid independent solution as considering the fine grid (717k cells) will be computationally costly with no significant improvements to the result. Thus, the intermediate grid was considered sufficient enough to be used for the remainder of this work.

The effect of different drag models (i.e. Schiller and Naumann, 1935; Tomiyama et al. 1999 and Universal drag) is being studied by comparing the predicted mean axial liquid velocities with experimental measured data obtained from LDA as shown in Figure 5. The lift model and the turbulent dispersion were enabled for this case study. It can be seen Schiller and Naumann (1935), a drag coefficient as a function of Reynolds number performed poorly in the overall results. It only performs fairly at height, $Y_1$ = 0.20 m which is the region most nearest to the sparger. Beyond that, the discrepancy became even larger with increasing height. It is worth noting that Schiller and Naumann (1935) model assumes spherical bubbles which neglects deformation of the bubble surface. On the other hand, both drag coefficients by Tomiyama et al. (1999) and Universal drag governed by the Eötvös number and different flow regimes, respectively were in closer agreement with experimental data obtained from LDA.

The Tomiyama et al. (1999) included the bubble Reynolds number and an additional Eötvös number in its expression whereby it considers the terminal velocities that are affected by gravity and surface tensions due to bubble deformation. The Tomiyama et al. (1999) drag model is expected to be more accurate when paired up with Tomiyama et al. (2002) lift model as the lift correlation was developed while accounting the Tomiyama drag coefficient [44]. However, the result especially in the downcomer in our study shows otherwise.

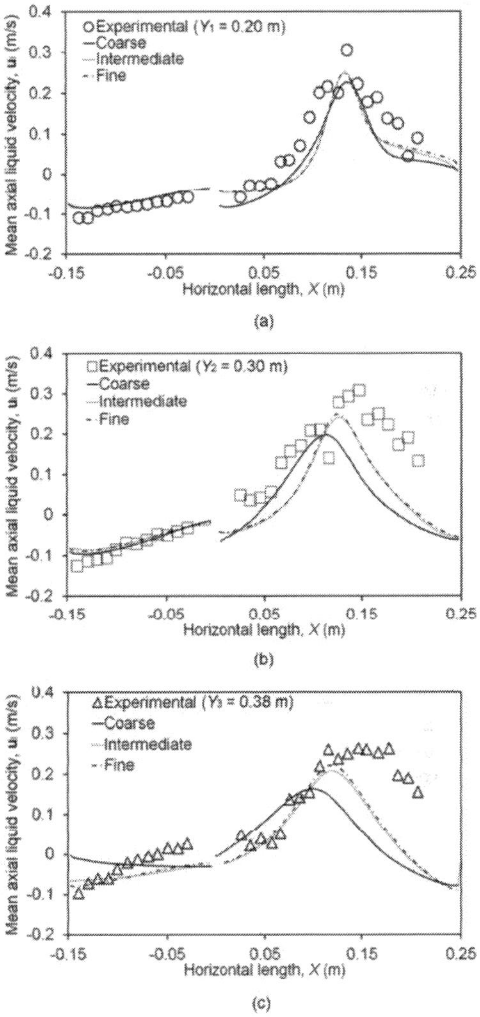


**Fig. 4.** Results on grid sensitivity study on the mean axial liquid velocity at three different heights, (a) $Y_1$ = 0.20 m, (b) $Y_2$ = 0.30 m and (c) $Y_3$ = 0.38 m.

Amongst the drag models, the Universal drag is best fitted with the experimental results in both the riser and downcomer although it only gives a slight improvement in accuracy in comparison to the Tomiyama et al. (1999) model. This result suggests that the drag model as a function of Rayleigh-Taylor instability is more accurate. Hence, the Universal drag model is being implemented

DOI: 10.1051/matecconf/201711101002

for the evaluation on the effect of considering the lift model.

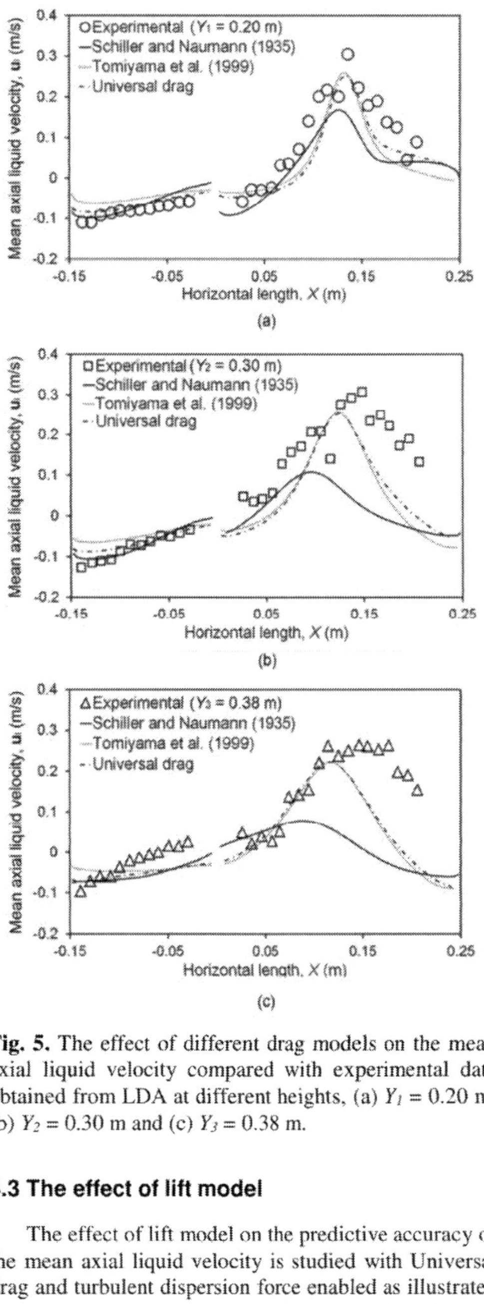


**Fig. 5.** The effect of different drag models on the mean axial liquid velocity compared with experimental data obtained from LDA at different heights, (a) $Y_1 = 0.20$ m, (b) $Y_2 = 0.30$ m and (c) $Y_3 = 0.38$ m.

### 4.3 The effect of lift model

The effect of lift model on the predictive accuracy of the mean axial liquid velocity is studied with Universal drag and turbulent dispersion force enabled as illustrated in Figure 6. When the lift effect is neglected, it is observed that the discrepancy between the simulation results with experimental result is minimal. This indicates that drag force plays a dominant role amongst the interfacial momentum forces. From the figure, the radial gas profile peaks are much higher, when lift force is neglected.

When lift force is enabled, the radial gas profile peak subsides. This is also similarly reported in a two dimensional airlift reactor study where they observed flatter radial profiles on the mean axial liquid velocity in a bubble column at different superficial gas velocity [45]. This is attributed to the consideration of lift force, drifting the bubbles tangentially from its flow direction, heading towards the wall. From the results it can be seen the role of the lift force more prominent with increasing height across the airlift reactor. Although the difference in peaks here are not that substantial, it is worth noting that the effect of the lift model is deem more significant at higher superficial gas velocity as the bubble sizes are polydisperse in nature [46].

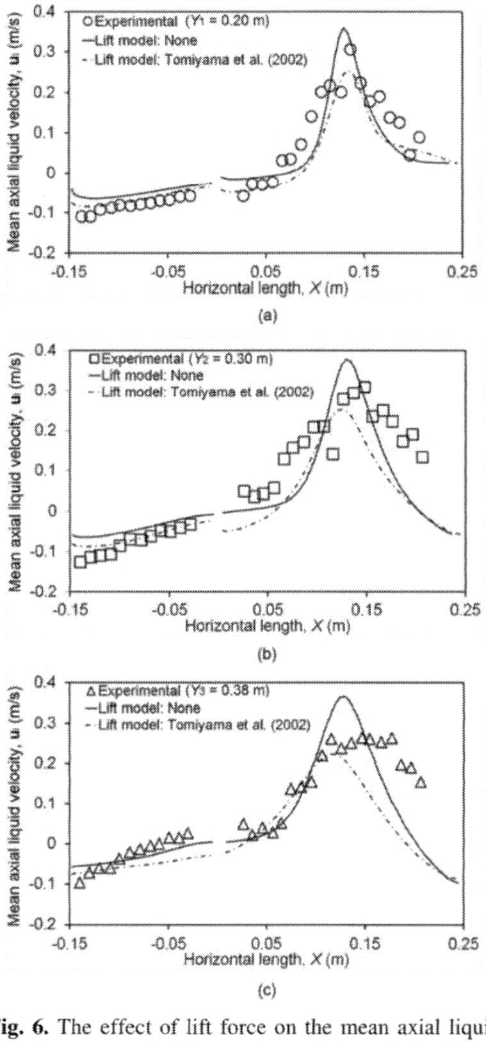


**Fig. 6.** The effect of lift force on the mean axial liquid velocity compared with experimental data obtained from LDA at different heights, (a) $Y_1 = 0.20$ m, (b) $Y_2 = 0.30$ m and (c) $Y_3 = 0.38$ m.

It is observable for all case studies that the riser at height $Y_2 = 0.30$ m and both the riser and downcomer height at $Y_3 = 0.38$ m, the prediction accuracy of the mean axial liquid velocity in the riser is far off although the trend is present. The combination of drag force, lift force and drift velocity leaves room for further improvement in the accuracy at those heights. Those are the position at which a swarm of bubbles are dispersed at. Hence, it is desired that in the future studies to include the bubble-induced model to improve the predictive accuracy. It is also interesting to observe the performance between the drag models, Tomiyama et al. (1999) and Universal drag model at different gas volume flow rate especially at higher gas flow rates. A study conducted in the bubble column have deem the Tomiyama et al. (1999) unsuitable at heterogeneous flow [47]. The study here indicates Tomiyama et al. (1999) to be on par with Universal drag model. Nevertheless, the results of this study will be useful to understand the roles of each interfacial forces especially the role of lift force at different heights across the reactor.

## 5 Conclusion

A simulation case study for a closure model was carried out in a gas-liquid internal-loop airlift reactor using two-fluid model approach. The study of the closure model consists of the evaluation of the effects between different drag models and the consideration of the lift model on the predictive accuracy. The simulation results were validated with experimental local measurements taken using LDA, obtained across three different heights ($Y$ = 0.20 m, 0.30 m and 0.38 m) of the internal-loop airlift reactor.

Results suggest that Schiller and Naumann (1935) drag model is poor at predicting the mean axial liquid velocity, giving major discrepancy with increasing height. The Universal drag performed the best in comparison to the Tomiyama et al. (1999), with closer agreement with experimental data. The Universal drag model was then implemented to evaluate the significance of considering the lift force in the gas-liquid internal-loop airlift simulation. The outcome reflects that the lift model subsides mean axial liquid velocity peaks, giving better improvements to the accuracy of the simulation results. The role of the lift force is seen more prominent with increasing height across the airlift reactor. This work reflects the importance of the closure model for better accuracy in the gas-liquid simulation results and LDA's capability at capturing the local hydrodynamics in an internal-loop airlift reactor. Future works in heterogeneous flow will allow further evaluation on the potential of this closure model.

**Acknowledgements**
S.Y. Liew thanks Ministry of Education Malaysia for MyMaster scholarship. We also acknowledge funding from University Malaysia Pahang (No. GRS160329) and the provision of a Master scholarship via Graduate Research Scheme.

## References

1. E. Kadic, T.J. Heindel, Bioreactor hydrodynamics and gas-liquid mass transfer (Wiley, Hoboken, New Jersey, 2014)
2. M.Y. Chisti, M. Moo-Young, Chem. Eng. Comm., **60**, 195-242 (1987)
3. J. Hills, Trans. Inst. Chem. Eng. **52**, 1-9 (1974)
4. R.A. Bello, C.W. Robinson, M. Moo-Young, Can. J. Chem. Eng., **62**, 573-577 (1984)
5. M.Y. Chisti, B. Halard, M. Moo-Young, Chem. Eng. Sci., **43**, 451-457 (1988)
6. R.F. Mudde, J.S. Groen, H.E.A. Van den Akker, Nucl. Eng. Des., **184**, 329-338 (1998)
7. V.V. Ranade, M. Perrard, C. Xuereb, N. Le Sauze, J. Bertrand, Chem. Eng. Res. Des., **79**, 957-964 (2001)
8. M. Raffel, C.E. Willert, J. Kompenhans, Particle image velocimetry: a practical guide (Springer, 2013)
9. R.F. Mudde, J.S. Groen, H.E.A. Van den Akker, Chem. Eng. Sci., **52**, 4217-4224 (1997)
10. J.J. Derksen, M.S. Doelman, H.E.A. Van den Akker, Exp. Fluids., **27**, 522-532 (1999)
11. A.A. Kulkarni, K. Ekambara, J.B. Joshi, Chem. Eng. Sci., **62**, 1049-1072 (2007)
12. V.V. Ranade, V.P. Mishra, V.S. Saraph, G.B. Deshpande, J.B. Joshi, Ind. Eng. Chem. Res., **31**, 2370-2379 (1992)
13. A.A. Kulkarni, J.B. Joshi, V.R. Kumar, B.D. Kulkarni, Chem. Eng. Sci., **56**, 5037-5048 (2001)
14. W.S. Tan, G.C. Dai, W. Ye, J.P. Shen, Chem. Eng. J. Biochem. Eng. J., **57**, B31-B36 (1995)
15. C. Vial, S. Poncin, G. Wild, N. Midoux, Chem. Eng. Sci., **57**, 4745-4762 (2002)
16. D. Law, F. Battaglia, T.J. Heindel, *Mechanical Engineering Conference Presentations, Papers, and Proceedings*, **134** (2008)
17. T. Zhang, C. Wei, C. Feng, J. Zhu, Bioresource Technol., **104**, 600-607 (2012)
18. A. Soman, Y. Shastri, Appl. Energy., **140**, 246-255 (2014)
19. S. Talvy, A. Cockx, A. Liné, AIChE J., **53**, 335-353 (2007)
20. M. Mohajerani, M. Mehrvar, F. Ein-Mozaffari, **9999**, 1-20 (2012)
21. X. Jiang, N. Yang, B. Yang, Particuology, **27**, 95-101 (2016)
22. J. Liao, T. Ziegenhein, R. Rzehak, J. Chem. Technol. Biotechnol., (2016)
23. R.S. Oey, R.F. Mudde, H.E.A. van den Akker, Fluid Mech. Transp. Phenom., **49**, 1621-1636 (2003)
24. K.M. Dhanasekharan, J. Sanyal, A. Jain, A. Haidari, Chem. Eng. Sci., **60**, 213-218 (2005)
25. R.F. Mudde, H.E.A. van den Akker, Chem. Eng. Sci., **56**, 6351-6358 (2001)
26. E. Olmos, C. Gentric, N. Midoux, Chem. Eng. Sci., **58**, 2113-2121 (2003)
27. M.R. Bhole, J.B. Joshi, D. Ramkrishna, Chem. Eng. Sci., **63**, 2267-2282 (2008)
28. A. Gupta, S. Roy, Chem. Eng. J., **225**, 818-836 (2013)
29. R. Rzehak, E. Krepper, Nucl. Eng. Des., **258**, 57-65 (2013)
30. O. Simonin, P.L. Viollet, Phenomena Multiphase Flows, 259-269 (1990)

31. N.G. Deen, T. Solberg, B.H. Hjertager, *Proceedings of 14th International Congress of Chemical and Process Engineering: CHISA* (2000)
32. J.S. Groen, R.F. Mudde, H.E.A. van den Akker, Exp. Fluids., **27**, 435-449 (1999)
33. P.M. Wilkinson, A. van schayk, J.P.M. Spronken, Chem. Eng. Sci., 48, 1213-1226 (1993)
34. D. Law, *Computational modeling and simulations of hydrodynamics for air-water external loop airlift reactors* (Virginia Tech, 2010)
35. R. Bannari, A. Bannari, B. Selma, P. Proulx, Chem. Eng. Sci., **66**, 2057-2067 (2011)
36. N.I. Kolev, *Multiphase Flow Dynamics 2: Thermal and Mechanical Interactions* (Springer, Berlin, Germany, 2005)
37. I. Zun, Int. J. Multiphase Flow, **6**, 583-588 (1980)
38. T.R. Auton, J. Fluid Mech., **183**, 199-218 (1987)
39. D. Drew, J.R.T. Lahey, Int. J. Multiphase Flow, **13**, 113-121 (1987)
40. S. Takagi, Y. Matumoto, *Proceedings of the Third International Conference on Multiphase Flow*, ICMF **98** (1998)
41. M.T. Dhotre, B. Niceno, B.L. Smith, M. Simiano, Chem. Eng. Sci., **64**, 2692-2704 (2009)
42. D. Lucas, A. Tomiyama, Int. J. Multiphase Flow, **37**, 1178-1190 (2011)
43. R.M. Wellek, A.L. Agrawal, A.H.P. Skelland, AIChE J., **12**, 854-862 (1966)
44. G. Montoya, D. Lucas, E. Baglietto, Y. Liao, Chem. Eng. Sci., **141**, 86-103 (2016)
45. A.E. Amooghin, S. Jafari, H. Sanaeepur, A. Kargari, Appl. Math. Model., **39**, 1616-1642 (2014)
46. M.V. Tabib, A.R. Swarnendu, J.B. Joshi, Chem. Eng. J., **139**, 589-614 (2008)
47. X.F. Liang, H. Pan, Y.H. Su, Z.H. Luo, Chem. Eng. Res. Des., **112**, 88-102 (2016)
48. L. Schiller, M. Naumann, Z. Ver. Deutsch. Ing., **77**, 318-320 (1935)
49. D.G. Karamanev, L.N. Nikolov, AIChE J., **38**, 1843-1846 (1992)
50. W. Dijkhuizen, I. Roghatr, M.V.S. Annaland, J.A.M. Kuipers, Chem. Eng. Sci., **65**, 1415-1426 (2010)
51. T. Takamasa, A. Tomiyama, *Proceedings of the 1999 NURETH-9 Conference* (1999)
52. A. Tomiyama, H. Tarnai, I. Zun, S. Hosokama, Chem. Eng. Sci., **57**, 1849-1858 (2002)

# Study on the fluid dynamics of nitrogen and hydrogen gases subjected to wires element in monolithic channel

*Amirah* Rashidi[1,*], *Zamri* Abdullah[1], and *Salman* Kashif[1]

[1]Chemical Engineering Department, Universiti Teknologi PETRONAS, 32610 Bandar Seri Iskandar, Perak Darul Ridzuan, Malaysia

**Abstract.** Ammonia has a very significant value in the fertilizer industry where it was being synthesized via Haber-Bosch process in the early 19th century. As the process utilize high operating conditions, it imposes high capital cost and is an energy-consuming process. Due to this unsustainable process, researchers have initiated an alternative to overcome this drawback by performing a simulation in microfluidic environment using ambient temperature and pressure (25°C and 1 atm). Wires element configured in a 50 mm x 10 mm, (L x D) dimension monolithic channel with different spacing and number of wires, arranged axially in 60° pitch have been introduced to investigate the dynamic mixing of nitrogen and hydrogen for ammonia synthesis. As the wires are configured in a different manner, the results show dissimilar volume fraction profile, contours and mixing index. Creating suitable obstruction with larger obstruction space enhanced the mixing. Reducing spacing from 2 mm to 1.5 mm illustrates fluctuating velocity at the centre of the channel causing the flow velocity become less than the set velocity 0.05ms$^{-1}$. By substituting from 19 wires to 13 wires to the flow, chaotic advection occurs lead to the increased of mixing index up to 94%.

**Keywords.** monolithic channel / computational fluid dynamic / dynamic mixing / ANSYS CFX / ammonia

## 1 Introduction

Ammonia ($NH_3$) is well-known in the industry for various applications with a high percentage, of 80% approximately from the total production, has been applied in synthetic fertilizer, therefore causing a great market demand in the industry business [1]. Consumption of fertilizer increased significantly in 2010 and is expected to grow in a stabilized way during the following years of the forecast period. World demand for total fertilizer is estimated to grow at 2.0% per annum from 2011 to 2015 [2].

Regrettably, the current production is capital and energy intensive but can only yield 10-20% of $NH_3$ [3]. This process of producing $NH_3$ through reaction of nitrogen ($N_2$) and hydrogen ($H_2$) which is known as Haber-Bosch process obliges to operate under a high operating condition temperature (400-500°C) and pressure (150-300 bar) where such circumstances requires a very careful handling as long with increasing the hazards to the operators and the plant facilities.

A new promising method to attain the high yield of $NH_3$ at ambient condition in microfluidic environment [4] is discovered by using wires element in monolithic channel. This method is seen as an alternative to improvise the efficiency of $NH_3$ production by exposing the raw materials, $H_2$ and $N_2$ to high surface area to volume ratio which have been proven previously by having parallel and perpendicular obstruction to enhance the dynamic mixing. The parallel and perpendicular placement leads to a favorable condition where the mixing index attains approximately 90% with the desired ratio for the raw materials to react [5]. As compared to simple square pitched arrangement, the possibility of having wires placed at 60° pitch has been explored in this work to elevate the yield of $NH_3$.

The usage of smaller dimension channel is proven to be more effective to obtain great reaction yield and product selectivity [6]. This reactor enables chemical reactions to be performed in reaction space several orders of magnitude smaller than conventional batch reactor [7]. The downscaling of devices has brought attractive features to the channel such as equipment installation time and time taken for the reaction to reach steady state has been reduced. The hazards and process safety risks may be reduced since the quantity of substances used in the channel is much more less.

Mixing is hardly achieved in minute system without modification of geometry. Mixing plays a significant role to increase the conversion of $N_2$ and $H_2$ to form $NH_3$. Mainly the importance is to reduce inhomogeneity that leads to secondary effects such as undesired reaction and change in properties [8]. Furthermore, great flow dynamic is preferred in a channel if intimate mixing is desired

 DOI: 10.1051/matecconf/201711101003

between two streams [9]. Several patterns of channel were investigated formerly to address the significant of mixing such as serpentine shaped channel [10], linked twist map barriers [11], floor-grooved channel [12] and rhombic micromixer [13]. In this project, the simulation is done through configuring the wires assembly in parallel arrangement inside the channel that will induce dynamic mixing via chaotic advection to increase the efficiency of ammonia production.

## 2 Simulation methodologies

Simulation is performed using ANSYS CFX software. Typical methodologies in CFD apply such as creation of geometry, development of mesh, and post-solver analysis as display in Figure 1. The project emphasizes on the 'synthesis of ammonia'. However, for the scope of this work, only flow dynamics without the reaction element will be simulated. It is important to create a geometry that will generate a great flow for the mixing.

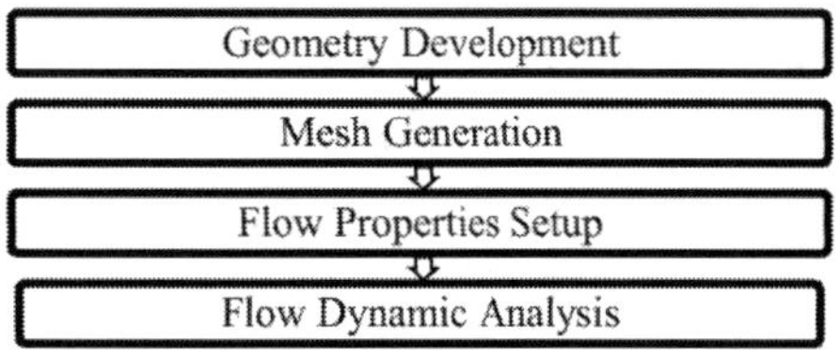


**Fig. 1.** Flow of simulation methodology.

### 2.1 Geometry development

As shown in Figure 2 the channel dimension is 10mm (D) x 50mm (L) while the diameters of the wires are 1mm. The wires assembly started at a distance on 3mm from the inlet and ended 3mm before the outlet as a mean to induce pre-mixing upon flowing inside which are constant for all three designs. The wires are arranged parallel to the channel with different spacing of wires and different number of wires aimed to create higher mixing efficiency.

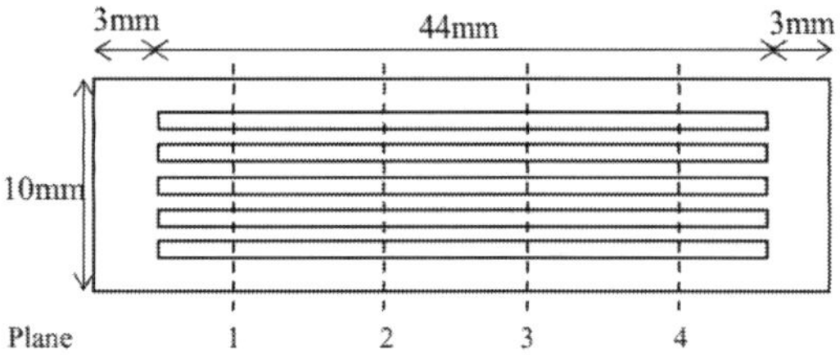


**Fig. 2.** General dimension of a monolithic channel.

Table 1 illustrates three designs that were developed in this project and the flow dynamics of the nitrogen and hydrogen gases throughout the monolithic channel will be observed and recorded. As it is arranged in parallel order, there is no intersection of wires and no change in wires path.

**Table 1.** Configuration of channel in radial view.

| | |
|---|---|
| 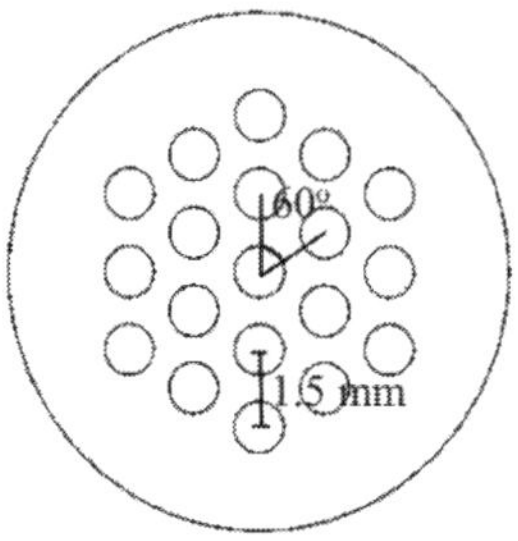 | Geometry A |
| | Number of wires: 19 |
| | Spacing: 1.5 mm |
| 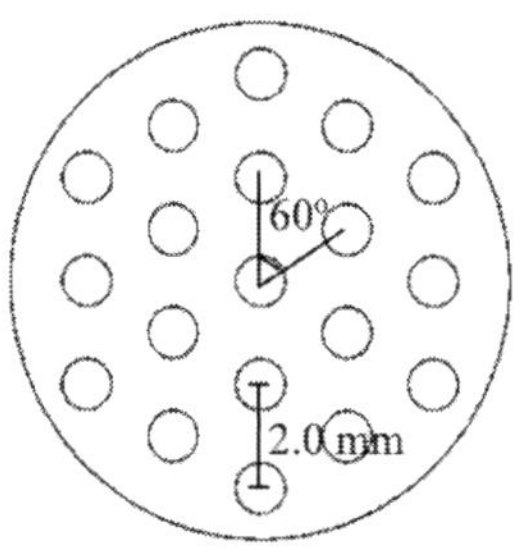 | Geometry B |
| | Number of wires: 19 |
| | Spacing: 2.0 mm |
| 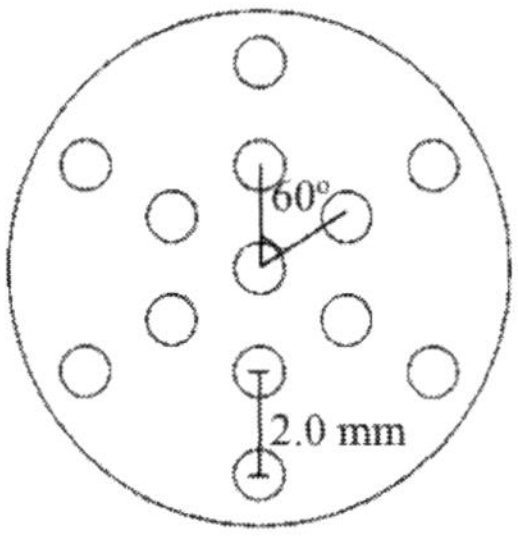 | Geometry C |
| | Number of wires: 13 |
| | Spacing: 2.0 mm |

### 2.2 Mesh Generation

The mesh properties for geometry B were set up to fine, coarse and medium. Per the orthogonal factor, mixing index and time taken to run the program, coarse mesh was selected to be the relevance centre for all design since there was not much difference in mixing index curve making mesh finer as shown in Figure 3.

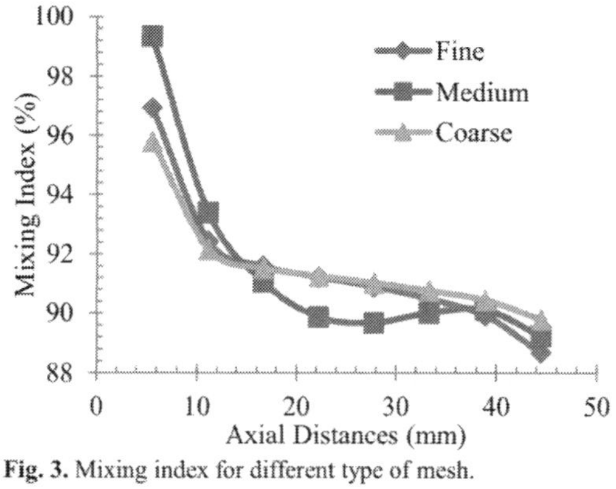


**Fig. 3.** Mixing index for different type of mesh.

* Corresponding author: nuramirahrashidi@gmail.com

### 2.3 Flow properties setup

Once the geometry designs for the configurations and the meshing are accomplished, several pre-setup inputs are required by ANSYS before stimulating the design. The physics are listed in the Table 2.

**Table 2.** Flow parameters setup.

| Parameter | Flow Properties |
|---|---|
| Gas Inlet Velocity | 0.05 $ms^{-1}$ |
| Volume Fraction | $H_2 : N_2 = 0.75 : 0.25$ |
| Temperature | 25°C |
| Pressure | 1 atm |
| Heat Transfer Model | Isothermal |
| Turbulence Model | K-Epsilon |
| Simulation Mode | Steady State |

### 2.4 Mixing index

The main indicator to measure the proximity of the flow ratio of both gasses is by using mixing index. With the intention of quantifying mixing efficiency, volume fraction of hydrogen data from CFD simulation is extracted from each line that have been constructed in the monolithic channel. 8 samples that are taken along the channel length as shown in Figure 4 are used to calculate the mixing index.

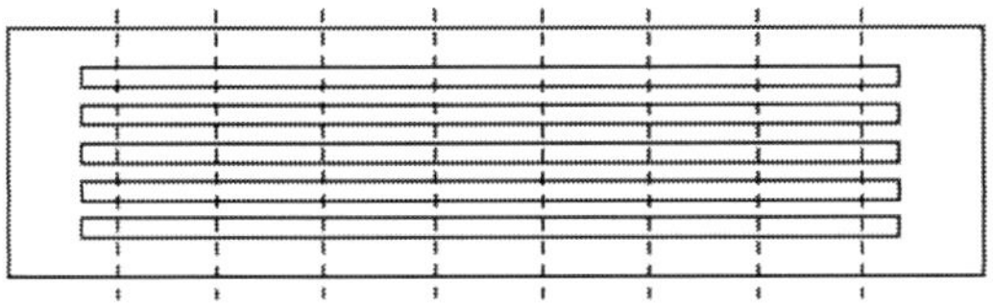

5.6 11.1 16.7 22.2 27.8 33.3 38.9 44.4

**Fig. 4.** Distance in mm of sample taken along monolithic channel for mixing index calculation

Based on the points taken, mixing index graph will be created using equation 1 and 2.

$$\gamma = \sqrt{\frac{1}{N}\sum(C_i - \overline{C_m})^2} \quad (1)$$

$$M = 1 - \sqrt{\frac{\gamma^2}{\gamma_{max}^2}} \quad (2)$$

Nomenclature

- $\gamma$ Variance
- $N$ Number of sampling point
- $C_i$ Mole fraction at point/node *i*
- $C_m$ Optimal mixing fraction
- $M$ Mixing index
- $\gamma^2$ Actual variance
- $\gamma_{max}^2$ Maximum variance from optimal mixing fraction

* Corresponding author: nuramirahrashidi@gmail.com

## 3 Results and discussion

The volume fraction and velocity of both nitrogen and hydrogen gases are extracted from the simulation in the form of contour plots in radial and axial manner. Each axial arrangement consists of four planes which are measured equally throughout the channel to observe the mixing of both gasses.

Mixing index for each geometry are calculated and presented in line chart to express which geometry create greater mixing to produce ammonia. The range of velocity profile is standardized from 0 $ms^{-1}$ until 0.09 $ms^{-1}$ while the range for volume fraction is from 0 to 1 which is illustrated at the legend on the right side of the tables. The results are taken from CFD Post.

Based on the results obtained from the simulation, effect of spacing between wires and effect of number of wires are investigated. Geometry A and B comprises the same geometry with different spacing of wires, 1.5 mm and 2.0 mm. Geometry B and C are studied with different number of wires, 19 wires and 13 wires respectively.

### 3.1 Effect of spacing of wires

Increasing the parallel spacing between the wires lead to a significant effect on mixing index.

Rendering to the XY-axis velocity contour of Figure 5, nitrogen particles in Geometry B are having more stable velocity while Geometry A shows inconsistent velocity at the center of the channel as the arrangement of the wires is closely packed at that position providing less space for mixing which leads to unfavorable outcome.

Velocity contour of Geometry B presented in Figure 6 demonstrates a fast moving of hydrogen gas at 0.063 $ms^{-1}$ to 0.072 $ms^{-1}$ throughout the channel. The axial obstruction of Geometry A generates higher velocity of hydrogen at the inner wall of channel which can cause it hard for the comparatively heavier nitrogen gas to have the similar pace of the hydrogen gas in that region.

However, as shown in Figure 7, Geometry B is not able to produce consistent volume fraction of nitrogen and hydrogen ratio, 3:1 to enhance the collision. Geometry A possess volume fraction of nitrogen that is lower than 0.2 in the centre area at the inlet of the channel.

Referring to Figure 8, the centreline in YZ-axis of Geometry B is seen to have steady velocity of nitrogen while the hydrogen possesses higher velocity than the set velocity. For Geometry A, velocity of hydrogen gas is lower at the inlet part than the outlet of the channel.

As the surface area for mixing of raw materials of Geometry B is higher, the mixing index for this design elevated better than Geometry A as displayed in Figure 13 [10].

Therefore, Geometry C is created using 2 mm as the spacing between wires same as Geometry B as the results produce good convergence among the designs.

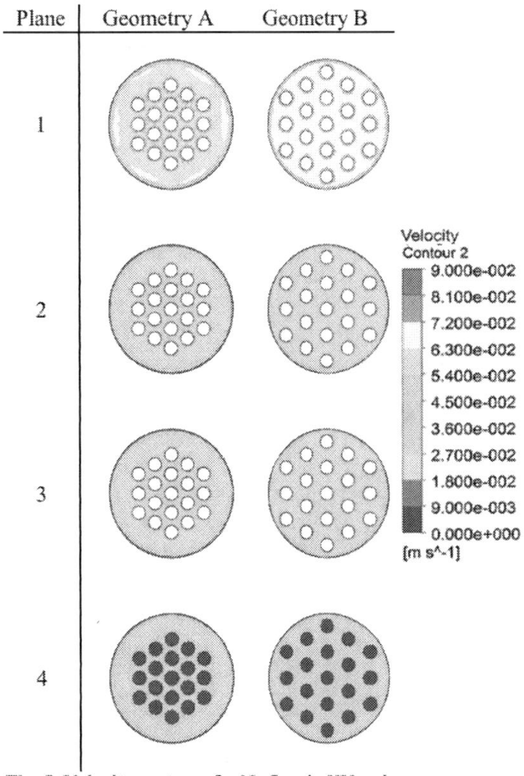


**Fig. 5.** Velocity contours for $N_2$ flow in XY-axis.

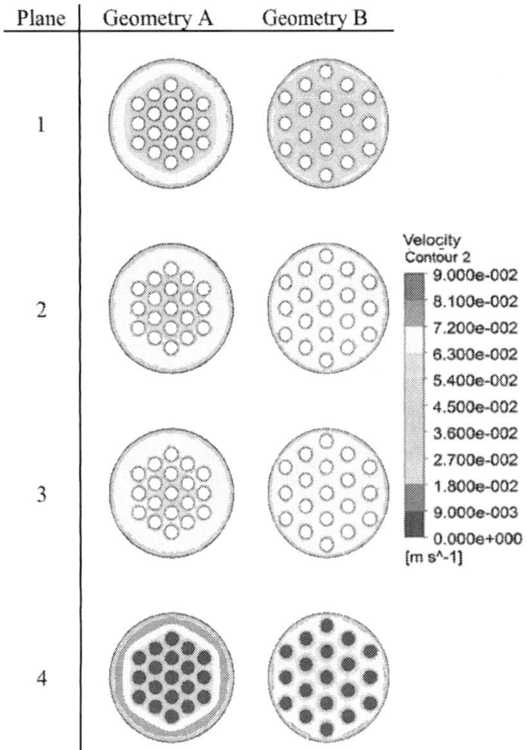


**Fig. 6.** Velocity contours for $H_2$ flow in XY-axis.

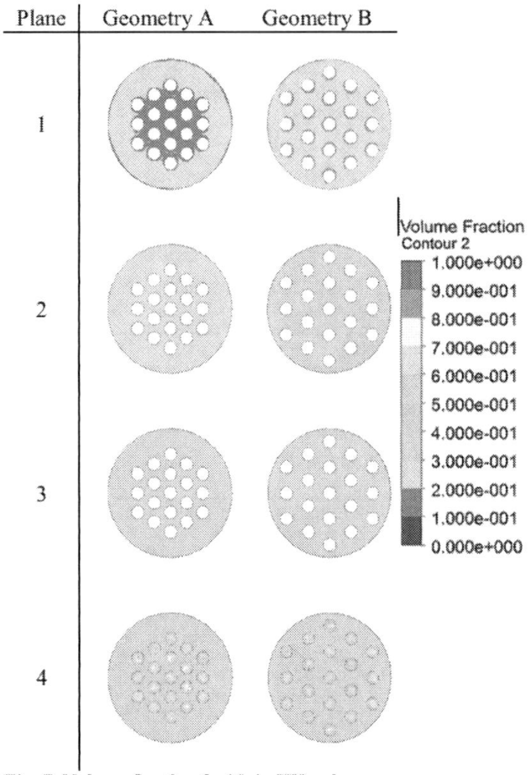


**Fig. 7.** Volume fraction for $N_2$ in XY-axis.

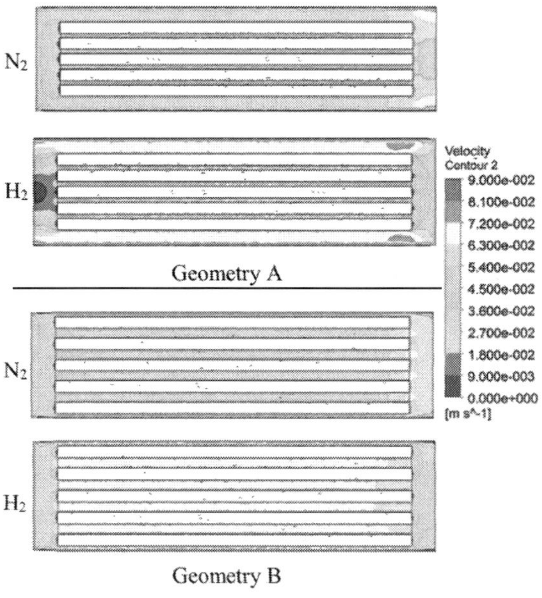


**Fig. 8.** Velocity contours at centreline in YZ-axis.

### 3.2 Effect of numbers of wires

From the results presented in Figure 9, Geometry C creates more uniform velocity except for plane 1 while Geometry B possess velocity of nitrogen flow approximately at 0.065 $ms^{-1}$ at the first plane which is

* Corresponding author: nuramirahrashidi@gmail.com

DOI: 10.1051/matecconf/201711101003

higher than the set velocity. An active dynamic mixing was expected when the tone of the contour for both gases achieve similarities. Based on the set velocity in Figure 10, hydrogen gas flows at faster speed for Geometry C where the velocity is between 0.054 $ms^{-1}$ and 0.063 $ms^{-1}$ from plane 1 to 4 while Geometry B retain velocity between 0.063 $ms^{-1}$ and 0.072 $ms^{-1}$. Since the velocity of nitrogen and hydrogen for Geometry C is nearly similar, it will lead to more homogenous mixing.

Plane 1 and 2 in Figure 11 for Geometry C showed suitable 0.25:0.75 volume fraction ratio while Geometry B achieve the appropriate volume fraction only in Plane 1. Therefore, lesser or higher volume fraction of the gases will result to lower the yield of ammonia when the actual reaction occurs in the channel.

Velocity of nitrogen and hydrogen can be visualized in Figure 12 at an altered angle where the contour is obtained at the point Z = 0 on YZ-axis, showing the radial centerline of the movement for both gases. Similar hue of contour for nitrogen and hydrogen contours along the channel indicating good mixing among the gases. The velocity contours at centreline for Geometry C illustrated better shifting of gasses when number of wires is lower.

As for the mixing index result, the Figure 13 illustrated that Geometry C achieved greater dynamic mixing than Geometry B. Hence, we can conclude that lower number of wires, 13 contributes to better mixing efficiency than 19 wires [15] at four different axial distance distributed evenly throughout the monolithic channel.

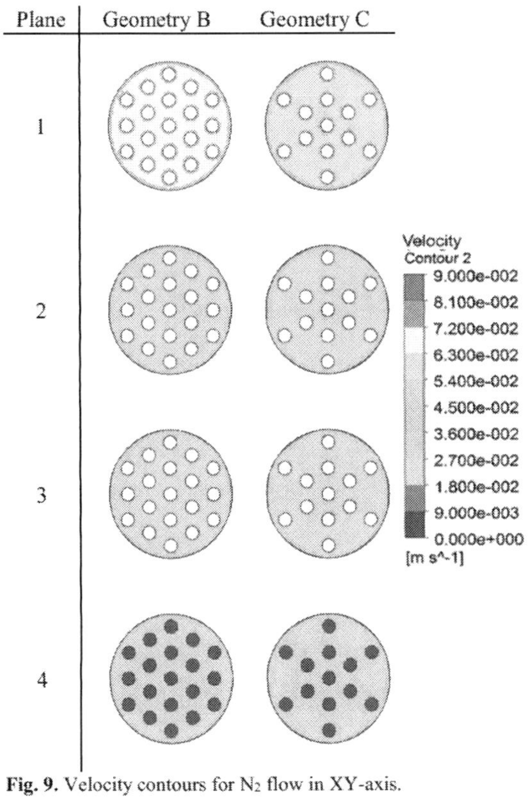


**Fig. 9.** Velocity contours for $N_2$ flow in XY-axis.

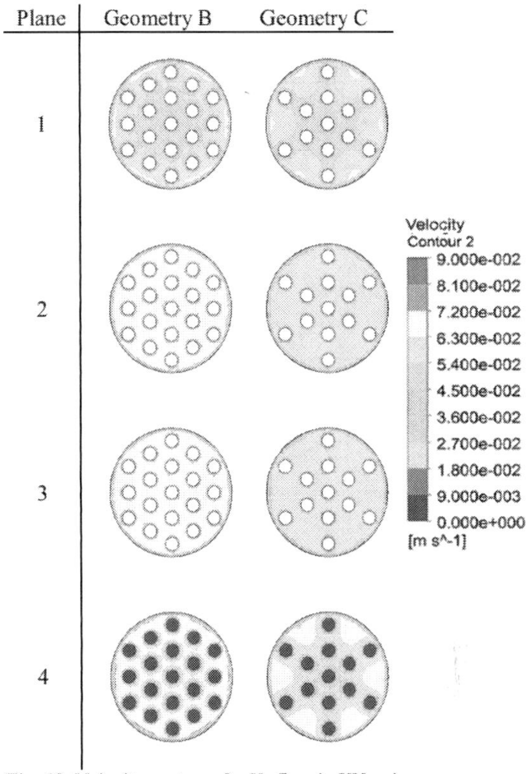


**Fig. 10.** Velocity contours for $H_2$ flow in XY-axis.

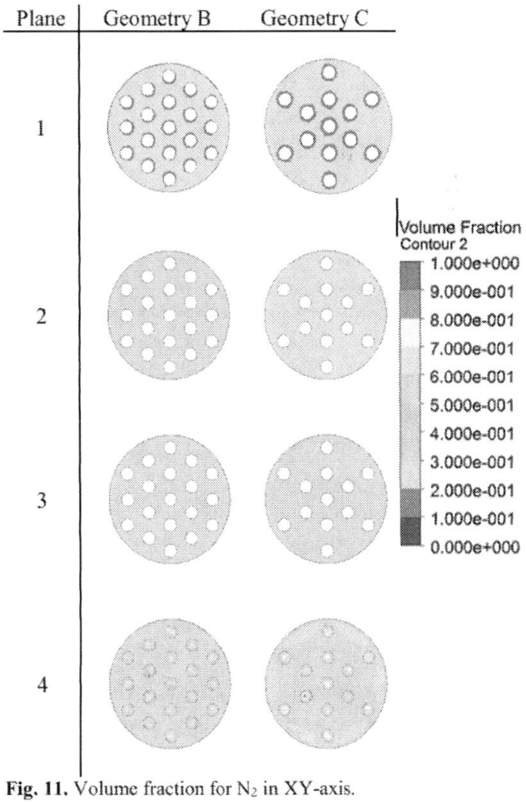


**Fig. 11.** Volume fraction for $N_2$ in XY-axis.

* Corresponding author: nuramirahrashidi@gmail.com

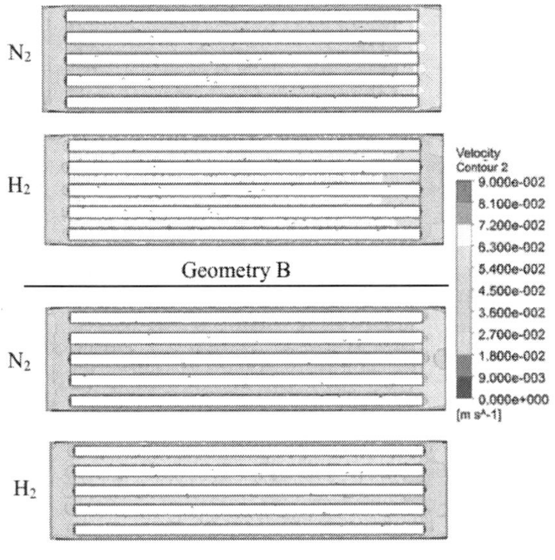


**Fig. 12.** Velocity contours at centreline in YZ-axis.

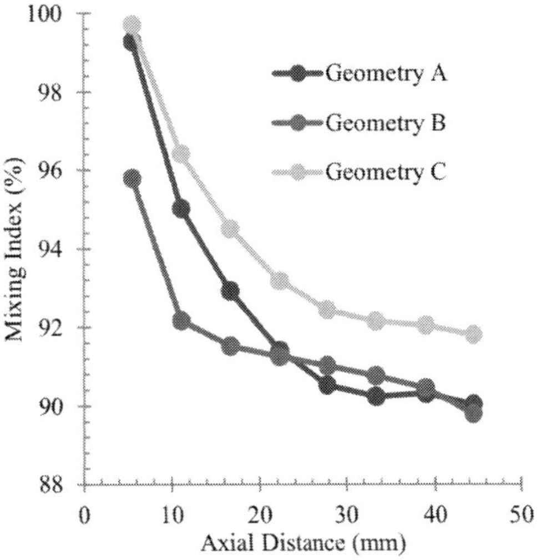


**Fig. 13.** Mixing index of geometries.

## 4 Conclusions

Enhancing the configuration of wires by increasing spacing and lowering number of wires causing a huge impact on inducing the dynamic mixing of gasses. Geometry C showed the best mixing result with uniform speed and volume fraction as well as generating high mixing index throughout the monolithic channel. By introducing proper structure of obstructions or wires to the monolithic channel, chaotic advection occurs and the flow direction of the gas stream altered laterally. It is proven that constructing obstruction with larger obstruction space boosted the mixing and rate of reaction between hydrogen and nitrogen gasses to synthesis ammonia. Therefore, the objectives of this project are accomplished successfully with the appropriate methods that have been used. In future, this research will consider the adding of wire supports to hold the wires element in place and inputting magnetic induction into the simulation section to provoke the mixing. Iron based nanocatalyst will be embedded on the wires element configured inside the monolithic channel. The mixing efficiency of the three geometries will be validated using Vibrating Sample Magnetometer (VSM) with various magnetic induction readings from range of 1 to 2.5 Tesla.

The authors gratefully acknowledge Universiti Teknologi PETRONAS for providing CFD computing facilities.

## References

[1] C. Egenhofer, L. Schrefler, *The Case of The Chemical Industry – Ammonia* (2014)
[2] FAO, *Food and Agriculture Organization of the United Nations Rome* (2011)
[3] P. Puspitasari, J. A. Razak, N. Yahya, *American Institute of Physics* (2012)
[4] N. Yahya, P. Puspitasari, K. Koziol, N. A. M. Zabidi, M. F. Othman, *International Journal of Basic and Applied Science*, **10** 60-64 (2010)
[5] K. Salman, A. Zamri, K. Kuzilati, *Procedia Engineering*, **148** 1266-1273 (2016)
[6] W. Wang, S. Zhao, T. Shao, M. Zhang, Y. Jin, Y. Cheng, *Chemical Engineering Science*, **84** 148-154 (2012)
[7] Y. Asano, S. Togashi, H. Tsudome, S. Murakami, *International Society for Pharmaceutical Engineering*, **30** (2010)
[8] N. Nguyen, Z. Wu, *Journal of Micromechanics and Microengineering*, **15** (2005)
[9] P. L. Mills, D. J. Quiram, J. F. Ryley, *Chemical Engineering Science*, **62** 6692-7010 (2007)
[10] R. B. Liu, M. A. Stremler, K. V. Sharp, M. G. Olsen, J. G. Santiago, R. J. Adrian, H. Aref, D. J. Beebe, *Journal of Microelectromechanical Systems*, **9** 190-197 (2000)
[11] R. Wang, J. Lin, H. Li, *Chaos, Solitons and Fractals*, **33** 1362-1366 (2007)
[12] Z. Zhang, C. Yim, M. Lin, X. Cao, *Biomicrofluidics*, **2** (2008)
[13] C. K. Chung, T. R. Shih, *Microfluid Nanofluid*, **4** 419-425 (2007)
[14] W. W.-F. Leung, Y. Ren, *International Journal of Heat and Mass Transfer*, **64** 457-467 (2013)
[15] H. Shou-Shing, H. Yi-Cheng, *Journal of Micromechanics and Microengineering*, **18** (2008)

* Corresponding author: nuramirahrashidi@gmail.com

# On Newton-Raphson formulation and algorithm for displacement based structural dynamics problem with quadratic damping nonlinearity

*Kim Jie* Koh[1,*], and *Airil Yasreen Mohd* Yassin[1]

[1]Department of Civil and Environmental Engineering, Universiti Teknologi PETRONAS, 32610 Seri Iskandar, Perak, Malaysia

**Abstract.** Quadratic damping nonlinearity is challenging for displacement based structural dynamics problem as the problem is nonlinear in time derivative of the primitive variable. For such nonlinearity, the formulation of tangent stiffness matrix is not lucid in the literature. Consequently, ambiguity related to kinematics update arises when implementing the time integration-iterative algorithm. In present work, an Euler-Bernoulli beam vibration problem with quadratic damping nonlinearity is addressed as the main source of quadratic damping nonlinearity arises from drag force estimation, which is generally valid only for slender structures. Employing Newton-Raphson formulation, tangent stiffness components associated with quadratic damping nonlinearity requires velocity input for evaluation purpose. For this reason, two mathematically equivalent algorithm structures with different kinematics arrangement are tested. Both algorithm structures result in the same accuracy and convergence characteristic of solution.

## 1 Introduction

In finite element analysis of time dependent problems, time integration is of primary importance. Other than satisfying accuracy and numerical stability, the selected time integration scheme has to be practical in term of computational effort. In general, there are implicit and explicit time integration schemes. From a computational perspective, the selection of a time integration scheme depends on the type of problem and analysis interest. When the desired overall time span is extensive, it is natural to adopt a time step that is as large as possible to save computing time but small enough to ensure an accurate solution. Such situation is the case for low speed dynamics such as structural dynamics where the structural response can be characterized by a few low frequency modes with sufficient accuracy [1]. Therefore, implicit time integration is normally preferred over explicit time integration in structural dynamics since a relatively larger time step can be used.

When applying implicit time integration onto nonlinear problem, an iterative procedure is required within each time step [1]. Since convergence is a must for each time step, proper implementation of the nonlinear algorithm is crucial. Misunderstanding of the nonlinear formulation can lead to a theoretically incompatible algorithm. Under certain circumstances, theoretically incompatible algorithm may ironically yield a converged solution. However, in most cases such algorithm will yield an inaccurate solution or give rise to numerical instability. Discussion on the implementation of nonlinear algorithm is vast in the literature of finite element analysis [1, 2, 3]. The numerical behavior of various nonlinear solvers is not within present work's scope. Rather, the full Newton-Raphson method which is widely practised in finite element analysis, is focused. On the basis of a full Newton-Raphson method, a specific source of nonlinearity arised from quadratic damping, also known as quadratic velocity or hydrodynamic damping, is deemed challenging for a few reasons as explained in the forthcoming paragraph.

For quadratic damping nonlinearity, the nonlinear source arises from the presence of quadratic structural velocity. In other words, the problem is nonlinear in term of the time derivative of structural displacement. In a standard displacement based finite element formulation, the primitive variable is structural displacement. Thus, such nonlinearity appears to be more complicated than the standard case which is nonlinear in term of structural displacement. A representative example of nonlinear quadratic damping exists in the estimation of fluid drag force. In particular, drag component of the Morison equation [4, 5] widely used in offshore applications contains the quadratic structural velocity. Many researchers in the past had presented the solution for structural analysis involving Morison equation, to name a few [6, 7, 8, 9]. Owing to the diverse focuses and objectives, details on the contribution of nonlinear quadratic damping in the Morison equation towards the formulation of tangent stiffness matrix are often omitted in these works. In ref. [10], the iterative procedure was neglected within each time step when dealing with quadratic damping nonlinearity. To address the lack of lucidity in the Newton-Raphson formulation, the main

* Corresponding author: kimjie24@gmail.com

objective in present work is to show how the quadratic damping nonlinearity contributes to the tangent stiffness matrix. As quadratic damping nonlinearity involves time derivative of displacement, the time integration-iterative algorithm appears to be challenging. In particular, a question arises such that should acceleration and/ or velocity be iterated simultaneously with displacement? To the best knowledge of the authors, such confusion involving quadratic damping nonlinearity has not been addressed before. Therefore, in present work the stated question about nonlinear algorithm is addressed.

The finite element discretization involved in present work is based on a semi-discrete formulation. For structural dynamics, following spatial discretization the semi-discrete form is of a set of hyperbolic ordinary differential equations [11]. A variety of implicit time integration schemes are available for temporal discretization and their implementations had been investigated rigorously in the past, to name a few [12, 13, 14, 15]. Similarly, it is not within the scope of the present work to compare and contrast the various time integration schemes. Rather, the trapezoidal rule or Newmark's constant average acceleration method [1] which is unconditionally stable in linear analysis is adopted. However, the nonlinear formulation can be analogously applied for other implicit time integration schemes.

As mentioned in the third paragraph of section 1, the lack of lucidity in the Newton-Raphson formulation may be due to the diverse focuses and objectives of works in the past. Thus, the authors learn that the best approach to achieve clarity is to present a specific structural dynamics problem where quadratic damping nonlinearity is involved. A suitable choice is the Euler-Bernoulli beam vibration problem because in general, slender structures are prone to bending thus can be modelled as Euler-Bernoulli beam. Meanwhile, drag force estimation which exhibits quadratic damping nonlinearity is normally valid only for slender structures [5]. Therefore, in present work finite element formulation for the Euler-Bernoulli beam vibration problem with quadratic damping nonlinearity is illustrated.

The present work is organized as such. In section 2, a specific Euler-Bernoulli beam vibration problem involving quadratic damping is described. In section 3-4, spatial discretization using Galerkin weighted residual method is performed followed by temporal discretization using trapezoidal rule. In section 5, the Newton-Raphson formulation is given with emphasis on the contribution of quadratic damping nonlinearity towards formulation of the tangent stiffness matrix. Following that, in section 6, discussion of the nonlinear algorithm is given such that the stated question about iterating acceleration and/ or velocity is addressed. Lastly, in section 7, the finite element solution is verified with an established semi-analytical solution, while the convergence characteristic of the nonlinear algorithm is discussed.

## 2 Euler-Bernoulli beam vibration with quadratic damping nonlinearity

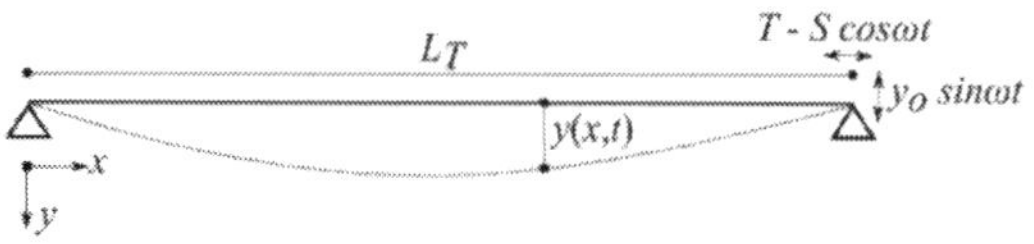


**Fig. 1.** 1D Euler-Bernoulli beam subjected to harmonic support excitation and harmonic tension fluctuation.

The presented example is taken from Patel and Park [16]. In their work, a marine tether subjected to a harmonic support excitation (forced excitation) and a harmonic tension fluctuation (parametric excitation) is idealized as a 1D Euler-Bernoulli beam with constant axial effect as shown in Fig. 1. The quadratic damping nonlinearity arises from the hydrodynamic damping. Patel and Park [16] had solved the problem using a semi-analytical approach. Herein, the same problem is solved using finite element method. The chosen problem is deemed representative because the only source of nonlinearity comes from the hydrodynamic damping. The partial differential equation is given as

$$m\frac{\partial^2 y}{\partial t^2} + EI\frac{\partial^4 y}{\partial x^4} - (T - Scos\omega t)\frac{\partial^2 y}{\partial x^2} + \frac{1}{2}C_D\rho_w D_o\left|\frac{\partial y}{\partial t}\right|\frac{\partial y}{\partial t} = 0; \quad 0 \le x \le L_T; t \ge 0 \qquad (1)$$

where $y$ is the displacement variable, a function of 1D space $x$ and time $t$, $m$ is the mass per unit length of beam, $EI$ is the flexural rigidity of beam, $T$ is the constant axial tension, $S$ and $\omega$ are the amplitude and angular frequency of tension fluctuation respectively, $C_D$ is the drag coefficient, $\rho_w$ is the external fluid density, $D_o$ is the outer diameter of an annular cross-section and $L_T$ is the total beam length.

Similar to the original work [16], only a pinned-pinned boundary condition is studied. Thus, the Dirichlet boundary conditions and Neumann boundary conditions are given respectively as

$$y(0,t) = 0; y(L_T,t) = y_o sin\omega t \qquad (2)$$

$$EI\frac{\partial^2 y}{\partial x^2}(0,t) = 0; EI\frac{\partial^2 y}{\partial x^2}(L_T,t) = 0 \qquad (3)$$

where $y_o$ and $\omega$ are the amplitude and angular frequency of support excitation respectively. Due to the nonhomogeneous Dirichlet boundary conditions, the initial displacement and initial velocity of the beam are not necessarily zero. For example, in this particular problem the initial displacement is zero but the initial velocity is not. For the finite element solution, the initial displacement and initial velocity are calculated based on the semi-analytical solution [16] at $t = 0$.

## 3 Spatial discretization using Galerkin weighted residual method

The beam's domain is discretized into multiple 2-noded Hermitian beam elements. With respect to the elemental

 DOI: 10.1051/matecconf/201711101004

domain, eq. (1) is weighted using the Bubnov-Galerkin method. Following integration by parts and substitution of cubic Hermitian interpolation function for the displacement variable, the elemental semi-discrete form is given in index notation as

$$\begin{aligned}&\left(m\int_L N_iN_j\,dx\right)\frac{\partial^2 u_j}{\partial t^2}\\&+\left(EI\int_L \frac{d^2N_i}{dx^2}\frac{d^2N_j}{dx^2}dx\right.\\&\left.+(T-Scos\omega t)\int_L \frac{dN_i}{dx}\frac{dN_j}{dx}dx\right)u_j\\&=-\left(\frac{1}{2}C_D\rho_wD_o\int_L N_i\left|N_m\frac{\partial u_m}{\partial t}\right|N_k\frac{\partial u_k}{\partial t}dx\right)\\&-\left\{N_i\left[EI\frac{\partial^3 y}{\partial x^3}-(T-Scos\omega t)\frac{\partial y}{\partial x}\right]\right\}_0^L\\&-\left\{\left(\frac{-dN_i}{dx}\right)EI\frac{\partial^2 y}{\partial x^2}\right\}_0^L\end{aligned} \tag{4}$$

where $L$ is the element length, $N$ is the shape function for a Hermitian beam element and $u$ is the nodal degree of freedom. The Neumann boundary conditions at the elemental level can be identified more clearly by rewriting eq. (4) as eq. (5).

$$\begin{aligned}&\left(m\int_L N_iN_j\,dx\right)\frac{\partial^2 u_j}{\partial t^2}\\&+\left(EI\int_L \frac{d^2N_i}{dx^2}\frac{d^2N_j}{dx^2}dx\right.\\&\left.+(T-Scos\omega t)\int_L \frac{dN_i}{dx}\frac{dN_j}{dx}dx\right)u_j\\&=-\left(\frac{1}{2}C_D\rho_wD_o\int_L N_i\left|N_m\frac{\partial u_m}{\partial t}\right|N_k\frac{\partial u_k}{\partial t}dx\right)\\&+N_i|_{x=0}b_1+N_i|_{x=L}b_3-\left.\frac{dN_i}{dx}\right|_{x=0}b_2-\left.\frac{dN_i}{dx}\right|_{x=L}b_4\end{aligned} \tag{5}$$

where

$$\begin{aligned}b_1&=\left.\left(EI\frac{\partial^3 y}{\partial x^3}-(T-Scos\omega t)\frac{\partial y}{\partial x}\right)\right|_{x=0};\\b_2&=\left.EI\frac{\partial^2 y}{\partial x^2}\right|_{x=0};\\b_3&=\left.\left(-EI\frac{\partial^3 y}{\partial x^3}+(T-Scos\omega t)\frac{\partial y}{\partial x}\right)\right|_{x=L};\\b_4&=-\left.EI\frac{\partial^2 y}{\partial x^2}\right|_{x=L}\end{aligned} \tag{6}$$

Equivalently, eq. (5) can be written in matrix form as

$$[M]\{\ddot{u}\}+[K]\{u\}=-\{F(\dot{u})\}+\{h\} \tag{7}$$

where $[M]$ is the mass matrix, $[K]$ is the stiffness matrix, $\{F(\dot{u})\}$ is the hydrodynamic damping force vector which is nonlinear in velocity $\dot{u}$, $\{h\}$ is the Neumann boundary load vector, $\{u\}$ is the nodal displacement vector and $\{\ddot{u}\}$ is the nodal acceleration vector. Components for the matrices and vectors are given as

$$\begin{aligned}M_{ij}&=m\int_L N_iN_j\,dx\\K_{ij}&=EI\int_L\frac{d^2N_i}{dx^2}\frac{d^2N_j}{dx^2}dx\\&\quad+(T-Scos\omega t)\int_L\frac{dN_i}{dx}\frac{dN_j}{dx}dx\\F_i(\dot{u})&=\frac{1}{2}C_D\rho_wD_o\int_L N_i\left|N_m\frac{\partial u_m}{\partial t}\right|N_k\frac{\partial u_k}{\partial t}dx\\h_i&=N_i|_{x=0}b_1+N_i|_{x=L}b_3-\left.\frac{dN_i}{dx}\right|_{x=0}b_2\\&\quad-\left.\frac{dN_i}{dx}\right|_{x=L}b_4\end{aligned} \tag{8}$$

In this particular problem, the Neumann boundary load vector $\{h\}$ in eq. (7) vanishes as the forcing effect is due to the nonhomogeneous Dirichlet boundary condition as stated in eq. (2) and the harmonic tension fluctuation. Hence $\{h\}$ is omitted hereinafter. Note that although the stiffness matrix $[K]$ is time-dependent, the same temporal discretization and Newton-Raphson formulation hold as in the case of a zero tension fluctuation i.e. when $S = 0$.

## 4 Temporal discretization using trapezoidal rule

To proceed, the equation of motion in eq. (7) is expressed at time $t + \Delta t$ which is to be solved as

$$[\mathrm{M}]_{t+\Delta t}\{\ddot{u}\}_{t+\Delta t}+[\mathrm{K}]_{t+\Delta t}\{u\}_{t+\Delta t}=-\{F(\dot{u})\}_{t+\Delta t} \tag{9}$$

Eq. (9) signifies that equilibrium condition should hold at any time and is mathematically equivalent to the notion that the external load vector is equal to the internal load vector at any time [1]. For temporal discretization using trapezoidal rule, the acceleration and velocity at time $t + \Delta t$ are given implicitly with respect to time $t + \Delta t$ and previous time $t$ as

$$\{\ddot{u}\}_{t+\Delta t}=\frac{4}{(\Delta t)^2}(\{u\}_{t+\Delta t}-\{u\}_t)-\frac{4}{\Delta t}\{\dot{u}\}_t-\{\ddot{u}\}_t \tag{10}$$

$$\{\dot{u}\}_{t+\Delta t}=\frac{2}{\Delta t}(\{u\}_{t+\Delta t}-\{u\}_t)-\{\dot{u}\}_t \tag{11}$$

where $\Delta t$ is the time step used. Substituting eq. (10) and eq. (11) to eq. (9) and rearranging, one obtains

$$\begin{aligned}&\left(\frac{4}{(\Delta t)^2}[\mathrm{M}]_{t+\Delta t}+[\mathrm{K}]_{t+\Delta t}\right)\{u\}_{t+\Delta t}\\&=-\{F(\dot{u})\}_{t+\Delta t}\\&+[\mathrm{M}]_{t+\Delta t}\left(\frac{4}{(\Delta t)^2}\{u\}_t+\frac{4}{\Delta t}\{\dot{u}\}_t\right.\\&\left.+\{\ddot{u}\}_t\right)\end{aligned} \tag{12}$$

Apparently, eq. (12) is nonlinear in velocity hence an iterative procedure is required within each time step.

## 5 Newton-Raphson formulation

All terms in eq. (9) or equivalently eq. (12) can be collected on one side such that the residual function desired at present iteration $r$ is defined as

$$\{R(\dot{u})\}_{t+\Delta t}^{r} = 0 \tag{13}$$

In Newton-Raphson formulation, the residual function at present iteration $r$ is expanded using Taylor series with respect to previous iteration $r-1$. Terms involving second order function and above are truncated.

$$\{R(\dot{u})\}_{t+\Delta t}^{r-1} + \frac{\partial\{R(\dot{u})\}_{t+\Delta t}^{r-1}}{\partial\{u\}_{t+\Delta t}^{r-1}}\left(\{u\}_{t+\Delta t}^{r} - \{u\}_{t+\Delta t}^{r-1}\right) \approx 0 \tag{14}$$

Thus, the tangent stiffness equations are obtained as

$$[T]_{t+\Delta t}^{r-1}\{\Delta u\}_{t+\Delta t} = -\{R(\dot{u})\}_{t+\Delta t}^{r-1} \tag{15}$$

where $[T]_{t+\Delta t}^{r-1}$ is the tangent stiffness matrix and $\{\Delta u\}_{t+\Delta t} = \{u\}_{t+\Delta t}^{r} - \{u\}_{t+\Delta t}^{r-1}$. Both tangent stiffness matrix and residual function are expressed at the known iteration $r-1$ hence eq. (15) can be readily solved in an iterative manner.

In particular, components of the tangent stiffness matrix can be evaluated as

$$\begin{aligned} T_{ij} &= \left.\frac{\partial R_i(\dot{u}_k)}{\partial u_j}\right|_{t+\Delta t}^{r-1} \\ &= \left.\frac{\partial(M_{im}\ddot{u}_m + K_{im}u_m + F_i(\dot{u}_k))}{\partial u_j}\right|_{t+\Delta t}^{r-1} \end{aligned} \tag{16}$$

Utilizing relations given in eq. (10) and eq. (11), eq. (16) becomes

$$\begin{aligned} T_{ij} = \left.\frac{\partial R_i(\dot{u}_k)}{\partial u_j}\right|_{t+\Delta t}^{r-1} = \Bigg[&\frac{4}{(\Delta t)^2}M_{ij} + K_{ij} \\ &+ \frac{\partial F_i(\dot{u}_k)}{\partial u_j}\Bigg]\Bigg|_{t+\Delta t}^{r-1} \end{aligned} \tag{17}$$

The third term on the right hand side of eq. (17) is elaborated as

$$Let\ B_{ij}(\dot{u}_k) = \left.\frac{\partial F_i(\dot{u}_k)}{\partial u_j}\right|_{t+\Delta t}^{r-1};$$

$$\begin{aligned} &B_{ij}(\dot{u}_k) \\ &= \left.\frac{\partial\left(\frac{1}{2}C_D\rho_w D_o \int_L N_i|N_m\dot{u}_m|N_k\dot{u}_k dx\right)}{\partial u_j}\right|_{t+\Delta t}^{r-1} \end{aligned} \tag{18}$$

By differentiating using product rule in eq. (18), it can be shown that

$$\begin{aligned} \frac{\partial|N_m\dot{u}_m|N_k\dot{u}_k}{\partial u_j} = \left(\frac{2}{\Delta t}\right)&N_j[(|N_m\dot{u}_m|) \\ &+ (N_k\dot{u}_k)(signum(N_m\dot{u}_m))] \end{aligned} \tag{19}$$

where

$$signum(N_m\dot{u}_m) = \begin{cases} -1, & N_m\dot{u}_m < 0 \\ 0, & N_m\dot{u}_m = 0 \\ 1, & N_m\dot{u}_m > 0 \end{cases} \tag{20}$$

Thus, eq. (18) can be rewritten as

$$\begin{aligned} B_{ij}(\dot{u}_k) = \left(\frac{2}{\Delta t}\right)\Bigg(\frac{1}{2}C_D\rho_w D_o \int_L\ &N_iN_j[(|N_m\dot{u}_m|) + \\ &(N_k\dot{u}_k)(signum(N_k\dot{u}_k))]dx\Big|_{t+\Delta t}^{r-1}\Bigg) \end{aligned} \tag{21}$$

Accordingly, components of the tangent stiffness matrix can be finalized as

$$T_{ij} = \left[\frac{4}{(\Delta t)^2}M_{ij} + K_{ij} + B_{ij}(\dot{u}_k)\right]\Bigg|_{t+\Delta t}^{r-1} \tag{22}$$

The spatial discretization, temporal discretization and Newton-Raphson formulation shown above are operated on the elemental level. However, this is immaterial for the global level operation as global matrices and vectors can be constructed using direct stiffness method.

## 6 Implementation of nonlinear algorithm

In solving nonlinear time dependent problems using implicit time integration, there are essentially two loops in the algorithm structure. The outer loop is for time integration whereas the inner loop is for Newton-Raphson iteration. Since input of velocity is required for evaluation of the nonlinear hydrodynamic damping force vector, an important question herein is that should the acceleration and/ or velocity be updated in the outer loop or the inner loop? The confusion may be further demonstrated through literature for the standard case where the problem is nonlinear in displacement. For instance, update in the outer loop is seen in *section 16.3.3* of ref. [17] whereas update in the inner loop is seen in *section 6.3.7* of ref. [18] and ref. [19].

In present work, both algorithm structures had been tested for quadratic damping nonlinearity. To clarify, this section shows that both algorithm structures are mathematically equivalent thus either algorithm structure can be implemented. However, when dealing with quadratic damping nonlinearity the velocity should be iterated in the inner loop for either algorithm structure to ensure convergence. Otherwise, the nonlinear hydrodynamic damping vector may be inappropriately evaluated thus producing divergence of solution.

In eq. (15), both the tangent stiffness matrix as given in eq. (9) or eq. (12) and residual function as given in eq. (22) are expressed at the known iteration $r-1$. If the next iteration is required, the matrices and vectors are naturally expressed at the succeeding iteration. For the residual function, the general picture is clearer by rewriting eq. (9) and eq. (12) respectively with the iteration superscript included as

$$[\mathrm{M}]_{t+\Delta t}\{\ddot{u}\}_{t+\Delta t}^{r-1} + [\mathrm{K}]_{t+\Delta t}\{u\}_{t+\Delta t}^{r-1} = -\{F(\dot{u})\}_{t+\Delta t}^{r-1} \tag{23}$$

 DOI: 10.1051/matecconf/201711101004

$$\left(\frac{4}{(\Delta t)^2}[\mathrm{M}]_{t+\Delta t} + [\mathrm{K}]_{t+\Delta t}\right)\{u\}_{t+\Delta t}^{r-1} = -\{F(\dot{u})\}_{t+\Delta t}^{r-1} + [\mathrm{M}]_{t+\Delta t}\left(\frac{4}{(\Delta t)^2}\{u\}_t + \frac{4}{\Delta t}\{\dot{u}\}_t + \{\ddot{u}\}_t\right) \quad (24)$$

Accordingly, for the standard case where the problem is nonlinear in displacement, the following statements shall apply. If the acceleration and velocity were updated in the outer loop, this indicates that the perspective is set such that eq. (22) and eq. (24) are meant. On the other hand, the perspective for the case where acceleration and velocity are updated in the inner loop is set such that eq. (22) and eq. (23) are meant. The latter can be implemented as following.

By substituting $\{u\}_{t+\Delta t}^{r} = \{u\}_{t+\Delta t}^{r-1} + \{\Delta u\}_{t+\Delta t}$ into eq. (10) and eq. (11) at the desired iteration $r$ [1], the acceleration and velocity to be iterated in the inner loop are given respectively as

$$\{\ddot{u}\}_{t+\Delta t}^{r} = \frac{4}{(\Delta t)^2}\left(\{u\}_{t+\Delta t}^{r-1} - \{u\}_t + \{\Delta u\}_{t+\Delta t}\right) - \frac{4}{\Delta t}\{\dot{u}\}_t - \{\ddot{u}\}_t$$

or $\quad (25)$

$$\{\ddot{u}\}_{t+\Delta t}^{r} = \{\ddot{u}\}_{t+\Delta t}^{r-1} + \frac{4}{(\Delta t)^2}\{\Delta u\}_{t+\Delta t}$$

$$\{\dot{u}\}_{t+\Delta t}^{r} = \frac{2}{\Delta t}\left(\{u\}_{t+\Delta t}^{r-1} - \{u\}_t + \{\Delta u\}_{t+\Delta t}\right) - \{\dot{u}\}_t$$

or $\quad (26)$

$$\{\dot{u}\}_{t+\Delta t}^{r} = \{\dot{u}\}_{t+\Delta t}^{r-1} + \frac{2}{\Delta t}\{\Delta u\}_{t+\Delta t}$$

As shown in eq. (25) and eq. (26), once $\{\Delta u\}_{t+\Delta t}$ is solved, acceleration and velocity can be updated in the inner loop. Note that in this particular problem however, the velocity should be iterated in the inner loop regardless of the algorithm structure used. For this particular problem, both algorithm structures are given in Table 1 and Table 2.

**Table 1.** Algorithm structure with update in the outer loop

| | |
|---|---|
| - | Based on initial displacement, $\{u\}_{\bar{t}=0}$ and initial velocity, $\{\dot{u}\}_{\bar{t}=0}$ , calculate initial acceleration, $\{\ddot{u}\}_{\bar{t}=0}$ from equation of motion i.e. eq. (9). |
| **{{OUT}}** | *Outer loop*: time integration, start with $t = 0$ |
| **OUT-1** | Initialization for kinematics at time $t + \Delta t$<br>$\{u\}_{\bar{t}=t+\Delta t} = \{u\}_{\bar{t}=t}$ ; $\{\dot{u}\}_{\bar{t}=t+\Delta t} = \{\dot{u}\}_{\bar{t}=t}$ ;<br>$\{\ddot{u}\}_{\bar{t}=t+\Delta t} = \{\ddot{u}\}_{\bar{t}=t}$ |
| **OUT-2** | Impose Dirichlet boundary conditions on $\{u\}_{\bar{t}=t+\Delta t}$ |
| **OUT-3** | Initialization for displacement at time $t + \Delta t$ to be iterated in the inner loop<br>$\{u\}_{\bar{t}=t+\Delta t}^{\bar{r}=1} = \{u\}_{\bar{t}=t+\Delta t}$ |
| **{IN}** | *Inner loop*:<br>Newton-Raphson iteration, start with $r = 1$ |
| **IN-1** | Calculate velocity at time $t + \Delta t$ with iteration $r$<br>$\{\dot{u}\}_{\bar{t}=t+\Delta t}^{\bar{r}=r} = \frac{2}{\Delta t}\left(\{u\}_{\bar{t}=t+\Delta t}^{\bar{r}=r} - \{u\}_{\bar{t}=t}\right) - \{\dot{u}\}_{\bar{t}=t}$ |
| **IN-2** | Impose Dirichlet boundary conditions on $\{\dot{u}\}_{\bar{t}=t+\Delta t}^{\bar{r}=r}$ |
| **IN-3** | Evaluate tangent stiffness matrix and residual function using eq. (22) and eq. (24) respectively and assemble into global matrices |
| **IN-4** | Impose stationary condition on $\{\Delta u\}_{t+\Delta t}$* for all known DOFs associated with Dirichlet boundary conditions |
| **IN-5** | Solve the tangent stiffness matrix equations i.e. eq. (15) for $\{\Delta u\}_{t+\Delta t}$ |
| **IN-6** | Check if tolerance is acceptable while iterating:<br>• If pass: go to **OUT-4**<br>• If fail: set current iteration as $r - 1$ and perform $\{u\}_{\bar{t}=t+\Delta t}^{\bar{r}=r} = \{u\}_{\bar{t}=t+\Delta t}^{\bar{r}=r-1} + \{\Delta u\}_{t+\Delta t}$ , go to **IN-1** |
| **OUT-4** | Calculate acceleration at time $t + \Delta t$<br>$\{\ddot{u}\}_{t+\Delta t} = \frac{4}{(\Delta t)^2}(\{u\}_{t+\Delta t} - \{u\}_t) - \frac{4}{\Delta t}\{\dot{u}\}_t - \{\ddot{u}\}_t$ |
| **OUT-5** | Impose Dirichlet boundary conditions on $\{\ddot{u}\}_{t+\Delta t}$ |
| **OUT-6** | Set current time as $t$ and go to **OUT-1** |
| * For any iteration in the same time instant $t + \Delta t$, the known DOFs associated with the nonhomogeneous Dirichlet boundary conditions do not change. | |

**Table 2.** Algorithm structure with update in the inner loop

| | |
|---|---|
| - | Based on initial displacement, $\{u\}_{\bar{t}=0}$ and initial velocity, $\{\dot{u}\}_{\bar{t}=0}$ , calculate initial acceleration, $\{\ddot{u}\}_{\bar{t}=0}$ from equation of motion i.e. eq. (9). |
| **{{OUT}}** | *Outer loop*: time integration, start with $t = 0$ |
| **OUT-1** | Initialization for kinematics at time $t + \Delta t$<br>$\{u\}_{\bar{t}=t+\Delta t} = \{u\}_{\bar{t}=t}$ ;<br>$\{\dot{u}\}_{\bar{t}=t+\Delta t} = \{\dot{u}\}_{\bar{t}=t}$ ;<br>$\{\ddot{u}\}_{\bar{t}=t+\Delta t} = \{\ddot{u}\}_{\bar{t}=t}$ |
| **OUT-2** | Impose Dirichlet boundary conditions on $\{u\}_{\bar{t}=t+\Delta t}$ |
| **OUT-3** | Initialization for kinematics at time $t + \Delta t$ to be iterated in the inner loop<br>$\{u\}_{\bar{t}=t+\Delta t}^{\bar{r}=1} = \{u\}_{\bar{t}=t+\Delta t}$<br>$\{\dot{u}\}_{\bar{t}=t+\Delta t}^{\bar{r}=1} = \frac{2}{\Delta t}\left(\{u\}_{\bar{t}=t+\Delta t}^{\bar{r}=1} - \{u\}_{\bar{t}=t}\right) - \{\dot{u}\}_{\bar{t}=t}$<br>$\{\ddot{u}\}_{\bar{t}=t+\Delta t}^{\bar{r}=1} = \frac{4}{(\Delta t)^2}\left(\{u\}_{\bar{t}=t+\Delta t}^{\bar{r}=1} - \{u\}_{\bar{t}=t}\right) - \frac{4}{\Delta t}\{\dot{u}\}_{\bar{t}=t} - \{\ddot{u}\}_{\bar{t}=t}$ |
| **OUT-4** | Impose Dirichlet boundary conditions on $\{\ddot{u}\}_{\bar{t}=t+\Delta t}^{\bar{r}=1}$ and $\{\dot{u}\}_{\bar{t}=t+\Delta t}^{\bar{r}=1}$ |
| **{IN}** | *Inner loop*:<br>Newton-Raphson iteration, start with $r = 1$ |
| **IN-1** | Evaluate tangent stiffness matrix and residual function using eq. (22) and eq. (23) respectively and assemble into global matrices |

 DOI: 10.1051/matecconf/201711101004

| | |
|---|---|
| **IN-2** | Impose stationary condition on $\{\Delta u\}_{t+\Delta t}$* for all known DOFs associated with Dirichlet boundary conditions |
| **IN-3** | Solve the tangent stiffness matrix equations i.e. eq. (15) for $\{\Delta u\}_{t+\Delta t}$ |
| **IN-4** | Check if tolerance is acceptable while iterating: <br>• If pass: go to **OUT-3** <br>• If fail: set current iteration as $r-1$ and perform <br>$\{u\}_{\bar{t}=t+\Delta t}^{\bar{r}=r} = \{u\}_{\bar{t}=t+\Delta t}^{\bar{r}=r-1} + \{\Delta u\}_{t+\Delta t}$ ; <br>$\{\dot{u}\}_{\bar{t}=t+\Delta t}^{\bar{r}=r} = \{\dot{u}\}_{\bar{t}=t+\Delta t}^{\bar{r}=r-1} + \frac{2}{\Delta t}\{\Delta u\}_{t+\Delta t}$ ; <br>$\{\ddot{u}\}_{\bar{t}=t+\Delta t}^{\bar{r}=r} = \{u\}_{\bar{t}=t+\Delta t}^{\bar{r}=r-1} + \frac{4}{(\Delta t)^2}\{\Delta u\}_{t+\Delta t}$, <br>go to **IN-1** |
| **OUT-4** | Set current time as $t$ and go to **OUT-1** |
| * For any iteration in the same time instant $t+\Delta t$, the known DOFs associated with the nonhomogeneous Dirichlet boundary conditions do not change. | |

## 7 Numerical example

To verify the nonlinear formulation, both finite element solution given in present work and semi-analytical solution given by Patel and Park [16] are computed for comparison. In the semi-analytical solution, the solution is expressed in the form of Fourier sine series. To obtain the unknown Fourier coefficients, eq. (1) is reduced to a system of 2nd order nonlinear ordinary differential equations using Galerkin mode decomposition method, which is subsequently solved using fourth order Runge-Kutta (RK4) method. With respect to Fig. 1, the problem parameters are given in Table 3. Similar to the original work [16], three cases are studied herein to illustrate the effect of different forcing excitations. Likewise, the solution parameters are given in Table 4. In the finite element solution, a rather strict tolerance criterion has been set as to test the feasibility of both algorithm structures in handling quadratic damping nonlinearity.

**Table 3.** Problem parameters

| | |
|---|---|
| Total length of beam, $L_T$ | 300 |
| Flexural rigidity, $EI$ (N m$^2$) | 1.5 $x$ 10$^9$ |
| Mass per unit length, $m$ (kg/m) | 200 |
| Constant axial tension, $T$ (N) | 1.5 $x$ 10$^6$ |
| Amplitude of tension fluctuation, $S$ (N) | ▪ 1.5 $x$ 10$^6$ (case I) <br>▪ 1.5 $x$ 10$^6$ (case II) <br>▪ 0 (case III) |
| Amplitude of support excitation, $y_o$ (m) | ▪ 3 (case I) <br>▪ 0 (case II) <br>▪ 3 (case III) |
| Angular frequency, $\omega$ (rad/s) | $2\pi/12$ |
| Drag coefficient, $C_D$ | 0.8 |
| Outer diameter of pipe, $D_o$ (m) | 0.75 |
| External fluid density, $\rho_w$ (kg/m$^3$) | 1025 |

**Table 4.** Solution parameters

| | |
|---|---|
| A. Finite element solution | |
| No. of Hermitian beam element | 20 (equally spaced) |
| Tolerance check | $\frac{\left\|\{\Delta u\}_{t+\Delta t}\right\|}{\left\|\{u\}_{\bar{t}=t+\Delta t}^{\bar{r}=r}\right\|} < 1E-12$ |
| Time step, $\Delta t$ (s) | 0.01 |
| Initial displacement & velocity | As per semi-analytical solution [16] at $t=0$ |
| B. Semi-analytical solution (for notation see [16]) | |
| Initial conditions for the system of 2nd order ODEs | $F_m(0) = 0.15$; $\frac{dF_m(0)}{d\tau} = 0$ |
| Time step, $\Delta t$ (s) for RK4 | 0.1 |
| No. of modes in Fourier sine series | 4 |

The displacement time histories of the beam at mid-length for all three cases are given in Fig. 2-4. As shown in the figures, both algorithm structures agree well with the semi-analytical solution. To further ascertain the feasibility of both algorithm structures, the number of iterations required and tolerance upon converged are analysed. As shown in Fig. 5-6, both algorithm structures used are effective in the numerical example with only three to four iterations required in each time step.

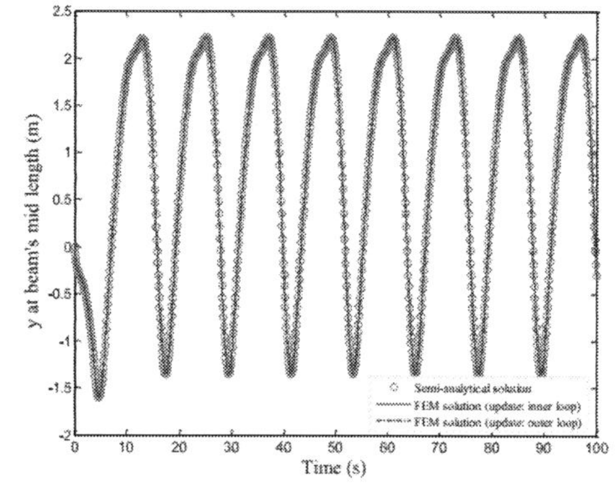


**Fig. 2.** Displacement time history of the pinned-pinned beam at mid-length, $x = 0.5L_T$ (case I: excitation due to both nonhomogeneous Dirichlet boundary condition at $x = L_T$ and harmonic tension fluctuation)

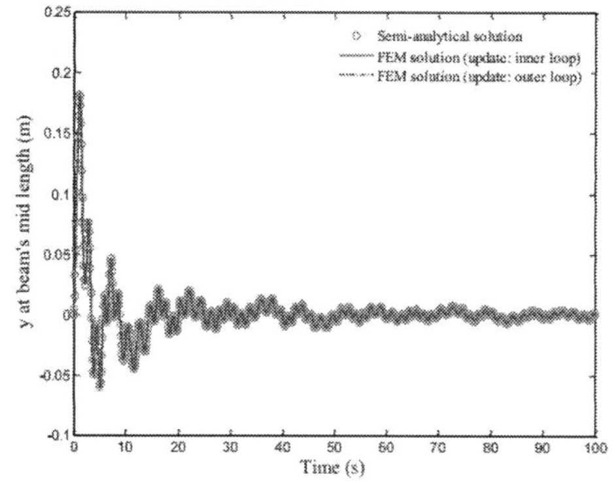


**Fig. 3.** Displacement time history of the pinned-pinned beam at mid-length, $x = 0.5L_T$ (case II: excitation due to harmonic tension fluctuation only)

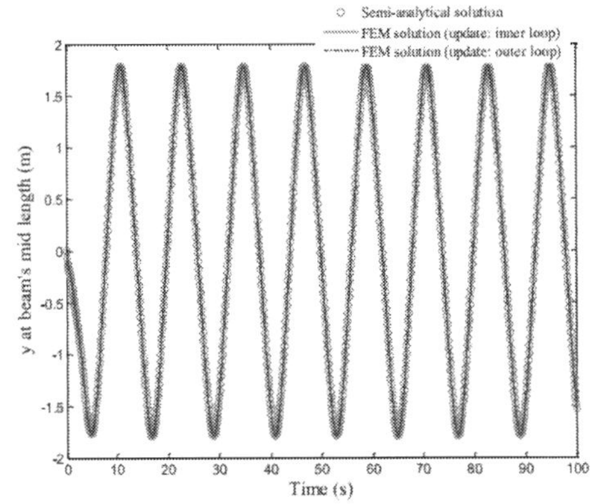


**Fig. 4.** Displacement time history of the pinned-pinned beam at mid-length, $x = 0.5L_T$ (case II: excitation due to nonhomogeneous Dirichlet boundary condition at $x = L_T$ only)

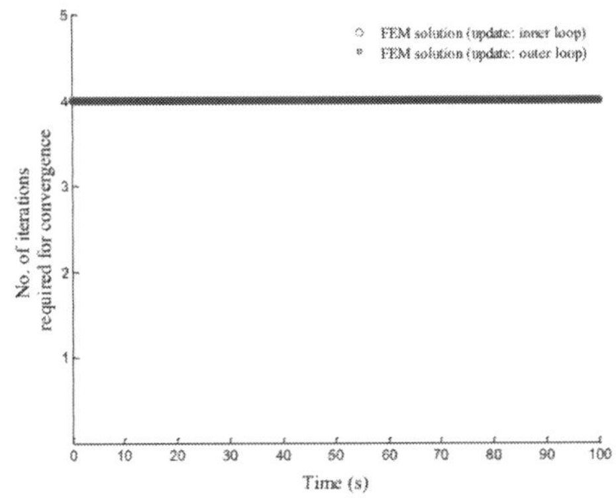


**Fig. 5.** Iterations required for each time step in case I & III

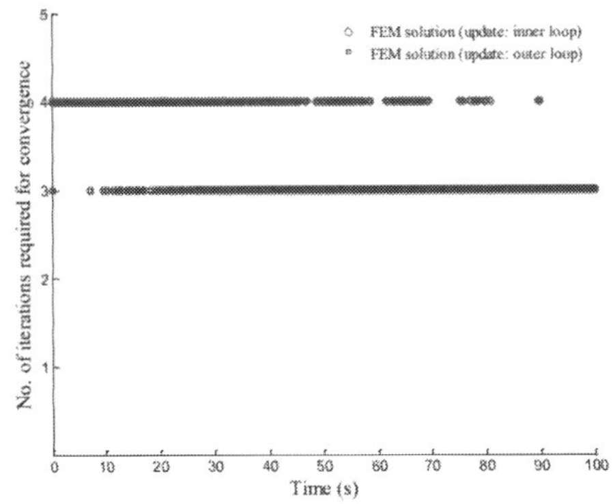


**Fig. 6.** Iterations required for each time step in case II

Furthermore, both algorithm structures are equivalent in term of number of iterations per time step. Another evidence that the algorithm structures are equivalent to each other is the resemblance of tolerance upon converged in each time step as shown in Fig. 7-9. The similarity of convergence properties in both algorithm structures is due to mathematical manipulations as explained earlier.

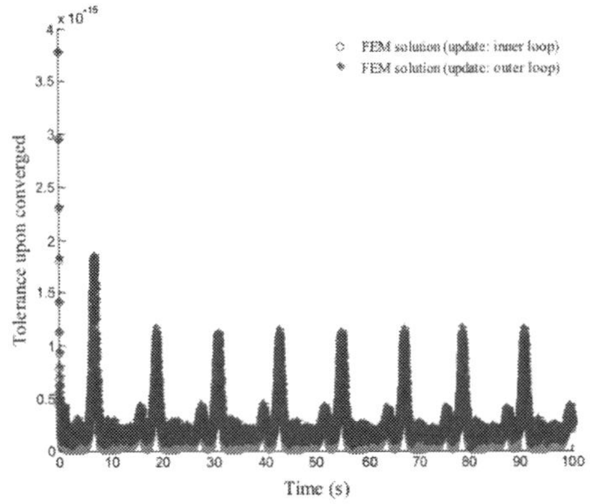


**Fig. 7.** Tolerance upon converged for each time step in case I

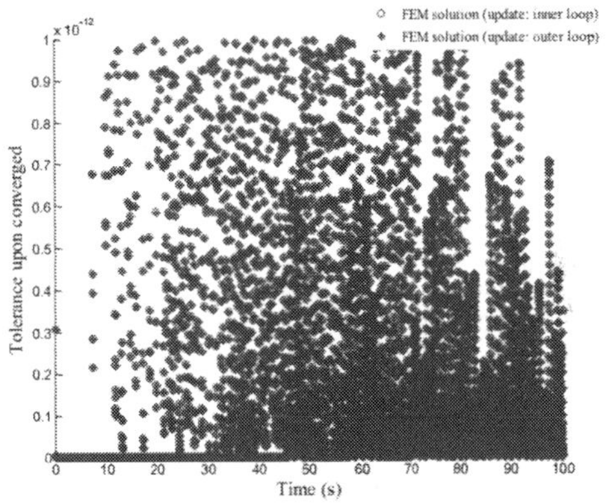


**Fig. 8.** Tolerance upon converged for each time step in case II

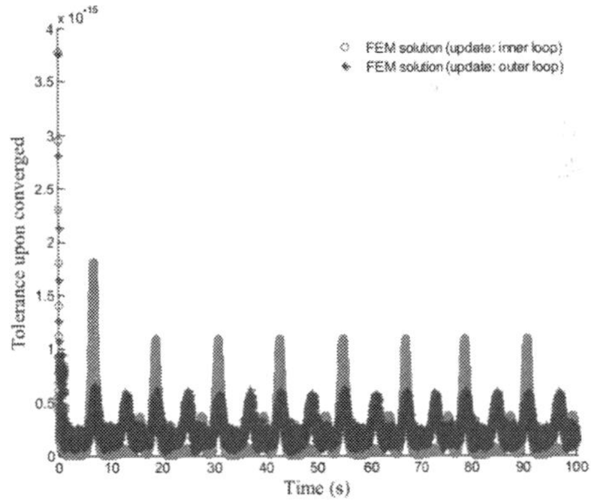


**Fig. 9.** Tolerance upon converged for each time step in case III

## 8 Conclusion

In present work, nonlinear formulation for an Euler-Bernoulli beam vibration problem with quadratic damping nonlinearity is demonstrated. The finite element solution shows a good agreement with the established semi-analytical solution by Patel and Park [16]. Through Newton-Raphson formulation, it is lucid that the tangent stiffness components associated with the quadratic damping nonlinearity require input of velocity for evaluation purpose in the iterative procedure. As a result, velocity has to be iterated in the inner (iterative) loop when implementing the time integration-iterative algorithm. The nonlinear formulation has been tested with two mathematically equivalent algorithm structures with different kinematics arrangement. The first algorithm structure which is termed "update in the outer loop", involves iteration of displacement and velocity, whereas the second algorithm which is termed "update in the inner loop", involves iteration of displacement, velocity and acceleration. As expected, both mathematically equivalent algorithm structures exhibit the same accuracy and convergence characteristic of solution.

## References

[1] K. J. Bathe, Finite element procedures, Klaus-Jurgen Bathe, 2006.

[2] K. J. Bathe and A. P. Cimento, "Some practical procedures for the solution of nonlinear finite element equations," *Computer methods in applied mechanics and engineering,* vol. 22, no. 1, pp. 59-85, 1980.

[3] K. J. Bathe and E. N. Dvorkin, "On the automatic solution of nonlinear finite element equations," *Computers & Structures,* vol. 17, no. 5, pp. 871-879, 1983.

[4] J. R. Morison, J. W. Johnson and S. A. Schaaf, "The force exerted by surface waves on piles," *Journal of Petroleum Technology,* vol. 2, no. 05, pp. 149-154, 1960.

[5] S. Chakrabarti, Handbook of Offshore Engineering (2-volume set), Elsevier, 2005.

[6] V. H. Safai, "Nonlinear dynamic analysis of deep water risers," *Applied Ocean Research,* vol. 5, no. 4, pp. 215-225, 1983.

[7] N. Haritos and D. T. He, "Modelling the response of cable elements in an ocean environment," *Finite Elements in Analysis and Design,* vol. 11, no. 1, pp. 19-32, 1992.

[8] J. I. Gobat and M. A. Grosenbaugh, "Time-domain numerical simulation of ocean cable structures," *Ocean Engineering,* vol. 33, no. 10, pp. 1373-1400, 2006.

[9] A. K. Jain, "Nonlinear coupled response of offshore tension leg platforms to regular wave forces," *Ocean Engineering,* vol. 24, no. 7, pp. 577-592, 1997.

[10] M. H. Patel, S. Sarohia and K. F. Ng, "Finite-element analysis of the marine riser," *Engineering Structures,* vol. 6, no. 3, pp. 175-184, 1984.

[11] J. N. Reddy, An Introduction to Nonlinear Finite element analysis: with applications to heat transfer, fluid mechanics, and solid mechanics., OUP Oxford, 2014.

[12] H. M. Hilber, T. J. Hughes and R. L. Taylor, "Improved numerical dissipation for time integration algorithms in structural dynamics," *Earthquake Engineering & Structural Dynamics,* vol. 5, no. 3, pp. 283-292, 1977.

[13] K. J. Bathe and M. M. I. Baig, "On a composite implicit time integration procedure for nonlinear dynamics," *Computers & Structures,* vol. 83, no. 31, pp. 2513-2524, 2005.

[14] K. J. Bathe, "Conserving energy and momentum in nonlinear dynamics: a simple implicit time integration scheme," *Computers & Structures,* vol. 85, no. 7, pp. 437-445, 2007.

[15] J. Chung and G. M. Hulbert, "A time integration algorithm for structural dynamics with improved numerical dissipation: the generalized-α method," *Journal of Applied Mechanics,* vol. 60, no. 2, pp. 371-375, 1993.

[16] M. H. Patel and H. I. Park, "Combined axial and lateral responses of tensioned buoyant platform tethers," *Engineering Structures,* vol. 17, no. 10, pp. 687-695, 1995.

[17] A. K. Chopra, Dynamics of structures: theory and applications to earthquake engineering (fourth edition), New Jersey: Prentice Hall, 2012.

[18] T. Belytschko, W. K. Liu, B. Moran and K. Elkhodary, Nonlinear finite elements for continua and structures, John Wiley & Sons, 2013.

[19] B. P. Jacob and N. F. F. Ebecken, "An optimized implementation of the Newmark/Newton-Raphson algorithm for the time integration of non-linear problems," *Communications in Numerical Methods in Engineering,* vol. 10, no. 12, pp. 983-992, 1994.

# Fluid-Structure Interaction (FSI) of Damped Oil Conveying Pipeline System by Finite Element Method

*Kah Soon* Fong[1,*] and *Airil Yasreen* Mohd Yassin[1]

[1] Department of Civil and Environmental Engineering, Universiti Teknologi PETRONAS, 32610 Seri Iskandar, Perak Darul Ridzuan, Malaysia

**Abstract.** The vibration of damped pipeline conveying fluid with the effect of fluid-structure interaction is known to pose challenging problems in oil and gas industry. In this study, the natural frequency of fluid-structure interaction in pipeline conveying fluid set on viscoelastic foundation is investigated by using finite element method. The governing partial differential equation is modelled based on Euler-Bernoulli beam theory. By applying Galerkin weighted residual method, the stiffness, damping, and mass matrices are obtained. For a given boundary condition which is simply supported, two components of the foundation (foundation stiffness and damping) which are influencing the damped natural frequency of the pipeline are studied for different fluid velocity. The results indicate that increasing the foundation stiffness from 10 kN/m$^3$ to 30 kN/m$^3$ increases the natural frequency of the pipeline, while increasing the foundation damping from 1 kN.s/m$^3$ to 3 kN.s/m$^3$ and fluid velocity decrease the natural frequency of the pipeline. The accuracy of the results obtained is validated against data from literature.

## 1 Introduction

Pipeline conveying fluid has a vital role in engineering fields particularly its application in marine engineering, oil and gas industry, oil transportation, and municipal water supply [1, 2]. In oil and gas industry, pipelines are commonly used to convey and transport crude oil. However, the interaction between the fluid and structure often causes vibration problems to the pipelines, which has resulted in machinery breakdown, pipeline damages, fatigue failures, fires, and explosions in refineries and petrochemical plants. Consequently, huge economic loss due to pipeline vibration is experienced even by advanced country every year. The damage has been estimated as high as $10 billion yearly [2]. Therefore, the research on the vibration characteristic of pipelines considering the effect of fluid-structure interaction such as natural frequency have gained the interest of researchers due to its great engineering and economic significance.

Pipeline conveying fluid is also considered as the simplest system of fluid–structure interaction problems, being able to describe significant dynamical behaviour of pipeline [3]. A century ago, the studies of pipeline conveying fluid were focused only on classic water hammer equation [4]. However, complex problems of pipeline conveying fluid were often encountered in various engineering fields. Therefore, it is necessary to consider the effect of fluid-structure interaction in the work due to its high development and demand in the field of engineering.

Important investigation in the study of vibration was made by Ashley and Haviland [5] which explained the vibration of the Trans-Arabian pipeline. The work involved a dynamic analysis of a simply supported pipe span using an approximate power series solution. However, the governing inertial force was not described adequately in the study as mentioned by Feodos'ev [6]. Therefore, Feodos'ev formulated a more complete equation by considering the governing inertial force in the study. The equation was then solved by Galerkin method. Housner [7] developed a more realistic equation by considering the transverse vibration of the Euler-Bernoulli pipeline conveying fluid. The work presented several major factors such as elastic restoring force, centrifugal force, Coriolis force, and inertia force affect the vibration behavior of pipeline conveying fluid. Later, Paidoussis and Issid [8], Wiggert, et al. [9], Lesmez, et al. [10], and Wiggert [11] considered the transverse motion of the pipe in the equation of motion of pipeline. However, the equation did not contain any fluid related term except for Paidoussis and Issid work.

In recent years, a finite element formulation was developed by Sreejith, et al. [12] for a fully coupled dynamic equations of motions to study the liquid-filled pipeline's behaviour subjected to valve closure excitation. The study showed the significant reduction in structural velocities when the effect of fluid structure interaction was considered. Chellapilla and Simha [13] presented the vibration and stability characteristics of the pipeline conveying fluid based on the effect of two parameter foundation model. Critical velocity of the pipeline was obtained by using Fourier series and Galerkin method. A transverse vibration model was presented by Huang, et al. [1]. The dynamic equation with fluid structure interaction

* Corresponding author: kahsoon0401@gmail.com

effect in the Euler-Bernoulli pipeline conveying fluid subjected to different boundary condition was solved using eliminated element-Galerkin method to obtain the natural frequency. The study showed that effect of the flow velocity on natural frequency can be estimated through its dimensionless fluid velocity and limit values. Another interesting finding from the study is that Coriolis effect can be neglected if the required precision of natural frequency is lower than 95. Huang, et al. [4] solved the same pipeline conveying fluid model for natural frequency analytically via separation of variables method and Ferrari's method..

In the present study, Galerkin finite element method is used to discretize the governing partial differential equation of one dimensional Euler Bernoulli pipeline conveying oil. The damped natural frequency of the pipeline is obtained through state space method with the imposed boundary condition of simply supported using the MATLAB program. The results are validated against the data reported by Mostafa [14].

## 2 Governing equation of pipeline

This study considers a horizontal pipeline rests on a viscoelastic medium that conveys oil as shown in Fig.1. The diameter of the pipeline is assumed to be comparatively smaller than the length of the pipeline. Therefore, the Euler-Bernoulli beam theory is applied to the pipeline model and the oil flow is approximated as plug flow model.

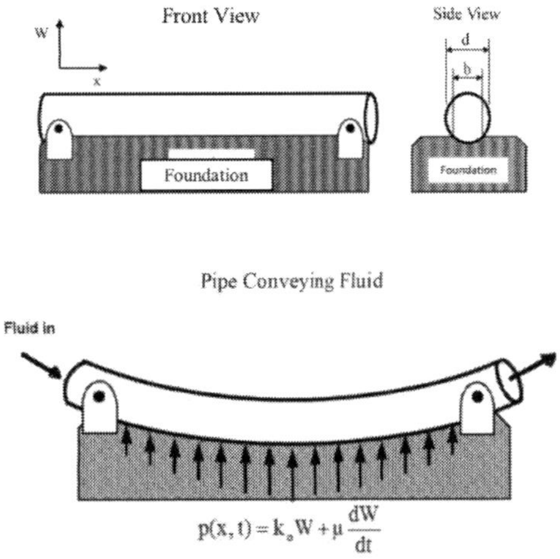


**Fig. 1.** Representation of modified Winkler foundation model[14]

The governing partial differential equation of motion that can represent the pipeline conveying oil on viscoelastic foundation based on the Euler-Bernoulli theory obtained from Mostafa [14] can be written as:

$$EI\frac{\partial^4 w}{\partial x^4}+k_0 w+\mu\frac{\partial w}{\partial t}+Mu^2\frac{\partial^2 w}{\partial x^2}+2MU\frac{\partial^2 w}{\partial x\partial t}$$

$$+(m+M)\frac{\partial^2 w}{\partial t^2}=0 \quad (1)$$

where $w(x,t)$ is the lateral pipeline displacement at position $x$, $EI$ is the bending stiffness of the pipeline, $u$ is the velocity of the fluid, $m$ is the pipe mass per unit length, $M$ is the conveying oil of mass per unit length, $k_0$ is the foundation stiffness coefficient per unit length and $\mu_0$ is the foundation damping coefficient per unit length.

The presence of Coriolis force is due to the relative motion of the fluid and is written as:

$$F_c = 2Mu(t)\frac{\partial^2 w}{\partial x\partial t} \quad (2)$$

The Coriolis force term $F_c$ is known as anti-symmetry whirligig "damping". Due to its effect, the finite element model becomes a complex eigenvalue problem.

## 3 Finite element formulation

The Galerkin weighted residual method is employed in the study as a tool for finite element formulation of pipeline governed by partial differential equation. The procedure is outlined in the following subsection.

### 3.1. Shape function

The guess solution of the lateral pipe displacement, $w(x,t)$ in an element can be expressed in term of Hermite cubic shape function as shown in Eq 3 or in form of component form as Eq 4 .

$$w(x,t) = N_1d_1+N_2d_2+N_3d_3+N_4d_4 \quad (3)$$

$$w(x,t) = [N(x)]\{d(t)\} = N_jd_j \text{ for j=1,2,3,4} \quad (4)$$

where $d_1$ and $d_3$ are nodal transverse degree of freedom while $d_2$ and $d_4$ are nodal rotational degree of freedom associated with node 1 and 2 respectively. The Hermite shape functions $N_1$, $N_2$, $N_3$, and $N_4$ are given as:

$$N_1 = 1-\frac{3x^2}{L^2}+\frac{2x^3}{L^3} \quad (5)$$

$$N_2 = x-\frac{2x^2}{L^2}+\frac{x^3}{L^2} \quad (6)$$

$$N_3 = \frac{3x^2}{L^2}-\frac{2x^3}{L^3} \quad (7)$$

$$N_4 = -\frac{x^2}{L}+\frac{x^3}{L^2} \quad (8)$$

### 3.2 Weak forms of differential equation

Having established shape functions and degree of freedom of the beam element, the partial differential equation in Eq 1 can be discretized by substituting Eq 4 into it. Since Bubnov-Galerkin formulation is used in this paper, the weight functions used are the shape functions. Therefore, integrating the equation with the multiplication of weight function $N_i$ and it yields the equation given by:

$$\int_0^l N_i\left(EI\frac{\partial^4(N_j d_j)}{\partial x^4}+k_0(N_j d_j)+\mu\frac{\partial(N_j d_j)}{\partial t}\right)dx$$

$$+\int_0^l N_i\left(Mu^2\frac{\partial^2(N_j d_j)}{\partial x^2}+2MU\frac{\partial^2(N_j d_j)}{\partial x\partial t}\right)dx$$

$$+\int_0^l N_i\left((m+M)\frac{\partial^2(N_j d_j)}{\partial t^2}\right)dx=0$$

(9)

By employing integration by part (IBP), the order of the original derivative terms is reduced. The final resulting weak form discrete finite element equation is written in the form of:

$$(m+M)[S_1]\{\ddot{d}\}+(\mu[S_1]+2MU[S_2])\{\dot{d}\}$$

$$+(EI[S_3]+k_0[S_1]+MU^2[S_4])\{d\}=0 \quad (10)$$

where,

$$[S_1]=\int_0^l[N_i][N_j],\qquad [S_2]=\int_0^l\left[\frac{dN_i}{dx}\right][N_j]$$

$$[S_3]=\int_0^l\left[\frac{d^2N_i}{dx^2}\right]\left[\frac{d^2N_j}{dx^2}\right],\quad [S_4]=\int_0^l\left[\frac{dN_i}{dx}\right]\left[\frac{dN_j}{dx}\right]$$

### 3.3 Finite element equations

These equations can be rewritten in matrix form as finite element equation for each element:

$$[m_1]\{\ddot{d}\}+([c_1]+[c_2])\{\dot{d}\}+([k_1]+[k_2]-[k_3])\{d\}=0 \quad (11)$$

where $[m_1]$ is element mass matrix, $[c_1]$ and $[c_2]$ the element damping matrix, $[k_1]$, $[k_2]$, and $[k_3]$ the element stiffness matrix, [4] the element displacement vector, $\{\dot{d}\}$ the element velocity vector, and $\{\ddot{d}\}$ the element acceleration vector. The respective element matrices are derived and shown below:

$$m_1=\frac{(m+M)l}{420}\begin{bmatrix}156 & 22l & 54 & -13l\\ 22l & 4l^2 & 13L & -3l^2\\ 54 & 13l & 156 & -22l\\ -13l & -3l^2 & -22l & 4l^2\end{bmatrix} \quad (12)$$

$$c_1=\frac{\mu l}{420}\begin{bmatrix}156 & 22l & 54 & -13l\\ 22l & 4l^2 & 13l & -3l^2\\ 54 & 13l & 156 & -22l\\ -13l & -3l^2 & -22l & 4l^2\end{bmatrix} \quad (13)$$

$$c_2=\frac{MU}{30}\begin{bmatrix}-30 & -6l & -30 & 6l\\ 6l & 0 & -6l & l^2\\ 30 & 6l & 30 & -6l\\ -6l & -l^2 & 6l & 0\end{bmatrix} \quad (14)$$

$$k_1=\frac{2EI}{l^3}\begin{bmatrix}6 & 3l & -6 & 3l\\ 3l & 2l^2 & -3l & l^2\\ -6 & -3l & 6 & -3l\\ 3l & l^2 & -3l & 2l^2\end{bmatrix} \quad (15)$$

$$k_2=\frac{k_o l}{420}\begin{bmatrix}156 & 22l & 54 & -13l\\ 22l & 4l^2 & 13l & -3l^2\\ 54 & 13l & 156 & -22l\\ -13l & -3l^2 & -22l & 4l^2\end{bmatrix} \quad (16)$$

$$k_3=\frac{MU^2}{30l}\begin{bmatrix}36 & 3l & -36 & 3l\\ 3l & 4l^2 & -3l & -l^2\\ -36 & -3l & 36 & -3l\\ 3l & -l^2 & -3l & 4l^2\end{bmatrix} \quad (17)$$

### 3.4 Dynamic eigenvalue of damped pipeline

These local matrices are then assembled into global matrices to represent the full length pipeline model. After imposing boundary condition (simply supported), the matrices are arranged into proper form to determine the dynamic characteristic of the structure. The asymmetrical damping matric resulting from the Coriolis force cause the problem to be complex eigenvalue problem. In order to obtain the eigenvalues of damped structure, the standard equation of motion in the finite element form Eq 18 is converted to first order differential equation Eq 19 by introducing the state space vector $y_1=d$ and $y_2=\dot{d}$.

$$[M]\{\ddot{d}\}+[C]\{\dot{d}\}+[K]\{d\}=0 \quad (18)$$

$$\dot{y}=\begin{bmatrix}\dot{y}_1\\ \dot{y}_2\end{bmatrix}=\begin{bmatrix}0 & I\\ -M^{-1}K & -M^{-1}C\end{bmatrix}\begin{bmatrix}y_1\\ y_2\end{bmatrix} \quad (19)$$

By solving the first order differential equation, the natural frequency or eigenvalues and mode shapes or eigenvectors can be obtained. The solution of eigenvalue problem yields the complex roots and the imaginary part of the root represent the natural frequency of the damped structure as written in Eq 20. The real part of the root represent the rate of decay of the vibration.

$$\lambda_i=-\xi_i\omega_i\pm\omega_i\sqrt{1-\xi_i^2}\,j \quad (20)$$

# 4 Numerical example

To investigate the effectiveness and accuracy of present formulation, the results are compared with the numerical study reported [14] of a uniform simply supported pipeline. The natural frequency of the model is determined and compared by using the MATLAB program. The pipeline, which is 2m long and made of steel with density 8000kg/m$^3$, conveys the fluid with the density of 1000kg/m$^3$. The thickness of the pipeline is of 0.001m thick and has the outer diameter of 0.01m. The elastic modulus of pipeline is 207 GPa and it is divided equally into 10 beam elements for the finite element analysis. The validated results using present formulation are shown and discussed in the following subsection.

## 4.1 Effect of fluid velocity on natural frequency at different foundation stiffness

This subsection investigates the natural frequency of a simply supported pipeline by varying the fluid velocity at different foundation stiffness. The present result is obtained by Galerkin finite element method while the result obtained by Mostafa [14] is based on energy finite element method. Both results are compared and excellent agreement between the results is achieved.

In Fig 2, the damped natural frequency of the pipeline with simply supported boundary condition for fluid velocity varying from 0 m/s to 80 m/s is illustrated. It is observed that the damped natural frequency decreases with an increase in fluid velocity. This is because higher fluid velocity weaken the structure's stiffness. In this example, the foundation viscous damping is set to be 100 N.s/m$^3$ and the parameters of foundation stiffness are equalled to 10 kN/m$^3$, 20 kN/m$^3$, and 30 kN/m$^3$. The relation between the foundation stiffness and damped natural frequency is plotted in Fig. 2. The damped natural frequency increases with an increase of foundation stiffness because it increases its overall structure stiffness.

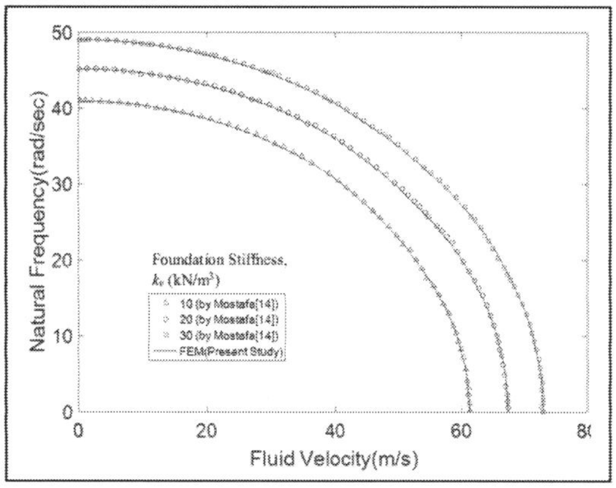


**Fig. 2.** Effect of fluid velocity on natural frequency at different foundation stiffness

## 4.2 Effect of fluid velocity on natural frequency at different foundation viscous damping

This subsection investigates the natural frequency of a simply supported pipeline by varying the fluid velocity at different foundation viscous damping. The present result is obtained by Galerkin finite element method while the result obtained by Mostafa [14] is based on energy finite element method. Both results are compared and excellent agreement between the results is achieved.

In Fig. 3, the damped natural frequency of the pipeline with simply supported boundary condition for fluid velocity varying from 0 m/s to 80 m/s is illustrated. Like previous discussion, the damped natural frequency decreases with an increase in fluid velocity. However, in this example, the parameter of foundation stiffness is set to be 20 kN/m$^3$ and the foundation viscous damping are equalled to 1 kN.s/m$^3$, 2 kN.s/m$^3$, and 3 kN.s/m$^3$. The relation between the foundation viscous damping and damped natural frequency is plotted in Fig. 3. Unlike previous result, the damped natural frequency decreases with an increase of foundation viscous damping. This is because the increment of foundation viscous damping leads to the reduction of dynamic properties of the pipe structure.

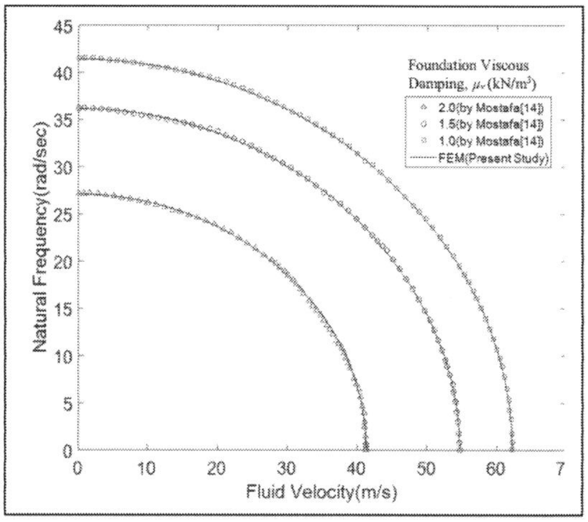


**Fig. 3.** Effect of fluid velocity on natural frequency at different foundation viscous damping

# 5 Conclusion

In this paper, the natural frequency of damped pipeline conveying oil is investigated and validated against the results of an existing work. Galerkin finite element method is employed to discretize the governing partial differential equation and the damped natural frequency of the structure are obtained though state space method. Several conclusions are drawn in this paper:

a) The presence of damping term, particularly external damping decreases the fluid critical velocity, while the increment of foundation stiffness increases the critical fluid velocity.

b) For the eigenvalue problem, the inclusion of Coriolis force resulted in complex roots. The imaginary part of the roots indicates the natural frequency of damped structure.

c) Large centrifugal force resulting from high fluid velocity has weakened the stiffness of the structure and hence decreases the fluid critical velocity.

## References

1. Y.-M. Huang, Y.-S. Liu, B.-H. Li, Y.-J. Li, and Z.-F. Yue, "Natural frequency analysis of fluid conveying pipeline with different boundary conditions," *Nuclear Engineering and Design,* vol. 240, pp. 461-467, 2010.
2. Y. Sun and Y. G. a. H. Xiong, "Studied and Their Application of Vibration Control Technologies," 2013.
3. H. Dai, L. Wang, and Q. Ni, "Dynamics of a fluid-conveying pipe composed of two different materials," *International Journal of Engineering Science,* vol. 73, pp. 67-76, 2013.
4. Y.-m. Huang, S. Ge, W. Wu, and J. He, "A direct method of natural frequency analysis on pipeline conveying fluid with both ends supported," *Nuclear Engineering and Design,* vol. 253, pp. 12-22, 2012.
5. H. Ashley and G. Haviland, "Bending vibrations of a pipe line containing flowing fluid," *Journal of Applied Mechanics-Transactions of the ASME,* vol. 17, pp. 229-232, 1950.
6. V. Feodos'ev, "Vibrations and stability of a pipe when liquid flows through it," *Inzhenernyi Sbornik,* vol. 10, pp. 169-170, 1951.
7. G. Housner, "Bending vibrations of a pipe line containing flowing fluid," *Journal of Applied Mechanics-Transactions of the ASME,* vol. 19, pp. 205-208, 1952.
8. M. P. Paidoussis and N. Issid, "Dynamic stability of pipes conveying fluid," *Journal of sound and vibration,* vol. 33, pp. 267-294, 1974.
9. D. Wiggert, F. Hatfield, and S. Stuckenbruck, "Analysis of liquid and structural transients in piping by the method of characteristics," *Journal of fluids engineering,* vol. 109, pp. 161-165, 1987.
10. M. W. Lesmez, D. Wiggert, and F. Hatfield, "Modal analysis of vibrations in liquid-filled piping systems," *Journal of Fluids Engineering,* vol. 112, pp. 311-318, 1990.
11. D. Wiggert, "Fluid transients in flexible piping systems," in *Hydraulic Machinery and Cavitation,* ed: Springer, 1996, pp. 58-67.
12. B. Sreejith, K. Jayaraj, N. Ganesan, C. Padmanabhan, P. Chellapandi, and P. Selvaraj, "Finite element analysis of fluid–structure interaction in pipeline systems," *Nuclear Engineering and Design,* vol. 227, pp. 313-322, 2004.
13. K. R. Chellapilla and H. Simha, "Critical velocity of fluid-conveying pipes resting on two-parameter foundation," *Journal of sound and vibration,* vol. 302, pp. 387-397, 2007.
14. N. H. Mostafa, "Effect of a Viscoelastic Foundation on the Dynamic Stability of a Fluid Conveying Pipe," *International Journal of Applied Science and Engineering 12 (1),* pp. 59-74, 2014.

# Mixed Finite Element Formulation for Magnetic Fluid Oil Flow in Electromagnetic Field

*Phey Hoon* Tan*[1] and *Airil Yasreen* Mohd Yassin[1]

[1]Department of Civil and Environmental Engineering, Universiti Teknologi PETRONAS,32610 Seri Iskandar, Perak Darul Ridzuan Malaysia.

**Abstract.** Pressure depletion and high viscosity of crude oil in oil reservoir are the main challenges in oil recovery process. A potential solution is to employ electromagnetic heating coupled with magnetic fluid injection. The present work delivers a fundamental study on the interaction between magnetic fluid flow with electromagnetic field. The two-dimensional, incompressible flow is solved numerically using mixed finite element method. The velocity fields, temperature and pressure are the variables of interest, to be obtained by solving mass, momentum and energy equations coupled with Maxwell' equations. The fluid stress arises simultaneously with the external magnetic force which mobilises and increases the temperature of the oil flow. Verification is made against available data obtained from different numerical method reported in literature. The results justify feasibility of the mixed finite element formulation as an alternative for the modelling of the magnetic fluid flow.

## 1 Introduction

Oil recovery in reservoir has become a mainstream of focus over the past decades. Various technologies associated with oil extraction have been developed to support energy source in the existing and future oil well. There are three extraction phases categorised as primary, secondary and tertiary recovery. Primary and secondary recovery uses conventional methods such as thermal heating, water flooding and carbon dioxide injection [1]. Because of the highly viscous properties of the crude oil flowing after the secondary recovery phase, the mobility of the fluid is greatly reduced and thus requires alternative treatment. As such, the tertiary recovery phase is introduced, also known as Enhance Oil Recovery (EOR) stage which utilises nanotechnology in oil production. The idea of EOR process is carried out by injecting the nanofluids into the reservoir channel to mobilize the oil residual by altering the surface tension between water and oil in the porous medium [2].

Nanofluid is a synthesized fluid, which exhibit superiors heat transfer performance and high adaptability in various oil and gas operations [1,2]. Ferrofluid, a type of nanofluids, is made up of non-magnetic bases fluids and dispersed iron nanoparticle of diameter lesser than 100 nm with a surfactant layer coat. According to [3], the layer contributes to the stability of the nanofluids in the enhancement of the thermal conductivity. As proposed by [4], these nanofluids contain metallic particle that give rise to significant thermal conductivity, which are approximately three times higher compared to the conventional fluids. Low concentration mixture of ferrofluids in crude oil permits the formation of efficient energy heat transfer medium and thus leads to a more cost-effective recovery process. [5] developed a model that describes the relationship between the energy dissipation of polarizable fluid and applied magnetic field for the enhancement of heat transfer. Their study highlights the importance of ferrofluids in industry applications. The magnetization of ferrofluids is also dependent on the temperature and in such property, it gives an effective energy conversion in heat transport system [6]. The inherent magnetization characteristic of the ferrofluids allows manoeuvre to take place via external magnetic field and. Hence interaction of nanofluids with electromagnetic occurs. In the absence of magnetic field the nanoparticles are scattered. Exposing ferrofluids to external magnetic field causes these nanoparticles to align in a direction and create a stronger bond, therefore the magnetization of ferrofluids is multiplied. This alignment provides a novel method to deliver thermal energy efficiently known as Neel relaxation [7]. The excited ferrofluids are aligned and used to supply heat to a specific location in reservoir. Such a relaxation will induce a higher heating rate in flowing fluids than static fluids. [8] believes that the excited ferrofluids has the potential to mobilize the traps oil in well under quasi static motion of fluids. [9] used single phase miscible ferrofluid to mobilize the underground flow using magnetic guidance. The numerical simulation taken into consideration of strong magnetic field. However, the interaction of the magnetic fluids with electromagnetic is not well addressed. The interaction of ferrofluids can further been explored with ferrohydrodynamic (FHD) numerical analysis.

* Corresponding author: suzannetan90@gmail.com

Numerous numerical procedures had been proposed in the past, such as Finite Volume Method(FVM), Finite Difference Method (FDM), Finite Element Method (FEM), and Lagrange multiplier. Among the methods, the most widely used is the FDM which is combined with stream function vorticity formulation due to the simplicity of algorithm. In early study of FHD principle, [10] developed a model which dealt with FHD coupled with uniform electromagnetic field. The mathematical model on a rectangular domain had been employ and solved using SIMPLE algorithm. In addition, spatially varying magnetic field had been studied using FDM and pressure-linked psuedotransient method on grid method [11-13]. As a result, dominant flow form was observed nearby the magnetic source with no evidence of secondary flow effect. [14] extended the work of [13] on hydrodynamic model, where a fully conservation system is taken into consideration, whereby they used Lorentz and magnetization force and solved using FVM. One of the major advantages of using FVM is that they are easily formulated to allow for unstructured mesh. However, in present study, Lorentz force is insignificant because of the magnetic fluids oil flow has low conductivity where it is only affected by magnetization. Improvement of the [14] model is carried out by adopting Boundary Element Method using Laplacian term, which automatically satisfy continuity term and resulting in a smaller sized algebraic equation [15]. This algorithm uses comparatively less computer resources. Computational results revealed a vortex enlarged formed near to the region of magnetic source and followed with an increase in magnetic intensity. Furthermore, the fluid flow temperature increases with the increase of intensity of magnetic field. High intensity magnetic field leads to the domination of the magnetic gradient term in the system and causes the solution instable. This turmoil can be overcome by FEM as it is a well-developed approach to achieve higher order accuracy with a fine meshing compared with FDM and FVM. The result of velocities and temperature are studied considering the influence of magnetic field.

Current study focuses on development of the numerical algorithm used to simulate the ferrohydrodynamic flow in a rectangular channel with the disturbance of external magnetic field. Present study has developed a hypothesis stating that by exciting the ferrofluids with external magnetic, additional displacement will be induced which is going to mobilize the magnetic oil flow and enhanced the recovery.

## 2 Governing Equations

Employing Boussinesq approximation, the governing equation for conservation of mass, momentum and energy in non-dimensional form for a steady laminar flow can be expressed as follows [11]:

*Continuity :*

$$\frac{\partial u}{\partial x}+\frac{\partial v}{\partial y}=0 \qquad (1)$$

*Momentum :*

$$u\frac{\partial u}{\partial x}+v\frac{\partial u}{\partial y}+\frac{\partial P}{\partial x}-\frac{1}{\mathrm{Re}}\left(\nabla^2 u\right)-Mn_F TH\frac{\partial H}{\partial x}=0 \qquad (2)$$

$$u\frac{\partial v}{\partial x}+v\frac{\partial v}{\partial y}+\frac{\partial P}{\partial y}-\frac{1}{\mathrm{Re}}\left(\nabla^2 v\right)-Mn_F TH\frac{\partial H}{\partial y}=0 \qquad (3)$$

*Energy :*

$$u\frac{\partial T}{\partial x}+v\frac{\partial T}{\partial y}-\frac{1}{\mathrm{Pr}\,\mathrm{Re}}\left(\nabla^2 T\right)$$
$$-Mn_F EcH(\varepsilon-T)\left(u\frac{\partial H}{\partial x}+v\frac{\partial H}{\partial y}\right)= \qquad (4)$$
$$\frac{E_c}{\mathrm{Re}}\left[2\left(\frac{\partial u}{\partial x}\right)^2+2\left(\frac{\partial v}{\partial y}\right)^2+\left(\frac{\partial v}{\partial x}+\frac{\partial u}{\partial y}\right)^2\right]$$

where $u,v,p,T$ are the $x$ and $y$ velocity components, pressure and temperature of the fluid, respectively. The magnitude of magnetic field intensity $H$ of the magnetic field is given by [12]

$$H(x,y)=\frac{b}{\sqrt{(x-a)^2+(y-b)^2}} \qquad (5)$$

where $a$ and $b$ are the position of the applied magnetic field.

The new dimensionless parameter is magnetic number $Mn_F$ can be written as:

$$Mn_F=\frac{h^2 BM}{\mu^2 \rho} \qquad (6)$$

where $B$ and $M$ are magnetic induction and magnetization, respectively.

### 2.1 Boundary Conditions

The inflow is assumed to be fully developed with parabolic profile in $x$-direction whereas the temperature has a linear profile. No slip condition is imposed for the velocities on upper and lower plate of channel. The pressure outlet and outflow conditions are normal derivatives taken to be zero. Fig. 1 shows the arrangement of boundary conditions and the magnetic device is assumed to be permanent magnet with nonuniform distribution.

DOI: 10.1051/matecconf/201711101006

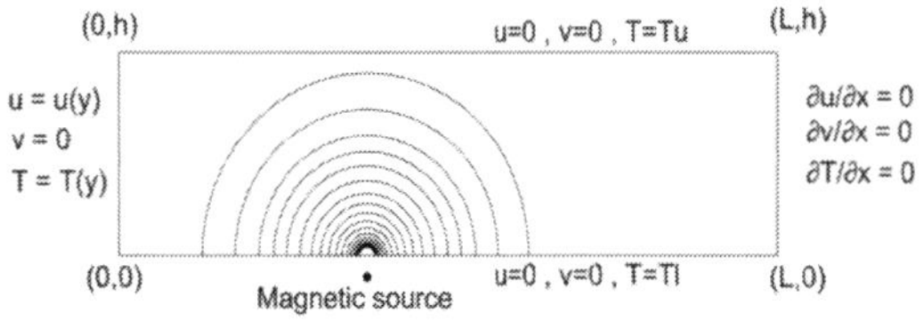


**Fig. 1.** Boundary conditions and magnetic field intensity.

## 3 Numerical Method

The system of partial differential equations (strong form) is discretized using Bubnov-Galerkin weighted residual method. The weighing function is taken to be the same as the interpolation function. This allows discretization of the nonlinear PDE's to obtain algebraic equations where is applicable to solve using computer programming. The stabilized formulation can be achieved by proper selection of interpolation function for velocity and pressure. The mixed interpolation means the order of the interpolation of velocity must higher than the pressure. Therefore, quadratic shape functions are employed in velocity whereas bilinear shape functions are used for the pressure. Standard triangular Taylor-Hood element was used to construct the domain with a denser mesh in between $x$=2 and $x$=3 as shown in Fig. 2. to capture the magnitude of the higher order of magnetic gradient. The point of evaluation for each element is shown at the Fig. 3. where a triangular element is having total six nodes at the vertices and 3 nodes at the midpoint.

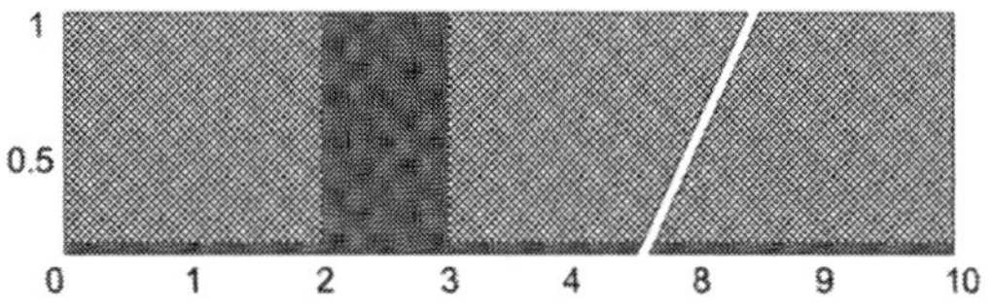


**Fig. 2.** Meshing of the flow domain.

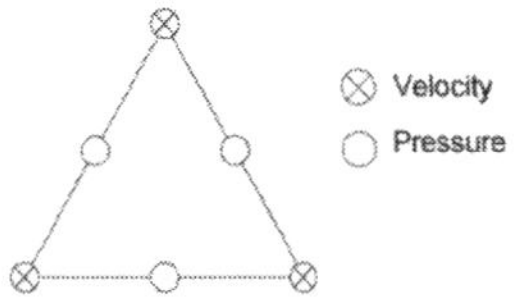


**Fig. 3.** Triangular Taylor-Hood element.

Non-structure grid was used to construct a non-uniform mesh. Mesh testing was conducted to ensure the grid independency of the solution. Subsequently, the total number of element used was 31,606 resulting 213,899 size of stiffness matrix. The FEM discretized system is solved iteratively with MATLAB package involving sparse solver.

### 3.1 Algorithm Efficiency

One of the main challenges in computation lies on solving massive matrix that is consuming computer resources and it can only be resolved using an efficient algorithm. Vector processing, which is a single instruction to operate multiple data gives a better performance in term of speed and memory usage. As shown in Fig. 4, different algorithm gives different efficiencies. Local imposition means impose boundary at local matrix has been found as able to execute the process more rapidly, which is 44 times faster than global imposition to complete the simulation.

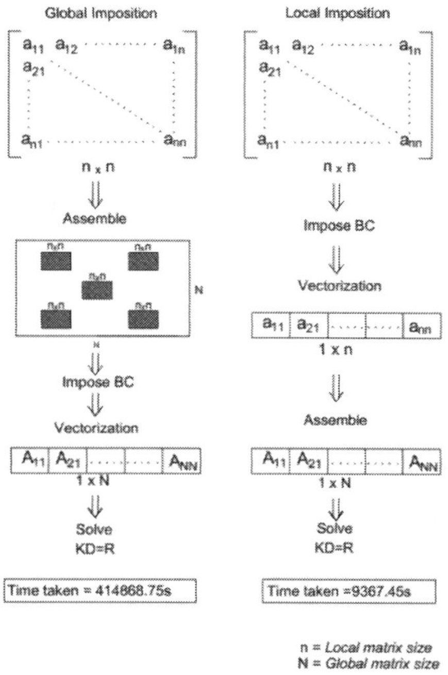


**Fig. 4.** Comparison of the flow model algorithms.

## 4 Numerical Results and Discussion

**Table 1.** Variables for Magnetic Fluid in Rectangular Channel.

| Variable | Value |
|---|---|
| Density, ρ | 1050 kg/m$^3$ |
| Viscosity, μ | 3.2e-3 kg/(ms) |
| a | 2.5m |
| b | -0.05m |
| Magnetization, M | 60 A/m |
| Reynold number, Re | 250 |
| Eckert number, Ec | 2.456x10$^{-6}$ |
| Prandlt number, Pr | 20 |
| Temperature number, ε | 8 |
| Magnetic number, $Mn_f$ | 315 |

DOI: 10.1051/matecconf/201711101006

Initially some verification calculations were performed for pure hydrodynamic case taken $Mn_f = 0$ and the results are well agreed with [16]. The purpose is to assess the stability and accuracy of the finite element algorithm.

Numerical solution for magnetic fluid is verified with [12]. The dimensionless axial velocity component is shown in Fig. 5 at various magnetic field strengths with Re=250. The velocity increases from 1.1 (2T), 1.3 (4T), 1.43 (6T) to 1.55 (8T). The increment of the velocity component is due to the additional energy transport from the stress arising from magnetic strength. At the upper channel, velocity increases as the magnetic strength increases whereas at lower edge the velocity is retarded. It is also observed that recirculation zone occurred at the lower part of channel where the back flow has the minimum value of -0.20. This physically shows that the flow has slowed down.

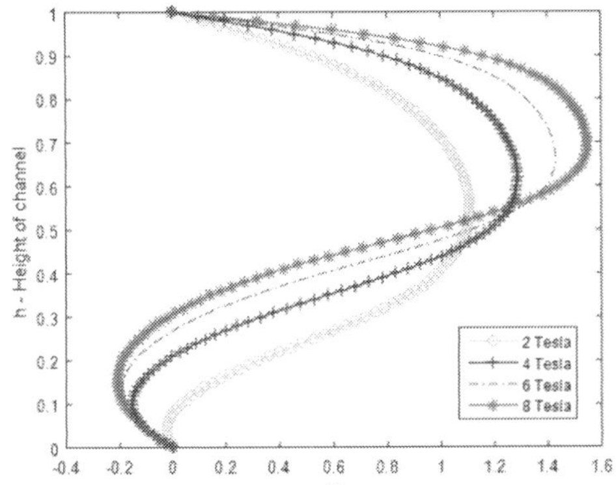


**Fig. 5.** Axial velocity profile in various magnetic field strengths.

Fig.6 shows the behaviour of the temperature profile along the middle of channel. The temperature increases rapidly in the region $x$=2.0 where it is close to magnetic source and plunged drastically to its minimum value at $x$=3.5 then raised steadily. From the point $x$=7.5 onward, temperature take its entrance value as it is approaching outlet. It can be seen that, the increment of $Mn_f$ leads to greater variation of the temperature value.

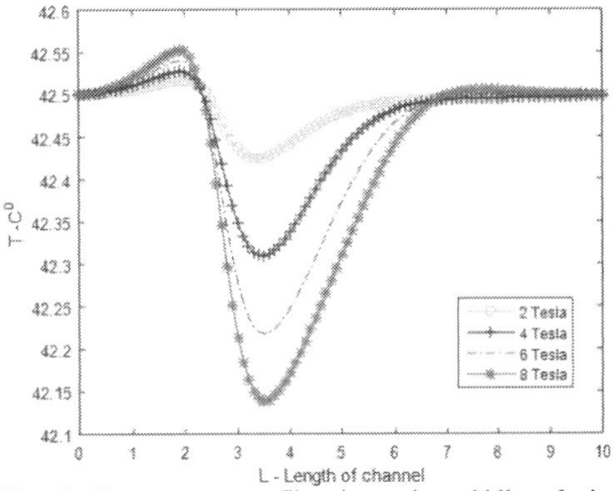


**Fig. 6.** Temperature profile along the middle of channel subjected to various magnetic strengths.

Fig.7 presents the velocity contours for 2T, 4T, 6T, and 8T magnetic strength corresponding to the magnetic number 78, 157, 236 and 315, respectively. The vortex grows near the source and extends in length and height nearby the magnetic source as the magnetic strength increase. At the exit of channel, the velocity gained back fully developed flow.

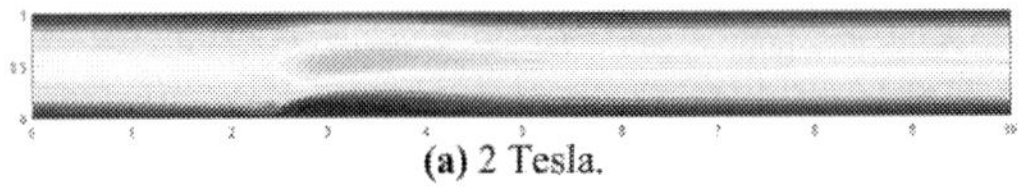

(a) 2 Tesla.

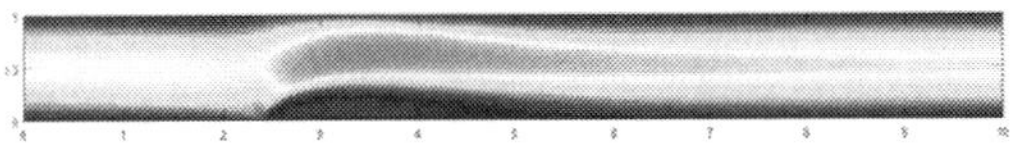

(b) 4 Tesla.

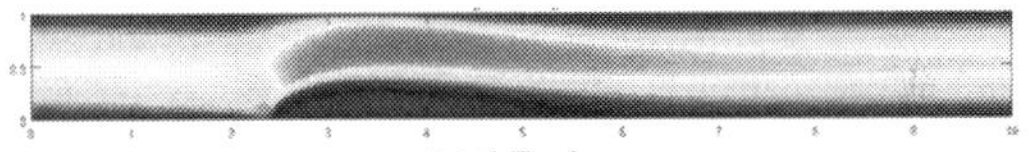

(c) 6 Tesla.

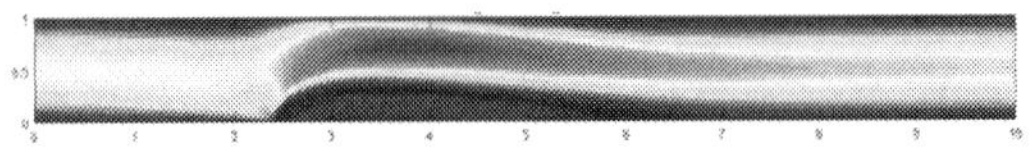

(d) 8 Tesla.

**Fig. 7.** Velocity contours for fluid flow model for magnetic field strength of (a) 2T, (b) 4T, (c) 6T and (d) 8T.

## 5 Concluding Remarks

A mixed finite element formulation for magnetic fluids is successfully developed up to 8 Tesla. The results indicate the feasibility of the mixed formulation as an alternative in the modelling of the magnetic fluid flow. The outcome of this study would provide an insight into the interaction of magnetic fluid with electromagnetics which can be important in various engineering fields especially in oil and gas extraction process.

## References


1. O. A. Alomair, K. M. Matar, and Y. H. Alsaeed, "Nanofluids Application for Heavy Oil Recovery," no. October, pp. 14–16, (2014).
2. A. Roustaei, S. Saffarzadeh, and M. Mohammadi, "An evaluation of modified silica nanoparticles' efficiency in enhancing oil recovery of light and intermediate oil reservoirs," *Egyptian Journal of Petroluem*, **vol. 22,** no. 3, pp. 427–433, (2013).
3. Q. Xue, "Model for effective thermal conductivity of nanofluids," *Physics Letters A*, **vol. 307,** no. 5–6, pp. 313–317, (2003).
4. J. a. Eastman, S. U. S. Choi, S. Li, L. J. Thompson, and S. Lee, "Enhanced Thermal Conductivity through the Development of Nanofluids," *MRS Proc.*, **vol. 457**, pp. 3–11, (1997).
5. R. E. E. Rosensweig, "Heating magnetic fluid with alternating magnetic field," *Journal of Magnetism and Magnetic Materials*, **vol. 252**, no. 0, pp. 370–374, (2002).
6. W. Lian, Y. Xuan, and Q. Li, "Characterization of miniature automatic energy transport devices based on the thermomagnetic effect," *Energy Conversion and Management,* **vol. 50**, no. 1, pp. 35–42, (2009).
7. A. Davidson, C. Huh, and S. L. Bryant, "Focused Magnetic Heating Utilizing Superparamagnetic Nanoparticles for Improved Oil Production Applications," in *SPE International Oilfield Nanotechnology Conference and Exhibition*, (2012).
8. F. S. D. A. Soares, M. Prodanovic, and C. Huh, "Excitable Nanoparticles for Trapped Oil Mobilization," *SPE Imroved Oil Recover. Simp.*, no. April, pp. 1–16, (2014).
9. C. Oldenburg, S. Borglin, and G. Moridis, "Numerical simulation of ferrofluid flow for subsurface environmental engineering applications," *Transport Porous Media*, **vol. 38** , pp. 319–344, (2000).
10. Y. Haik, V. Pai, and C. Chen, "Biomagnetic fluid dynamics," *Fluid dynamics at interfaces,* pp. 439-452, (1999).
11. E. Tzirtzilakis and V. Loukopoulos, "Biofluid flow in a channel under the action of a uniform localized magnetic field," *Computational Mechanics,* **vol. 36**, pp. 360-374, (2005).
12. Loukopoulos, V. C. and E. E. Tzirtzilakis, "Biomagnetic channel flow in spatially varying magnetic field," *International Journal of Engineering Science* **vol. 42,** no. 5: 571-590, (2004).
13. E. Tzirtzilakis, "A mathematical model for blood flow in magnetic field," *Physics of Fluids (1994-present),* **vol. 17**, p. 077103, (2005).
14. S. Kenjereš, "Numerical analysis of blood flow in realistic arteries subjected to strong non-uniform magnetic fields," *International Journal of Heat and Fluid Flow,* **vol. 29**, pp. 752-764, (2008).
15. M. Tezer-Sezgin, C. Bozkaya, and Ö. Türk, "BEM and FEM based numerical simulations for biomagnetic fluid flow," *Engineering Analysis with Boundary Elements,* **vol. 37**, pp. 1127-1135, (2013).
16. L. Q. Tang and T. T. Tsang, "A least-squares finite element method for time-dependent incompressible flows with thermal convection," *International Journal for Numerical Methods in Fluids,* **vol. 17**, pp. 271-289, (1993).

# Prediction of Stream Flow in Humid Tropical Rivers by Support Vector Machines

*Mohammed Seyam*[1,*], *Faridah Othman*[2] *and Ahmed El-Shafie*[2]

[1*]*Civil Engineering Programme., University College of Technology Sarawak, Sibu, Sarawak.* (a.seyam@ucts.edu.my)

[2]*Civil Engineering Dept., Faculty of Engineering, University of Malaya, Kuala Lumpur, Malaysia*

**Abstract.** Stream flow (SF) prediction is considered as a very complex due to the hydrological systems of surface water are complex and dynamic. The reliable prediction of stream flow (SF) can be performed by either conceptual or data-driven based models. In the modelling of hydrological processes, the support vector machine (SVM) is a novel, data-driven approach. Hence, six SVM-based models were generated in this study to predict real time hourly SF in the Selangor River Basin from the water level and rainfall of upstream stations. These models composed of six different combinations of input variables and were trained and tested under hourly records of SF, rainfall, and water level over one year (2011). Among the SVM-based models, SVM-M6, which has nine input variables, was the most effective. Under the training and testing data sets, its correlation coefficient and mean absolute error values were 0.992, 0.953, 0.061 and 0.253 respectively.

**Keywords:** stream flow, data-driven based models, support vector machine, prediction, hydrological modelling.

## 1 Introduction

To predict SF, various models have been established. These models can be classified into two main types: knowledge- driven and data-driven. Each type has specific advantages and disadvantages based on data availability and modelling condition [1, 2]. Knowledge-driven models are also known as physical or conceptual models. They are designed to simulate interior sub processes in prototypes, as well as physical mechanisms that dictate the natural process. These models use a mathematical structure that depends on basin features, such as the specific characteristics of rainfall (intensity and duration), the basin (area, shape, slope and land use, vegetation cover, and soil nature), and climate (temperature, humidity, and wind speed) to model and predict SF [3-5]. However, these models are too complex and demanding. In some cases, conceptual models cannot predict SF accurately and reliably given the lack of required data, especially in developing countries [6], furthermore, the physical process is complicated by the gathering of data on multiple model variables that vary spatially and temporally [7-10].

Data-driven models include those developed using artificial intelligence (AI) techniques, such as artificial neural networks, genetic algorithms, support vector machine (SVM), and fuzzy rule-based systems. These models are adequate alternatives in many hydrological applications, especially when data are inadequate to generate conceptual models [11-14].

This study mainly aims to develop SVM-based models to predict hourly SF of downstream area from the water level and rainfall records of upstream stations in the river basins of humid tropical regions. These models are generated based on hourly records of SF, rainfall, and water level throughout one year (2011). The performance of the models are assessed based on two criteria, namely, correlation coefficient ($R$) and mean absolute error (MAE).

## 2 Methodology

In developing SVM-based models for SF prediction, we primarily consider data collection and analyses, followed by the selection of adequate input and output variables for the model. In small basins, these variables depend completely on the estimated lag time between the upstream and downstream stations. Thereafter, we determine model structure. Finally, we assess the developed models according to the evaluation criteria to obtain the model that best predicts hourly SF.

### 2.1 Case study

In this study, we investigate the Selangor River Basin, which is one of the main rivers in Malaysia. It is located in the Selangor state and has an approximate area of 1960 km$^2$ [15]. From northeast to southwest, the Selangor River is approximately 110 km long [6, 16, 17]. Moreover, the Selangor River Basin provides approximately 50% of the water consumed in Selangor and Kuala Lumpur [18, 19]. Figure 1 presents the location map of the Selangor River Basin in Peninsular Malaysia, as well as its topography maps.

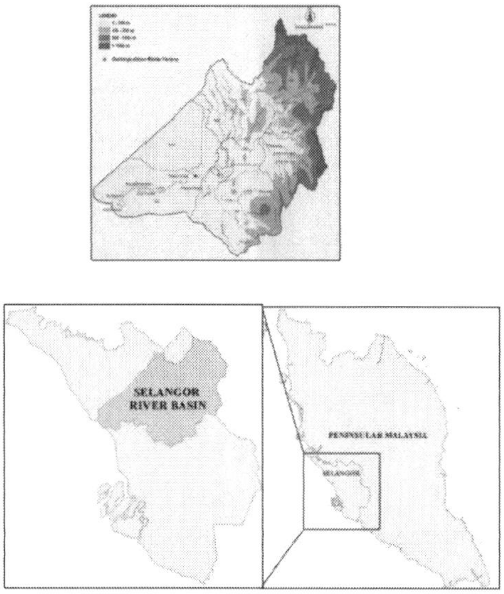


**Fig. 1** Location and topography map of the Selangor River Basin [20].

## 2.2 Data collection and analyses

The SF data of the downstream station were obtained from the Rantau Panjang gauging station, which is located in the downstream of the Selangor River. Before this station, all of the major tributaries of this river converge. Thus, the SF at the Rantau Panjang Station is the best indicator of the stream flow at the study area. Water level and rainfall data were obtained from four upstream stations. The study stations were selected based on data availability and modelling requirement. Moreover, the stations that gauge rainfall and water level are very close to one another. Figure 2 displays the location of the hydrological stations and the flow paths among them in the Selangor River Basin.

**Table 1.** Hydrological stations and the statistical characteristics of the data used.

| Station | Function | Mean | Min. | Max. | Std. Dev. |
|---|---|---|---|---|---|
| Rantau | SF (m$^3$/s) | 60.35 | 23.94 | 294.64 | 39.0 |
| Ulu Yam | RF (mm/h) | 32.24 | 30.56 | 35.49 | 0.49 |
| Batang Kali | RF (mm/h) | 32.42 | 27.03 | 34.71 | 0.78 |
| Kerling | RF (mm/h) | 44.18 | 43.93 | 45.61 | 0.12 |
| Ampang Pecah | RF (mm/h) | 50.16 | 49.61 | 50.89 | 0.15 |
| Ulu Yam | WL (m) | 0.16 | 0.00 | 19.33 | 0.73 |
| Batang Kali | WL (m) | 0.24 | 0.00 | 22.67 | 0.91 |
| Kerling | WL (m) | 0.25 | 0.00 | 25.33 | 1.06 |
| Ampang Pecah | WL (m) | 0.24 | 0.00 | 28.00 | 1.08 |

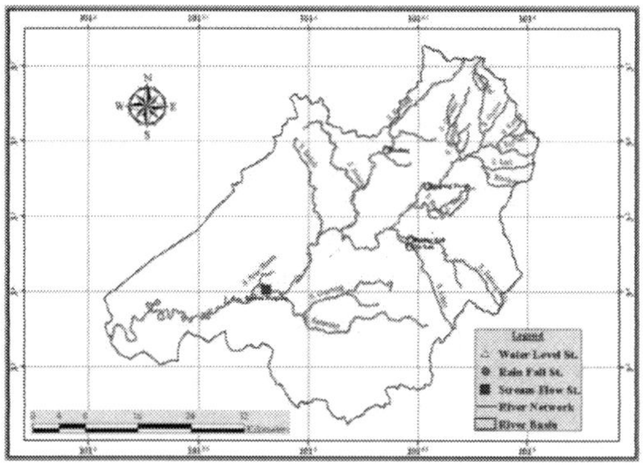

**Fig. 2** Location of the hydrological stations in the Selangor River Basin.

## 2.3 Models development

### *2.3.1 Determination of model variables*

In the development of AI-based models, determining the adequate input and output variables is a key issue. In models of SF prediction, model variables are commonly selected based on a priori knowledge of river basin hydrology, which provides initial indications of potential inputs and outputs [21]. The SF in tropical rivers can be characterized as the function of several influential variables, including rainfall, water level, and the physical characteristics of the river [22]. This study mainly aims to predict hourly SF of downstream area from the water level and rainfall records of upstream stations. Thus, we use the hourly records of water level and rainfall at the upstream stations as input variables and those of SF data in the downstream station as the output variable. Eq. 2 describe the relationship between SF and the influential variables:

$$Sf_t = f(X_{(t)}) + e \quad (2)$$

where $Sf_{(t)}$ represents the SF; $X(t)$ is the input vector that includes the input variables (i.e., rainfall and/or water level); and $e$ is the random error.

We consider three scenarios in selecting the input and output variables of the models. First, we apply the rainfall data of upstream stations as input variables. Second, we regard the water level data of these stations as inputs. Third, we utilize both water level and rainfall data from these stations as inputs.

In these three scenarios, we apply two input vectors. In the first, we use the single antecedent record of upstream stations. In the second, we obtain the average of these antecedent records. Given six input vectors, the single antecedent record of SF in the downstream station is considered another input variable that predicts the SF for a head period equal to the lag time between the upstream and downstream stations. The estimated lag time between these stations determines the final input variables for the six input vectors. Using these vectors, we generated six SVM-based models to predict hourly SF.

### 2.3.2 Model description

SVM is a new learning system that has been developed based on the statistical learning theory aiming at minimizing the generalized model error rather than just minimizing the training error, which consequently increases SVM generalization ability [23, 24].

SVM is a comparatively new AI modelling technique based on statistical learning theory introduced by Vapnik in the 1970s. SVM has been developed as a classification tool and it was applied successfully in a wide range of classification and clustering applications in. Recently, SVM have been successfully extended to apply in regression and prediction applications [11, 25, 26].

Figure 3 presents the Schematic diagram of SVM, where the K(xi,x) is the output of the ith hidden node for input vector x, it is a mapping of the input x and the support vector xi by selecting the kernel function (Chen & Yu, 2007).

SVM has been applied in the time-series prediction of river flow by Samsudin, Saad [6]; in SF prediction under multiple time scales by Asefa, Kemblowski [27]; in the real-time forecasting of flood stage by Yu, Chen [25]; in flood forecasting by [28]; in long-term discharge prediction by Lin, Cheng [29]; in the long-range forecast of SF by [30]; and in the monthly forecasting of SF by Guo, Zhou [31], Noori, Karbassi [32], Shabri and Suhartono [33], and Ch, Anand [34].

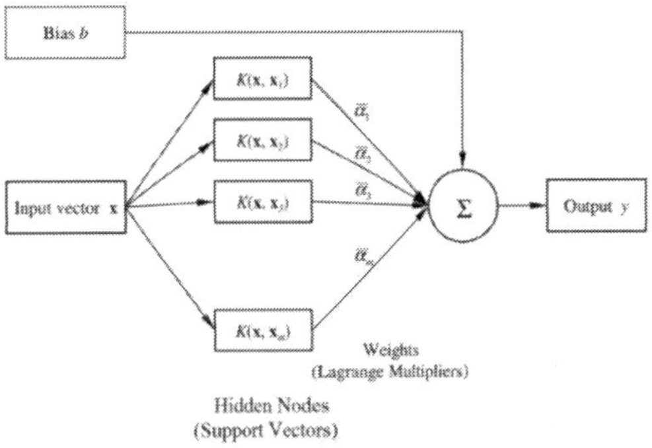


**Fig. 3:** Schematic diagram of SVM architecture

### 2.3.3 Model training

Once the structure of the SVM-based model has been determined, we set the conditions that halt the training processes. These conditions should be set prior to model training. Training is controlled by conditions such as the maximum number of iterations, maximum training time, the target performance that specifies the tolerance between the observed and predicted SF, and the minimum learning rate.

To generate the models, we determined approximately 8,753 patterns of hourly SF, water level, and rainfall records throughout a single year (2011). Table 1 lists the basic statistical characteristics of the hourly records obtained from the stations, such as minimum, maximum, mean, standard deviation, and skewness. The modelling data were divided into two data sets: 75% for training with 6,580 patterns and 25% for model testing with 2,193 patterns. The training data set is used to train the models, and the testing data set assesses the performance of the SVM-based models [35].

In this study, we are applying the SVM as modelling tool. The training process was performed internally using close-source programing which is available in AI toolbox in Statistica software. Hence, the best training algorithm was selected by the software using built-in optimization technique, where

 DOI: 10.1051/matecconf/201711101007

Levenberg-Marquardt technique is adopted as the training algorithm because it provides the best performance over other algorithms. The main attention of the this paper is maximize the correlation and minimize the error, regardless the details of training algorithms and techniques.

#### *2.3.4 Performance evaluation criteria*

The performance of the models was assessed based on two criteria: $R$ and MAE. $R$ is a statistical technique that indicates the strength and direction of a linear relationship between two variables [36, 37]. In this study, $R$ was used to validate the agreement between the observed and predicted hourly SFs. $R^2$ describes the variance between two variables as determined by the linear fit. $R$ can be calculated under different modes, but the most popular one is the Pearson $R$. This value is computed by dividing the covariance of the two variables by the product of their standard deviations, as described in the following equation:

$$R = \frac{\sum_{i=1}^{n}(x_i - \bar{x})\,(y_i - \bar{y})}{\sqrt{\sum_{i=1}^{n}(x_i - \bar{x})^2 \sum_{i=1}^{n}(y_i - \bar{y})^2}} \tag{3}$$

where $n$ is the number of data pairs and $x$ and $y$ are the variables.

In a perfectly increasing linear relationship, $R$ is +1. By contrast $R$ is −1 in a perfectly decreasing linear relationship. The $R$ values between +1 and −1 indicate the strength of the linear relationship between the variables. $R = 0$ signifies that the variables are not linearly related.

MAE evaluates the residual or the differences between the observed and predicted SF. Theoretically, its minimum value should be zero to indicate the perfect fit of the model. However, this value is difficult to obtain. Moreover, MAE has no maximum value and is calculated using the following equation:

$$MAE = \frac{\sum_{i=1}^{n}\left|X_{o,i} - X_{m,i}\right|}{n} \tag{4}$$

where $X_m$ represents the data predicted by a model and $X_o$ is the observed data.

## 3 Results and discussion

To predict hourly SF in the Rantau Panjang Station, we selected six AI-based models under different combinations of input variables. Table 2 presents the model structures. The six models were trained and developed by SVM to predict hourly SF. The performances of the models were assessed based on the training and testing data sets, as well as the overall performance of the data sets. The best fit model to predict hourly SF is thus determined according to the performance of the testing data sets.

SVM-M6 model displays the highest $R$ values (0.992 and 0.953) and the lowest MAE (0.061 and 0.253) in both the training and testing data sets, respectively. Figure 4 shows the correlation between the observed and predicted hourly SF in the SVM-M6 model given training and testing data sets. The observed and predicted hourly stream flow of the training and testing data sets, seem to be in good accord with $R^2$ 0.986 and 0.909 respectively. Figure 5 compares the observed and predicted hourly SF in SVM-M6 for the period of September 2013. These flows are highly consistent.

**Table 2.** Input and output variables of the AI models.

| **Model** | **Inputs** | **Output** | **No. of input Variables** |
|---|---|---|---|
| M1 | $Rf_{u(t)}$, $Rf_{b(t)}$, $Rf_{k(t)}$, $Rf_{a(t)}$, $Sf_{(t)}$ | $Sf_{(t+17)}$ | 5 |
| M2 | $Rf_{u(t)}$, $Rf_{b(t)}$, $Rf_{k(t)}$, $Rf_{a(t)}$, $Sf_{(t)}$ | $Sf_{(t+17)}$ | 5 |
| M3 | $Wl_{u(t)}$, $Wl_{b(t)}$, $Wl_{k(t)}$, $Wl_{a(t)}$, $Sf_{(t)}$ | $Sf_{(t+12)}$ | 5 |
| M4 | $Wl_{u(t)}$, $Wl_{b(t)}$, $Wl_{k(t)}$, $Wl_{a(t)}$, $Sf_{(t)}$ | $Sf_{(t+12)}$ | 5 |
| M5 | $Wl_{u(t)}$, $Wl_{b(t)}$, $Wl_{k(t)}$, $Wl_{a(t)}$, $Rf_{u(t-5)}$, $Rf_{b(t-5)}$, $Rf_{k(t-5)}$, $Rf_{a(t-5)}$, $Sf_{(t)}$ | $Sf_{(t+12)}$ | 9 |
| M6 | $Wl_{u(t)}$, $Wl_{b(t)}$, $Wl_{k(t)}$, $Wl_{a(t)}$, $Rf_{u(t-5)}$, $Rf_{b(t-5)}$, $Rf_{k(t-5)}$, $Rf_{a(t-5)}$, $Sf_{(t)}$ | $Sf_{(t+12)}$ | 9 |

DOI: 10.1051/matecconf/201711101007

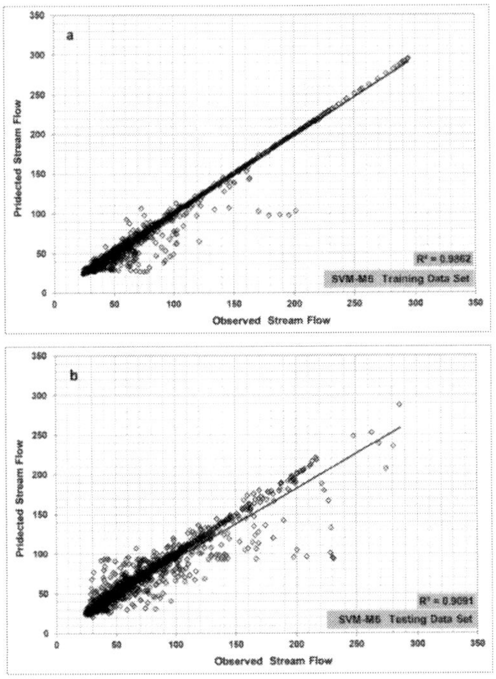


**Fig. 4** Correlation between the observed and predicted hourly stream flow in the SVM-M6 model: (a) training data set and (b) testing data set

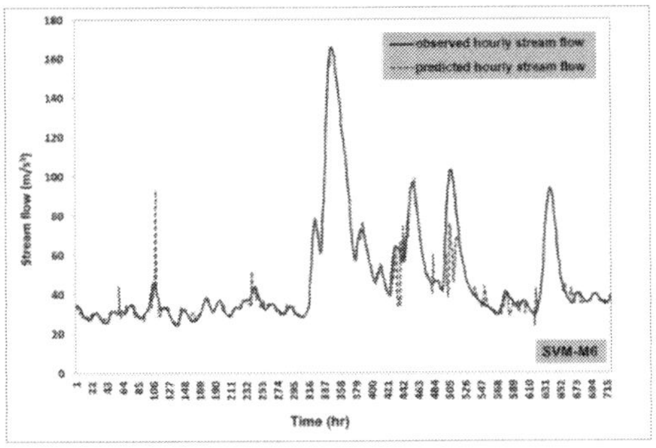


**Fig. 5** Comparison between the observed and predicted hourly stream flows in the SVM-M6 model

## 4 Conclusions

In this study, the ability of SVM to predict hourly SF in downstream area from the upstream water level and rainfall records in a humid tropical area was explored. In the process, we developed six SVM-based models with different combinations of input variables. The hourly records of SF, rainfall, and water level throughout one year (2011) were utilized to train and test the SVM models. The hourly water level and rainfall data in the upstream stations were considered input variables, and the hourly SF data in the downstream station were regarded as the output variable. The performance levels of the models were assessed based on two evaluation criteria, namely, *R* and MAE. The best fit model to predict hourly stream flow was determined based on the performance of the testing data sets. Among the SVM-based models, SVM-M6 model displays the highest *R* values (0.992 and 0.953) and the lowest MAE (0.061 and 0.253) in both the training and testing data sets, respectively. The ability of the three techniques of SVM SVM for SF prediction in the downstream area from the upstream WL and RF records in the Selangor River basin was explored and successfully achieved with high performance throughout two modelling phases. Although SVM performed well in real-time SF prediction, higher R in the Selangor River basin can be investigated by employing other AI techniques, such as FRBSs and GAs.

## References

1. Aqil, M., et al., *Neural Networks for Real Time Catchment Flow Modeling and Prediction.* Water Resources Management, 2007. **21**(10): p. 1781-1796.

2. Kentel, E., *Estimation of river flow by artificial neural networks and identification of input vectors susceptible to producing unreliable flow estimates.* Journal of Hydrology, 2009. **375**(3–4): p. 481-488.

3. Jain, A. and A.M. Kumar, *Hybrid neural network models for hydrologic time series forecasting.* Applied Soft Computing, 2007. **7**(2): p. 585-592.

4. Nilsson, P., C.B. Uvo, and R. Berndtsson, *Monthly runoff simulation: Comparing and combining conceptual and neural network models.* Journal of Hydrology, 2006. **321**(1–4): p. 344-363.

5. Kneis, D., *A lightweight framework for rapid development of object-based hydrological model engines.* Environmental Modelling & Software, 2015. **68**(0): p. 110-121.

6. Samsudin, R., P. Saad, and A. Shabri, *River flow time series using least squares support vector machines.* Hydrol. Earth Syst. Sci., 2011. **15**(6): p. 1835-1852.

7. Firat, M. and M.E. Turan, *Monthly river flow forecasting by an adaptive neuro-fuzzy inference system.* Water and Environment Journal, 2010. **24**(2): p. 116-125.

8. Akhtar, M.K., et al., *River flow forecasting with artificial neural networks using satellite observed precipitation pre-processed with flow length and travel time information: case study of the Ganges river basin.* Hydrol. Earth Syst. Sci., 2009. **13**(9): p. 1607-1618.

9. Nativi, S., et al., *Big Data challenges in building the Global Earth Observation System of Systems.* Environmental Modelling & Software, 2015. **68**(0): p. 1-26.

10. Seyam, M. and Y. Mogheir, *Application of artificial neural networks model as analytical tool for groundwater salinity.* Journal of Environmental Protection, 2011. **2**(01): p. 56.

11. Solomatine, D., L.M. See, and R.J. Abrahart, *Data-Driven Modelling: Concepts, Approaches and Experiences*, in *Practical Hydroinformatics*, R. Abrahart, L. See, and D. Solomatine, Editors. 2008, Springer Berlin Heidelberg. p. 17-30.

12. Kisi, O., et al., *Intermittent Streamflow Forecasting by Using Several Data Driven Techniques.* Water Resources Management, 2012. **26**(2): p. 457-474.

13. Daniel, E.B., et al., *Watershed Modeling and its Applications: A State-of-the-Art Review.* The Open Hydrology Journal, 2011. **5**: p. 26-50.

14. Kanevski, M., et al., *Environmental data mining and modeling based on machine learning algorithms and geostatistics.* Environmental Modelling & Software, 2004. **19**(9): p. 845-855.

15. Seyam, M. and F. Othman, *Long-term variation analysis of a tropical river's annual streamflow regime over a 50-year period.* Theoretical and Applied Climatology, 2015. **121**(1): p. 71-85.

16. Lee, C.M., *Master Plan Study on Flood Mitigation and River Management for Sg. Selangor River Basin.*, 2002, Drainage and Irrigation Department (DID) Malaysia.

17. Hassan, A.J., A.A. Ghani, and R. Abdullah, *Development Of Flood Risk Map Using GIS For Sg. Selangor Basin*, 2004, National Hydraulic Research Institute of Malaysia: Malaysia.

18. Subramaniam, V., *Managing Water Supply In Selangor And Kuala Lumpur*, in *BULETIN INGENIEUR*2004, THE BOARD OF ENGINEERS MALAYSIA: 50580 Kuala Lumpur, Malaysia. p. 12-20.

19. Seyam, M. and F. Othman, *The Influence of Accurate Lag Time Estimation on the Performance of Stream Flow Data-driven Based Models.* Water Resources Management, 2014. **28**(9): p. 2583-2597.

20. Seyam, M. and F. Othman, *Long-term variation analysis of a tropical river's annual streamflow regime over a 50-year period.* Theoretical and Applied Climatology, 2014.

21. Maier, H.R. and G.C. Dandy, *Neural networks for the prediction and forecasting of water resources variables: a review of modelling issues and applications.* Environmental Modelling & Software, 2000. **15**(1): p. 101-124.

22. Firat, M., *Artificial Intelligence Techniques for river flow forecasting in the Seyhan River Catchment, Turkey.* Hydrol. Earth Syst. Sci. Discuss., 2007. **4**(3): p. 1369-1406.

23. Behzad, M., K. Asghari, and E.A. Coppola Jr, *Comparative Study of SVMs and ANNs in Aquifer Water Level Prediction.* Journal of Computing in Civil Engineering, 2010. **24**: p. 408.

24. Asefa, T., et al., *Multi-time scale stream flow predictions: The support vector machines approach.* Journal of Hydrology, 2006. **318**(1-4): p. 7-16.

25. Yu, P.-S., S.-T. Chen, and I.F. Chang, *Support vector regression for real-time flood stage forecasting.* Journal of Hydrology, 2006. **328**(3–4): p. 704-716.

26. Wu, C.L., K.W. Chau, and Y.S. Li, *River stage prediction based on a distributed support vector regression.* Journal of Hydrology, 2008. **358**(1–2): p. 96-111.

27. Asefa, T., et al., *Multi-time scale stream flow predictions: The support vector machines approach.* Journal of Hydrology, 2006. **318**(1–4): p. 7-16.

28. Chen, S.-T. and P.-S. Yu, *Pruning of support vector networks on flood forecasting.* Journal of Hydrology, 2007. **347**(1–2): p. 67-78.

29. Lin, J.Y., C.T. Cheng, and K.W. Chau, *Using support vector machines for long-term discharge prediction.* Hydrological Sciences Journal, 2006. **51**(4): p. 599-612.

30. Basketfield, D. and N. She, *Long Range Forecast of Streamflow Using Support Vector Machine*, in *Impacts of Global Climate Change*. 2005. p. 1-9.

31. Guo, J., et al., *Monthly streamflow forecasting based on improved support vector machine model.* Expert Systems with Applications, 2011. **38**(10): p. 13073-13081.

32. Noori, R., et al., *Assessment of input variables determination on the SVM model performance using PCA, Gamma test, and forward selection techniques for monthly stream flow prediction.* Journal of Hydrology, 2011. **401**(3–4): p. 177-189.

33. Shabri, A. and Suhartono, *Streamflow forecasting using least-squares support vector machines.* Hydrological Sciences Journal, 2012. **57**(7): p. 1275-1293.

34. Ch, S., et al., *Streamflow forecasting by SVM with quantum behaved particle swarm optimization.* Neurocomputing, 2013. **101**(0): p. 18-23.

35. Tiwari, M.K. and C. Chatterjee, *Development of an accurate and reliable hourly flood forecasting model using wavelet–bootstrap–ANN (WBANN) hybrid approach.* Journal of Hydrology, 2010. **394**(3–4): p. 458-470.

36. Perugu, M., A. Singam, and C. Kamasani, *Multiple Linear Correlation Analysis of Daily Reference Evapotranspiration.* Water Resources Management, 2013. **27**(5): p. 1489-1500.

37. Seyam, M., F. Othman, and A. El-Shafie, *RBFNN Versus Empirical Models for Lag Time Prediction in Tropical Humid Rivers.* Water Resources Management, 2016.

# Numerical estimation of natural ventilation of cubical urban arrays with different packing density

*Saidatul Sharin* Shuhaimi[1], *Sheikh Ahmad Zaki*[1,*], *Mohamed Sukri* Mat Ali[1], and *Nurul Huda* Ahmad[2], *Fitri* Yakub[1]

[1]Department of Mechanical Precision Engineering, Malaysia-Japan International Institute of Technology, Universiti Teknologi Malaysia, 54100 Kuala Lumpur, Malaysia
[2]Engineering Department, UTM Razak School of Engineering and Advanced Technology, Universiti Teknologi Malaysia, Kuala Lumpur, 54100 Kuala Lumpur, Malaysia

**Abstract.** In an urban environment where buildings are closely packed, natural ventilation performance is undesirably disturbed by the effect of surrounding buildings. Cross-ventilation refers to the regulation of air within a building, which is essential in providing good air quality and thermal comfort for the occupants. Thus, this study focuses on the impact of packing density on ventilation rate of cross-ventilated buildings. The numerical estimation is performed by means of computational fluid dynamic (CFD) using the Reynolds Averaged Navier-Stokes (RANS) with RNG k – ε (RNG) turbulence model. Three configurations of simplified generic cubes which are regularly aligned with packing density of 25%, 35%, and 50% were considered. Velocity distribution around and inside the buildings as well as the ventilation rate are analysed. The case with packing density of 25% exhibits a reduction of 90% in the ventilation rate compared to the isolated case and continues to decrease as the packing density increase up to 35%. However, further increase of packing density up to 50%, slightly increases the ventilation rate. Hence, the result of this study imposed that surrounding buildings have a substantial influence on ventilation performance of cross-ventilated buildings.

## 1 Introduction

Living in a period of tremendous technological development and densely inhabited cities, exploitation of natural ventilation into built environment for the construction of new buildings like residential, commercial buildings and public facilities are substantial for a country like Malaysia. Natural ventilation is conceptually simple; however the design to fully utilize the source can be a challenge. Ventilation performance is affected by microclimate (wind speed, wind direction, temperature and humidity), building form (building shape, size of openings and orientation) and surrounding (topography and location). Great ventilation has positive impact on occupant for the liveability of city. As the density of a city increase, occupants experience thermal discomfort and require usage of cooling system commonly in the form of mechanical ventilation. As a result, the quality of air decreases and larger energy is consumed. Natural ventilated building, relying on difference in temperature and wind induced pressure is the most common form of natural ventilation [1].

The concept of natural ventilation is not contemporary. It has been exploited for many centuries long before mechanical ventilation is invented. In cross ventilation, air enters from windward façade moving through leeward façade by openings installed on opposite sides of the enclosed area by wind-induced pressure difference between the opposite openings. Previous studies demonstrated the benefits of cross ventilation compared to single sided ventilation [2,3]. An experiment on a cross ventilation model revealed that the discharge coefficient on inlet did not fluctuate significantly with direction of wind [4]. A series of wind tunnel experiment were conducted to examine the wind load on a low-rise structure within an array of buildings of various spacing [5]. On top of that, both experimental and numerical methods have been utilized to investigate the effect of opening configurations on cross ventilation rate [6,7]. Despite the exhaustive research in this field, the influence of surrounding buildings has only been explored in a limited manner. One can be notable is a study on the significance of separation distance and building disposition in enhancing the ventilation rate of a cross ventilated buildings [8]. An experimental study using wind-tunnel suggested that the pressure coefficient of a cube array with solid walls decrease as the packing density increase [9]. Whereas a study using large-eddy simulation for heterogeneous buildings revealed that as packing density increase, a larger sheltering effect is observed towards surrounding buildings [10]. However, a similar study on a building with opening is still inadequate to strengthen the hypothesis. Thus, this paper aims to reveal how packing density affects ventilation rate in an urban environment with the aid of CFD.

## 2 Methodology

* Corresponding author: sheikh.kl@utm.my

## 2.1 Model Validation

The building model used in this validation is shown in Figure 1. The building has a size of 0.3 m × 0.3 m × 0.15 m (width × length × height) with two openings of size 0.06 m × 0.03 m (width × height) located at the centre of both windward facade and leeward facade. The thickness of the wall is $10^{-6}$ m. It is a cross ventilated building replicated from the wind tunnel experiment [4].

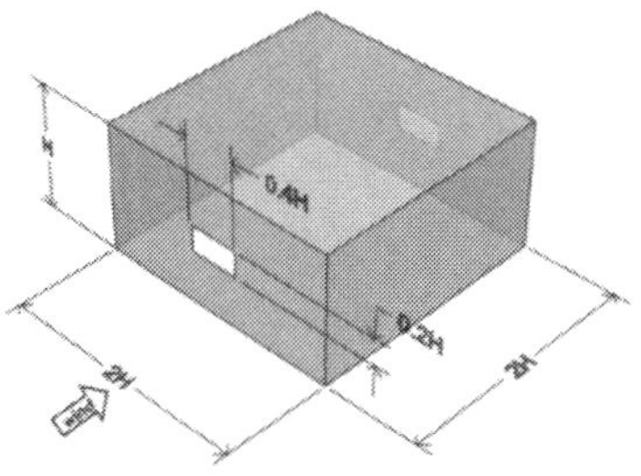

**Fig. 1.** Schematic view of building model (H =0.15m).

The inlet velocity profile is defined by the power law expression

$$U(z) = U_{ref}\left(\frac{z}{H}\right)^{n} \tag{1}$$

where $U_{ref}$ is the velocity at building height in m/s, H is the building height in metres, $z$ is the vertical location also in metres, and $n$ is the power coefficient with a value of 0.25. The inlet velocity profile of the RNG model used in the present study is plotted against $Z/H$ to fit the velocity profile used in the wind tunnel experiment [4], as shown in Figure 2.

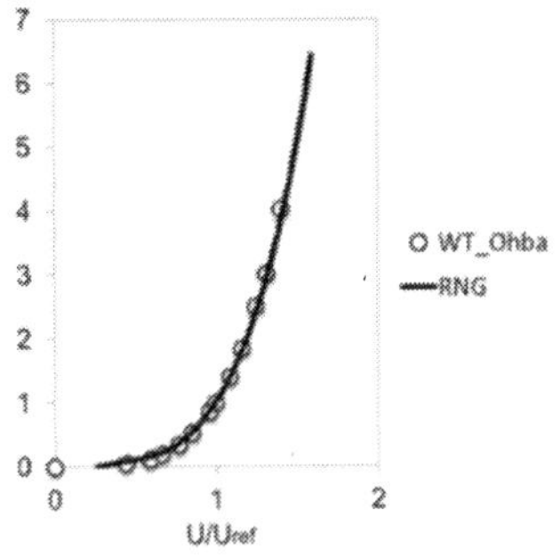


**Fig. 2.** Inlet velocity profile.

The dimension of the domain was determined based on the existing best practice guidelines [9], however the outflow region located at the leeward area of the domain was not determined according to the guidelines. The size of the computational domain is 17H × 12H × 6.5H.

The velocity inflow is prescribed at X = -7.5H in the upstream path perpendicular to the opening while the outflow boundary is placed at X = 7.5H downstream from the leeward building façade to the outlet plane. Width of the domain is set at Y = ± 5H from the sidewalls of the building, whereas the height of the domain is Z = 6.5H. Zero flux condition is applied on the lateral and upper boundaries meanwhile the building and the ground are set with wall functions.

Figure 3 (a) and (b) compares the velocity vector inside and around the building for simulation and experiment [4] respectively. Comparing both of these velocity vectors, the result shows a good agreement in general. The incoming flow is separated near the windward roof. However, a small discrepancy can still be observed near the edge of the roof where RNG under predict the separation flow compared to the experiment. A recirculating eddy can be observed on the upwind face on both RNG and wind tunnel case. The RNG model is able to capture the downward tendency of the flow passing through the windward opening, similar to that of wind tunnel. The flow inside the building shows similar behaviour as the experiment and the weakening flow right after the leeward openings are also captured within acceptable accuracy.

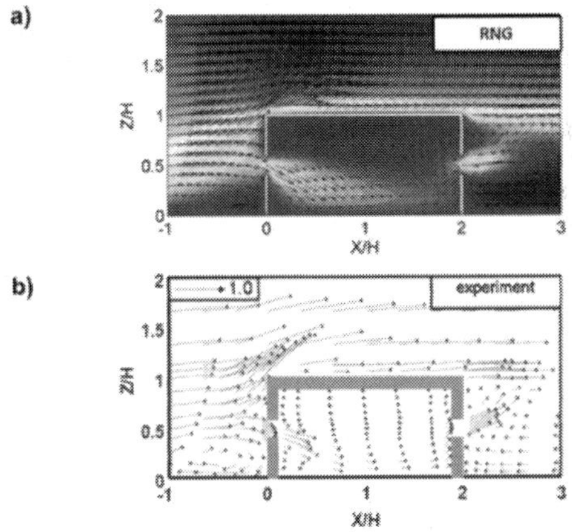


**Fig. 3.** Velocity vector and contour for a) RNG and b) Wind tunnel by Ohba et al.[4]

Ventilation rate in the present study is calculated directly from the airflow passing through the inlet of the openings, expressed as

$$Q = \overline{U} \times A \tag{2}$$

where Q is the ventilation rate in $m^3$/s, $\overline{U}$ is the mean velocity at opening in m/s, and A is the area of opening in $m^2$. The reference ventilation rate, $Q_{ref}$ are defined as

$$Q_{ref} = \overline{U}_{ref} \times A \tag{3}$$

where $Q_{ref}$ is the reference ventilation rate in $m^3$/s and $\overline{U}_{ref}$ is the mean velocity taken at building height.

Grid size located within the building area is refined up to H/64 to certify adequate degree of grid independence of the result. The normalized volumetric flow rates of three different grid sizes on the openings are compared with that of experiment in Table 1. The coarse grid size over predicts the ventilation rate by about 20% difference when compared to the experimental value. Hence the coarse grid size is not preferable in this study due to higher discrepancy. The medium grid size predicts the ventilation rate reasonably

DOI: 10.1051/matecconf/201711101008

with only 6% of discrepancy, while the fine grid size has a value almost identical to the experimental data with a discrepancy about 2% only. Hence the selection of fine grid size was used for the follow-up investigations.

**Table 2.** Normalized ventilation rate of different grid size.

| Grid Size | $Q/Q_{ref}$ |
|---|---|
| Coarse (H/12.5) | 0.547 |
| Medium (H/25) | 0.483 |
| Fine (H/64) | 0.465 |
| Wind tunnel | 0.456 |

## 2.2 Buildings Configurations and Computational Domain

Simplified generic cube was used to represent urban environment. The building model investigated in present study is a cross-ventilated building that is used in the validation exercise. The three dimensional computational domain composed of nine buildings, arranged in a regularly aligned manner. The building with openings is located in the centre of the arrangement. Three urban configurations are defined by varying the separation distance, W, between the buildings to represent packing density, $\lambda_p$ of 25% (PD25), 35% (PD35) and 50% (PD50) as in Figure 4. The size of the computational domain varies as packing density changes [11]. The wind direction is assumed to be normal to the opening of the building.

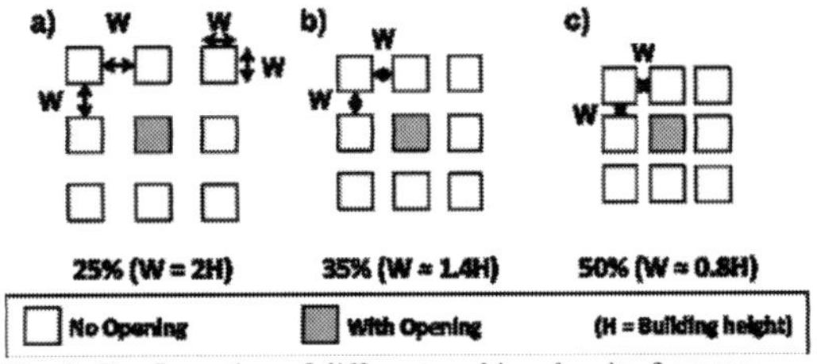


**Fig. 4.** Configuration of different packing density from top view. a) PD25 b) PD35 c) PD50.

## 2.3 Boundary Conditions and Turbulence Model

All the boundary conditions used in the validation process are applied onto the case study. Inlet velocity profiles of power law are expressed at the inlet boundary using Equation 3. All the walls and the ground of the buildings are treated with wall function. RNG turbulence model is applied due to its superior performance in validation exercise [12,13]. Open source software named OpenFOAM with version 2.3.1 is used to resolve the RANS equation of the RNG turbulence model considering its reliability in simulating airflow distribution in buildings [14]. Pressure velocity coupling is solved using the SIMPLE algorithm. Second order discretization scheme are used for the viscous terms of the governing equations. Convergence is obtained when the scaled residuals reached a minimum of $10^{-6}$ for pressure and $10^{-7}$ for dissipation rate, kinetic energy, and velocity.

# 3 Results and Discussion

## 3.1 Ventilation rate

A graph of normalized ventilation rate, $Q/Q_{ref}$ versus packing density in Figure 5 is plotted. The trend exhibited that ventilation rate on the openings decreases as packing density increases from 0% (isolated building) to 35%. The building experienced a reduction of almost 90% as it is surrounded by other buildings with no opening as in PD25. This figure is slightly higher than the one mentioned in previous study [15] which is most probably due to disparity in the building configuration and opening size. The normalized ventilation rate dropped to nearly 0 in PD35 signifying a very low ventilation performance in the building. However, the ventilation rate is slightly increased as the packing density is increased to 50% in PD50. The reduction of ventilation rate can be explained by the fact that smaller pressure difference is produced on building surfaces as the packing density increase [9].

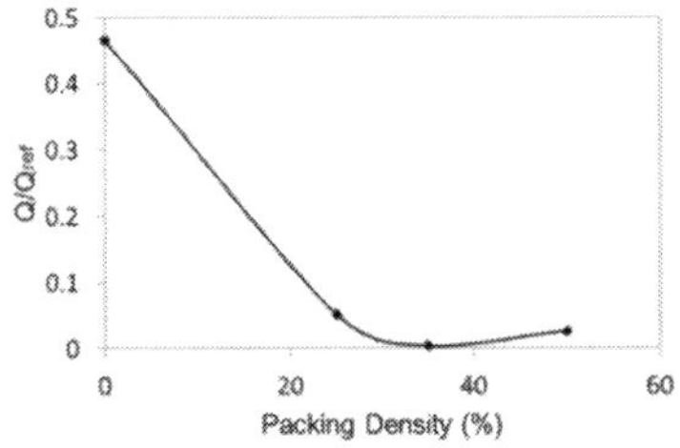


**Fig. 5.** Normalized ventilation rate for different packing density.

Figure 6 visualize the flow distribution of each case in terms of vectors and contours. A skimming flow regime [16] is observed inside the street canyon perpendicular to the wind direction in all cases. Nevertheless, the flow distribution is slightly modified as the distance between adjacent buildings is closer. PD25 shows the higher amount of flow entering the inlet compared to others. The flow inside building becomes weaker in case of PD35 and PD50. In contrast with PD25 and PD35, the flow in PD50 is observed to enter the building through the outlet on the leeward wall and escape through the windward opening. This indicates that ventilation behaviour depends on the characteristic of flow inside street canyon.

DOI: 10.1051/matecconf/201711101008

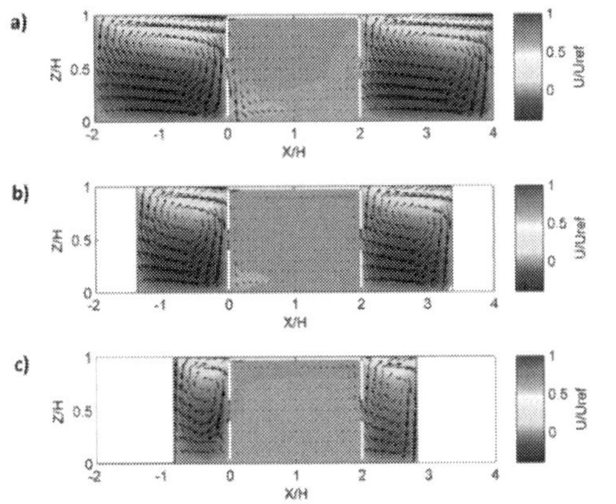


**Fig. 6.** Velocity vector and contour for a) PD25 b) PD35 and c) PD50

The negative value of $U/U_{ref}$ taken at the centre of the inlet as in Figure 7 justified the aforementioned statement regarding the direction of wind at the opening for case PD50.

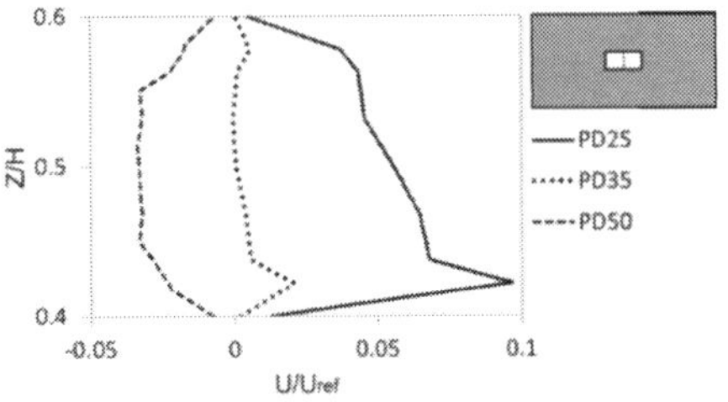


**Fig. 7.** $U/U_{ref}$ at the centre of the inlet opening.

## 4 Conclusions

One of the key elements in a sustainable environment is the exploitation of natural ventilation. Nevertheless, the utilization of natural ventilation is hampered due to rapid urbanization in the recent years.

CFD simulations in investigation of cross ventilation buildings with different packing densities are presented in this paper. 3D steady-state RANS approach is paired with the RNG k-ε model to provide closure for the turbulence model. Model validation is conducted using a reference case adopted from wind tunnel experiment, where it is replicated into computational model and the vertical velocity profile from both models are compared. The result shows a very good agreement with the experimental data; therefore proved the reliability of the RNG k-ε two-equation model in simulation building ventilation. Sensitivity study is conducted to choose the appropriate grid size for the study and to justify the selection. The verification study gives more insight on the effect of different grid size on the accuracy of the solution, where finer grid size exhibits superior accuracy compared to coarser grid size.

The case study focused on three different packing densities, particularly 25%, 35%, and 50%. The effect of packing density on ventilation rate are examined by calculating the normalize ventilation rate, $Q/Q_{ref}$ of hypothetical urban environment in regular alignment. The ventilation rate on the openings are analysed, calculated, and compared. Further analyses are done to attain more perspicacity of the behaviour of the flow.

For packing density of 25%, the ventilation rate dropped about 90% compared to the reference case. The ventilation rate becomes weaker when the packing density is increased to 35%. Contradictorily, a further increase in packing density up to 50% shows a slight increase with a negative sign in ventilation rate indicating the opposite direction of airflow compared to the ventilation rate of packing density 35%. Comparing all three cases simultaneously, it can be confirmed that the ventilation performance strongly depends on the area packing density as it affects the flow regimes inside the canyon. Hence, the fundamental relation between ventilation performances of cross-ventilated buildings to packing density is explored.

This research was financially supported by the AUNSEED-NET Grant (Vot 4B229) project of Universiti Teknologi Malaysia.

## References

1. R. Teppner, B. Langensteiner, W. Meile, G. Brenn, and S. Kerschbaumer, *Air change rates driven by the flow around and through a building storey with fully open or tilted windows: An experimental and numerical study*, Energy and Buildings. **76**:640-653 (2014).
2. K. Niachou, S. Hassid, M. Santamouris, and I. Livada, *Comparative monitoring of natural, hybrid and mechanical ventilation systems in urban canyons*, Energy and Buildings. **37(5)**: 503-513. (2005).
3. N.F.M. Kasim, S.A. Zaki, M.S.M Ali, N. Ikegaya, and A.A Razak, *Computational study on the influence of different opening position on wind-induced natural ventilation in urban building of cubical array*, Procedia Engineering. **169**:256-263 (2016).
4. M. Ohba, K. Irie, and T. Kurabuchi, *Study on airflow characteristic inside and outside a cross-ventilation model and ventilation flow rates using wind tunnel experiments*, Journal of Wind Engineering and Industrial Aerodynamics. **89 (14-15)**: 1513-1524 (2001).
5. B.E. Lee, M. Hussain, and B. Salman, *A method for the assessment of the wind-induced natural ventilation forces acting on low rise building arrays*, Building Services Engineering Research and Technology. **1(1)**: 35-48 (1980).
6. P. Karava, T. Stathopoulos, and A. Athienitis, *Airflow assessment in cross-ventilated buildings with operable façade elements*, Building and Environment. **46(1)**: 266-279 (2011).
7. J. Seifert, Y. Li, J. Axley, and M. Rösler, *Calculation of wind-driven cross ventilation in buildings with large openings*, Journal of Wind Engineering and Industrial Aerodynamics. **94(12)**: 925-947 (2006).

8. J. O. P. Cheung, and C. H. Liu, *CFD simulations of natural ventilation behavior in high-rise buildings in regular and staggered arrangements at various spacing*. Energy and Buildings. **43(5)**: 1149-1158 (2011).
9. S. A. Zaki, A. Hagishima, and J. Tanimoto, *Experimental study of wind-induced ventilation in urban building of cube arrays with various layouts,* Journal of Wind Engineering and Industrial Aerodynamics, **103**: 31-40 (2012).
10. A.F. Mohammad, S.A. Zaki, M.S.M Ali, H. Aya, A.A Razak, M. Shirakashi and N. Arai, *Large eddy simulation of wind pressure distribution on heterogenous buildings in idealised urban models*. **78**: 3055-3060 (2015).
11. J. Franke, A. Hellsten, H. Schlünzen, and B. Carissimo, *Guideline for the CFD simulation of flows in the urban environment: COST Action 732 Quality Assurance and Improvement of Microscale Meteorological*. 1-52 (2007).
12. A. D. Ferreira, A. C. M. Sousa, and D. X. Viegas, *Prediction of building interference effects on pedestrian level comfort*, Journal of Wind Engineering and Industrial Aerodynamics. **90(4-5)**: 305-319 (2002).
13. Q. Chen, *Ventilation performance prediction for buildings: A method overview and recent applications*, Building and Environments. **44(4)**: 848-858 (2009).
14. N. F. M. Kasim, S. A. Zaki, M.S. Mat Ali, A.F. Mohammad and A. Abd Razak, *A verification and validation study of CFD Simulation of wind-induced ventilation on building with single-sided opening*, Applied Mechanics and Materials. **554**: 696-700 (2014).
15. Y. Tominaga and B. Blocken, *Wind tunnel experiments on cross-ventilation flow of a generic building with contaminant dispersion in unsheltered and sheltered conditions*, Building and Environment, **92**:452-461 (2015).
16. T. R. Oke, *Street design and urban canopy layer climate*, Energy and Buildings. **11(1-3)**: 103-113 (1998).

 DOI: 10.1051/matecconf/201711101009

# Rapid Prototyping of Microfluidics Devices using Xurograhy Method

*Fiona* W.M. Ling[1]*, *Wafaa* K. Mahmood [1] *and Hayder* A. Abdulbari [1]

[1] Centre of Excellence for Advanced Research in Fluid Flow, Universiti Malaysia Pahang, Lebuhraya Tun Razak 26300 Kuantan, Pahang, Malaysia

**Abstract.** Rapid prototyping of microchannel gain lots of attention from researchers along with the rapid development of microfluidic technology. The conventional methods carried few disadvantages such as high cost, time consuming, required high operating pressure and temperature and involve expertise in operating the equipment. In this work, new method adapting xurography method is introduced to replace the conventional method of fabrication of microchannels. The novelty in this study is replacing the adhesion film with clear plastic film which was used to cut the design of the microchannel as the material is more suitable for fabricating more complex microchannel design. The microchannel was then mold using polymethyldisiloxane (PDMS) and bonded with a clean glass to produce a close microchannel. The microchannel produced had a clean edge indicating good master mold was produced using the cutting plotter and the bonding between the PDMS and glass was good where no leakage was observed. The materials used in this method is cheap and the total time consumed is less than 5 hours where this method is suitable for rapid prototyping of microchannel.

## 1 Introduction

Microfluidics technology involves the utilization of the devices that have the dimensions less than 1 mm which allow the precision in handling of fluids. This technology have been developed tremendously and have massive impact in wide variety of applications. Low sample and reagent volume is use due to the usage of micrometer-sized channel resulting in creating less waste and environmental problems. The miniaturization of analytical technique using microfluidic technology bring advantages such as reduce the equipment size, shorten the reaction or analysis time and the possibility of develop of portable devices.

The miniaturization system has provide a new platform to fluids manipulation components [1] such as pumps, valves, filters and mixing. Microfluidic devices were also widely used in reaction [2-4] and heat transfer [5, 6]. Microfluidic plays an important role in detection technique including separation [7] and food analysis [8, 9]. Besides, lab-on-chips were greatly incorporated in clinical [10-12], cell biology [13] and droplet generators [14, 15].

Rapid prototyping of microfluidic devices have been the concern of the researchers along with the instantaneous development of microfluidic technique nowadays. Several rapid prototyping methods to fabricate microchannels were introduced including wet etching, conventional machining [16], photolithography, soft lithography, hot embossing, injection molding, laser ablation, and plasma etching [17].

There are some disadvantages of these conventional methods. Machining itself involve combination of few techniques of microchannels fabrications such as wet etching, photolithography and others which also requires the clean room facilities thus increases the cost of fabrication [18-20]. Embossing process consumes lots of time [21] and it is not suitable for prototyping few microchannels for testing purposes. Due to the complexity of molding equipment and high operating temperature and pressure [22], injection molding is more suitable for industrial compare to research application.

In most of the laboratory environment, soft lithography is used as rapid prototyping due to the advantages such as low cost and less time consuming [23, 24]. The material use in soft lithography is polydimethylsiloxane (PDMS) which has high optical transparency and high gas permeability thus suitable even for cell and biology applications [25]. Due to the elastomeric properties of PDMS even complex 3D microchannels also can be fabricated [26]. PDMS also showed excellent bonding to itself or other materials [27]. However, soft lithography involving fabrication of special silicon master composed of photoresist with the design of microchannel on it. This process is much more consuming time and need special equipment also chemicals especially when changes of design required.

With the development of the cutting plotter nowadays, the resolution and alignment of cutting plotter have drastically improved thus lead to microscale prototyping [28]. Xurography is a method where a master mold is cut using a cutting plotter machine.

* Corresponding author: fionalingwm@gmail.com

Martínez-López et al. [29] and Neuville et al. [30] utilized adhesive film to cut the master mold. However, the produced master mold from adhesive film is suitable for simple design where it is hard to manipulate when complicated design is required. In order to solve the problems occur when using the conventional method, therefore in this study, new low-cost rapid fabrication of microchannel adapting xurography method is introduced. Clear plastic film is chosen in this study, replacing adhesive films, to cut the desired design.

## 2 Methodology

Semi-clogged heart blood vessels (Fig 1a) according to the angiogram of the real heart vessels (Fig 1b) was first drawn using Graphtec Studio. A clear plastic film with the thickness around 200 μm was used to cut the master mold using Graphtec cutting plotter CE6000-60 (Fig 2a). The design was peel off from the plastic film (Fig 2b) and pasted on glass (Fig 2c) using a thin layer of heat resistance glue. Glass was chosen due to its chemical resistance property.

Mixture of PDMS and curing agent at the ratio of 10:1 was prepared in a disposable plastic cup. 60 g of PDMS and 10 g of curing agent were weighed and then mixed using a mixing machine (model: Thinky mixer). The revolution balance dial was adjusted according to the total weight of the cup and the adapter of the mixer. The program was set to mix the mixture at 1000 rpm for 5 min and defoaming (degassing) at 400 rpm for another 5 min.

The degassed mixture was then poured on the design (Fig 2d) which placed in a petri dish which was then placed in a desiccator for 40 min to remove all the air bubbles that might trapped inside (Fig 2e). The PDMS was then placed in an oven at 60 °C for 2 and a half hours for soft baking process which can solidify the PDMS (Fig 2f).

The PDMS was cut along the edge of the glass using a scalpel (surgery knife) then peeled off carefully and slowly from the design using a tweezer (Fig 2g). The inlet and outlet of the microchannel were punched using a puncher (Fig 2h). Both the PDMS and glass slide were placed in plasma cleaner for 2 minutes (Fig 2i) and then put together where the design of microchannel facing the glass slide and pressed slightly to let them bond together (Fig 2j). A suitable length of tubes was cut, and all the inlet and outlets were connected using the tubes. The microchannel was ready for further testing.

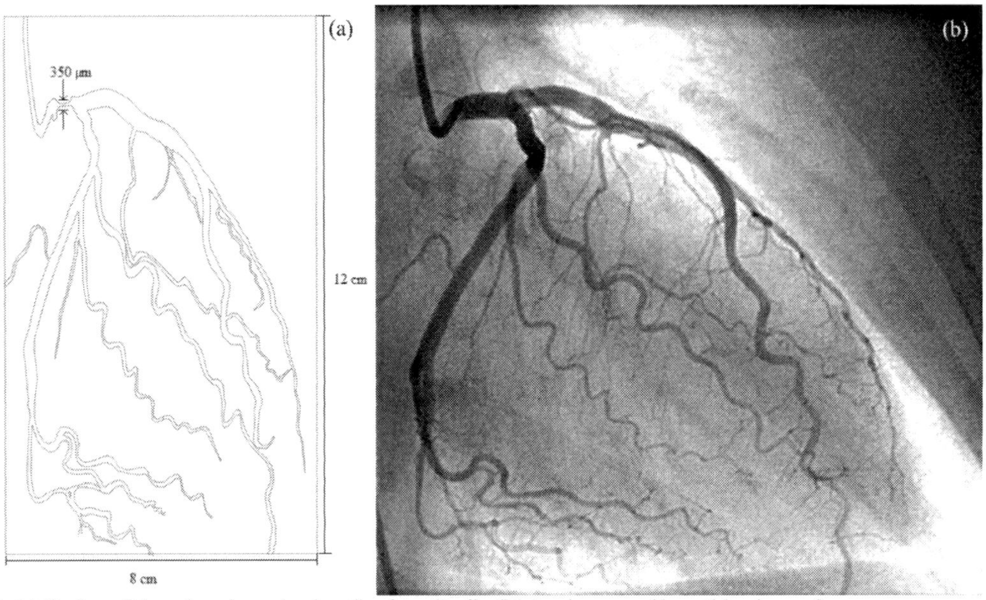


**Fig 1.** (a) Design of the microchannel using Graphtec Studio (b) Angiogram of heart blood vessels

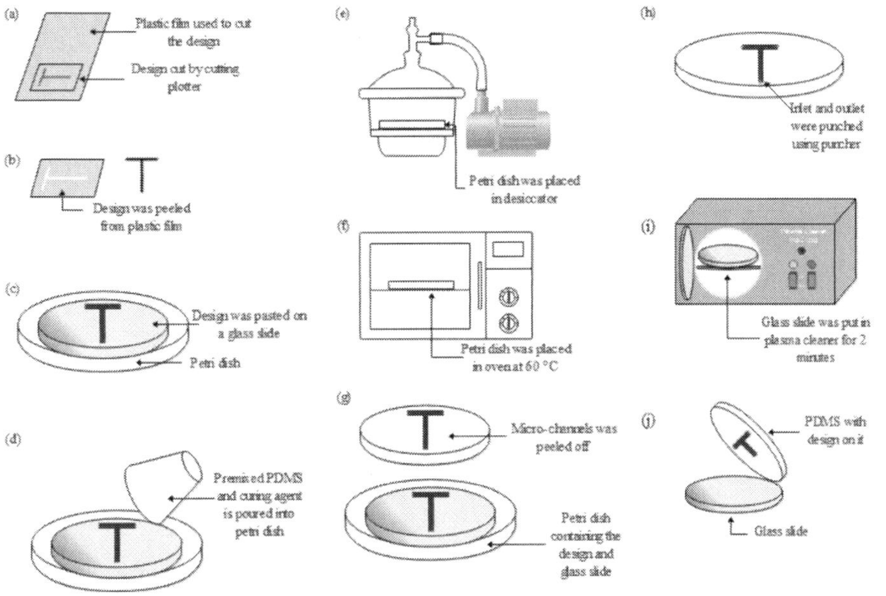


**Fig 2.** Schematic diagram depicting xurography method

## 3 Results and Discussions

The cutting quality of the cutting plotter is depended on some parameters such as the sharpness of cutting blade, cutting force, cutting speed, the condition of the cutting area and the material used. It is advisable to test cut for cutting quality and setting of the cutting plotter checking. For a more complicated design, it is always recommended that use a lower speed for a better cutting quality especially for a design having curvy edges. For a thicker material, the higher cut force will be needed. However, test cut start from a lower cut force is necessary to prevent damage to both cutting blade and cutting area of the cutting plotter where the penetration of the blade should not exceed the material itself. It is also suggested not to use the maximum cut force. The material use should always be compatible with the specification of the blade to longer the lifespan of the blade. In this case, a clean and high-quality cut is required. Thus the cut speed was set at a lower level in contrast to the higher level of cut force was set since the thick material was used.

Qualitative visualization of the cross-sectional of the microchannel was done to evaluate the quality of the produced master mold (clear plastic film containing design) which would directly affect the quality of microchannel produced. A small part of the straight microchannel was cut and brought to a scanning electron microscope (SEM) for inspection purpose. Straight microchannel was selected for clearer visualization. As shown in Fig 3, clean edge of microchannel can be observed at the cross-sectional area and along the entire straight channel which indicated that high quality of master mold was produced. This result also indicated that the resolution of the cutting plotter and the sharpness of the cutting blade were good enough to produce the master mold as it did not produce master mold with crooked edge which would then affect the quality of the produced microchannel. As a result, this method can be used in rapid prototyping of microchannel as high quality of master mold can be produced.

The complete microchannel was then brought for testing by flowing liquid across it. This process was done to test whether there is any leakage especially at inlet and outlet of the microchannel. Besides, by flowing liquid across the microchannel, checking for the adhesion of the PDMS on glass also can be done. Green food dye was used for the ease of visual inspection. The flow of the fluid was provided by the vacuum and pressure controller (model: Elveflow OB1) with a pressure set of 100-500 mbar to push the fluid out from the reservoir. Different pressure was set to check the maximum volume of fluid can be carried by the microchannel before backflow of liquid happen. In this study, the maximum pressure set is around 450 mbar where the flow rate of the fluid is around 3000 μl/min. From Fig 4, there was no any leakage at either the inlet, outlets or along the microchannel even when the pressure was set at higher pressure, in this case 500 mbar. The result indicated that the bonding between the PDMS and the glass was good. It is expected that the great bonding between these two surfaces can withstand higher pressure and flow rate in the channel without affecting the bonding. This quality is advantageous for researchers who required higher pressure and flow rate across the channel for their work.

By stimulating the real blood vessels, researchers can use the microchannels instead of animal testing which did not follow the ethics of life. Beside heart vessels, other blood vessels also can be stimulated using this method. By using microchannel, the time consuming for the experiment is believed to significantly reduced compared to animal testing. Thus, this method of fabrication of microchannel will play a significant role in the development of clinical and drug delivery studies.

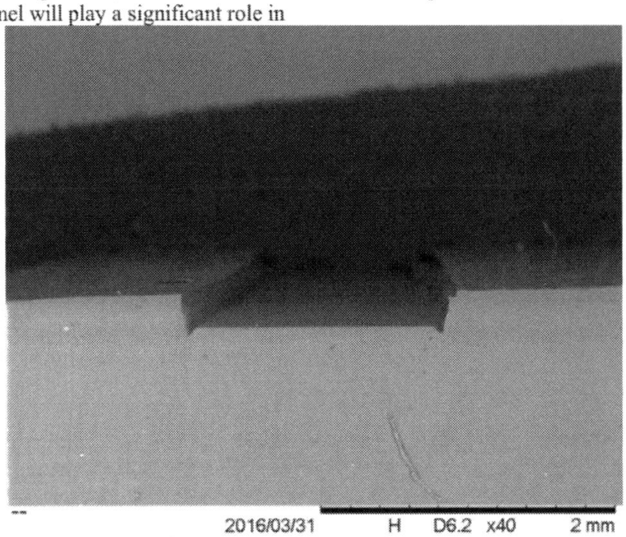


**Fig 3.** SEM image of cross-section of microchannel showing the straight channel. The edge of the microchannel was clean indicating high quality of the master mold was produced.

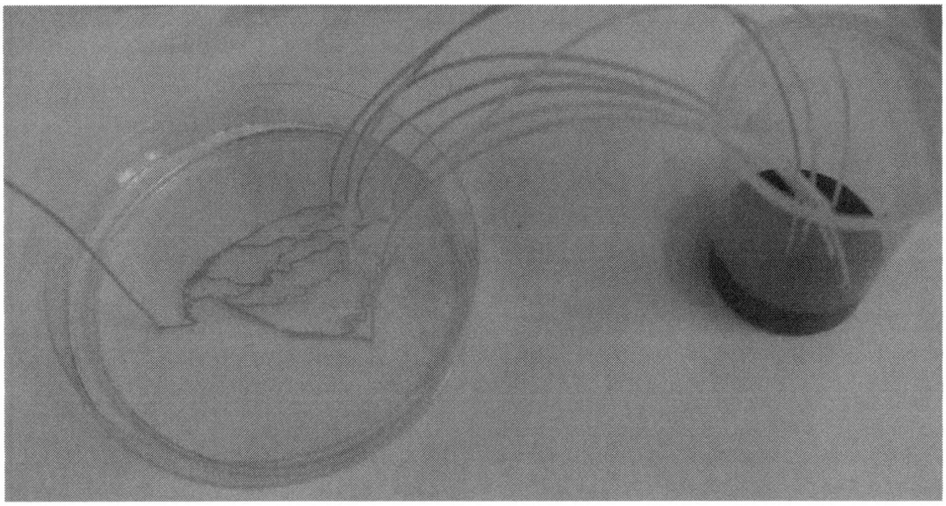

**Fig 4.** The complete microchannel was brought for flow testing. Food dye was used for visual inspection of leakage testing of the complete microchannel. The bonding of the PDMS and glass was good and no leakage was observed.

## 4 Conclusions

In this study presented a rapid and low-cost fabrication of microchannel using xurography method where the fabrication process only took less than 5 hours and there is no need expertise to operate all the equipment. The quality of the microchannel was depended on the cutting quality of the master mold where high quality of the cutting would produce a high-quality microchannel with clean edges. The thickness of the microchannel is depending on the paper material that used to cut the design using a cutting plotter. The current method is currently useful in producing complicated design such as blood streams that is useful in clinical researches thus replacing animal testing. The feasibility of using other materials and the ability to cut blade to cut even smaller dimension should be further explored to develop this method to other fields.

## References

1. Fiorini, G.S. and D.T. Chiu, *Disposable microfluidic devices: Fabrication, function, and application.* BioTechniques 2005. **38**: p. 429-446.
2. deMello, A.J., *Control and detection of chemical reactions in microfluidic systems.* Nature, 2006. **442**(7101): p. 394-402.

3. Truter, L.A., et al., *The application of palladium and zeolite incorporated chip-based microreactors.* Applied Catalysis A: General, 2016. **515**: p. 72-82.
4. Ying, X., et al., *Efficient Fischer–Tropsch microreactor with innovative aluminizing pretreatment on stainless steel substrate for Co/Al2O3 catalyst coating.* Fuel Processing Technology, 2016. **143**: p. 51-59.
5. Morgan, A.J.L., et al., *Efficient microwave heating of microfluidic systems.* Sensors and Actuators B: Chemical, 2013. **181**: p. 904-909.
6. Lee, S.H. and I. Mudawar, *Investigation of flow boiling in large micro-channel heat exchangers in a refrigeration loop for space applications.* International Journal of Heat and Mass Transfer, 2016. **97**: p. 110-129.
7. Nam, Y., M. Kim, and T. Kim, *Pneumatically controlled multi-level microchannel for separation and extraction of microparticles.* Sensors and Actuators B: Chemical, 2014. **190**: p. 86-92.
8. Guo, L., et al., *Application of microfluidic "lab-on-a-chip" for the detection of mycotoxins in foods.* Trends in Food Science & Technology, 2015. **46**(2, Part A): p. 252-263.
9. Muijlwijk, K., C. Berton-Carabin, and K. Schroën, *Cross-flow microfluidic emulsification from a food perspective.* Trends in Food Science & Technology, 2016. **49**: p. 51-63.
10. Boussommier-Calleja, A., et al., *Microfluidics: A new tool for modeling cancer–immune interactions.* Trends in Cancer, 2016. **2**(1): p. 6-19.
11. Kozminsky, M., Y. Wang, and S. Nagrath, *The incorporation of microfluidics into circulating tumor cell isolation for clinical applications.* Current Opinion in Chemical Engineering, 2016. **11**: p. 59-66.
12. Tay, A., et al., *Advances in microfluidics in combating infectious diseases.* Biotechnology Advances, 2016. **34**(4): p. 404-421.
13. Grünberger, A., W. Wiechert, and D. Kohlheyer, *Single-cell microfluidics: opportunity for bioprocess development.* Current Opinion in Biotechnology, 2014. **29**: p. 15-23.
14. Zhu, Y. and Q. Fang, *Analytical detection techniques for droplet microfluidics—A review.* Analytica Chimica Acta, 2013. **787**: p. 24-35.
15. Shah, R.K., et al., *Designer emulsions using microfluidics.* Materials Today, 2008. **11**(4): p. 18-27.
16. T., S., et al., *Microstructure grooves with a width of less than 50 [mu]m cut with ground hard metal micro end mills.* Precision Engineering, 1999. **23**: p. 229-235.
17. Rossier, J.S., et al., *Plasma etched polymer microelectrochemical systems.* Lab Chip, 2002. **2**(3): p. 145-50.
18. Martynova, L., et al., *Fabrication of plastic microfluid channels by imprinting methods.* Analytical Chemistry, 1997. **69**(23): p. 4783-9.
19. Xu, J., et al., *Room-temperature imprinting method for plastic microchannel fabrication.* Analytical Chemistry, 2000. **72**(8): p. 1930-3.
20. Rodriguez, I., et al., *Rapid prototyping of glass microchannels.* Analytica Chimica Acta, 2003. **496**(1–2): p. 205-215.
21. Tsao, C.-W., et al., *Rapid polymer microchannel fabrication by hot roller embossing process.* Microsystem Technologies, 2012. **18**(6): p. 713-722.
22. Çetin, B., A.K. Koska, and M. Erdal, *Warpage characterization of microchannels fabricated by injection molding.* Journal of Micro and Nano-Manufacturing, 2015. **3**(2): p. 021005-021005.
23. Friend, J. and L. Yeo, *Fabrication of Microfluidic Devices using Polydimethylsiloxane.* Biomicrofluidics, 2010. **4**.
24. Martin, C.R. and I.A. Aksay, *Microchannel molding: A soft lithography-inspired approach to micrometer-scale patterning.* Journal of Materials Research, 2005. **20**(8): p. 1995 - 2003.
25. Faustino, V., et al., *Biomedical microfluidic devices by using low-cost fabrication techniques: A review.* Journal of Biomechanics.
26. Wu, H., et al., *Fabrication of complex three-dimensional microchannel systems in PDMS.* J Am Chem Soc, 2003. **125**(2): p. 554-9.
27. Becker, H. and L.E. Locascio, *Polymer microfluidic devices.* Talanta, 2002. **56**(2): p. 267-287.
28. Bartholomeusz, D.A., R.W. Boutté, and J.D. Andrade, *Xurography: rapid prototyping of microstructures using a cutting plotter.* IEEE/ASME Journal of Microelectromechanical Systems, 2005. **14**(6): p. 1364-1374.
29. Martínez-López, J., et al., *Xurography as a rapid fabrication alternative for point-of-care devices: Assessment of passive micromixers.* Sensors, 2016. **16**(5): p. 705.
30. Neuville, A., et al., *Xurography for microfluidics on a reactive solid.* Lab on a Chip, 2017. **17**(2): p. 293-303.

# Factorial Analysis on Sulfide Removal from Petrochemical Industry Wastewater using Acclimatized Activated Sludge

*Norazwina* Zainol, *Akmal Arif B.* Mohamed Zaki, *Nur Hidayah* Mat Yasin, *Sunarti* Abd Rahman, *Azilah* Aziz, Chin *Sim Yee*

Faculty of Chemical Engineering and Natural Resources, University Malaysia Pahang, Lebuhraya Tun Razak, 26300 Kuantan, Pahang, Malaysia

**Abstract.** In wastewater treatment, sulfide or hydrogen sulfide is a major problem in refining process and can cause the pollution in the wastewater plant. In this study, acclimatized activated sludge was used to remove sulfide from the petrochemical wastewater. Five independent factors which affecting the degradation of sulfide were investigated in this paper - wastewater/acclimatized activated sludge (WW/AS) ratio, type of system, temperature, agitation and hydraulic retention time (HRT). Fractional factorial design was used to investigate the effect of independent factors as well as the interaction factors on the sulfide and COD removals (%). Some of the independent factors were shown to have significant effect on sulfide removals (%). The results obtained here showed that the order of significance in sulfide removal (%) is as follows: agitation > WW/AS ratio > type of system > temperature > HRT. Amongst of the operation factors in question, it was found that the interaction factor of temperature*WW/AS ratio had the strongest effect on the sulfide removal (%).

## 1 Introduction

Petroleum and petrochemical industry produced more than 2500 useful products from crude oil [1]. The petrochemical industry also includes the treatment of hydrocarbon such as sulfide from the petroleum refining industry and natural gas liquid in oil and gas industry [2]. Most sulfide in nature are produced biologically (under anaerobic conditions) and occurs as free hydrogen sulfide ($H_2S$). Even concentrations as low as 0.1 μg/L, the odor of hydrogen sulfide can be detected and may be offensive in the plant area and the surrounding communities [3].

Therefore, wastewater treatment is important in petrochemical industry to remove the contaminated wastewater which usually enriched with high sulfide concentrations. Various methods to remove sulfide from petrochemical industry are available such as microbial fuel cells (MFCs), anaerobic method and chemical reaction by using hydrogen peroxide [4]. However, up to date any of the mentioned methods are still not yet commercially used because of high cost of maintenance. Thus, the activated sludge process was used in this research as a strategy to control the problem. Activated sludge is a process for treating sewage and industrial wastewaters using air and a biological floc composed of bacteria and protozoa in that wastewater to trap the sulfide [5]. This process uses a mass of microorganism (bacteria) to aerobically treat wastewater [6].

In this industrial research, factorial analysis was performed by using the concept of two level factorial (TLF) via Design Expert software. TLF is a statistical method based on the multivariate non-linear model that is useful in studying interactions of various parameters affecting the process [7-8]. Factorial analysis was applied to determine the most important factors that affecting sulfide removal from petrochemical industry wastewater. Based on the requirement from industry, five selected factors on sulfide removal had been chosen which are wastewater to acclimatized activated sludge ratio (WW/AS), type of system, temperature, agitation and hydraulic retention time (HRT).

## 2 Materials and Methods

### 2.1 Synthetic Wastewater Preparation

The synthetic wastewater was prepared according to the sulfide and COD contents in the real petrochemical industry wastewater (PW) sample with the design basis of sulfide and COD were 30 mg/L and 150 mg/L, respectively. Sodium sulfide and glucose were added into the synthetic wastewater to achieve the design basis.

### 2.2 Acclimatization of Activated Sludge (AS)

This process was conducted for adaptation of the activated sludge to the sulfide content in PW. The acclimatization process was done in 10 L of batch reactor. Initially, 1 L of activated sludge was mixed with 528 mL synthetic wastewater. Later the same volume of synthetic wastewater was added into the reactor daily for three weeks.

### 2.3 Experimental Design

In this study, five factors e.g. WW/AS ratio, type of system, temperature, agitation and HRT were taken into account to investigate their effects on sulfide removal (%) using a $2^4$ fractional factorial design. The design of experiment was performed by Design Expert software where all the factors were randomized. Table 1 shows the design factors and levels were coded as -1 (low level) and +1 (high level) where low and high level indicates the lowest and highest range of the factors. Shake flask method was applied to study the reaction between

* Corresponding author: azwina@ump.edu.my

DOI: 10.1051/matecconf/201711102001

synthetic wastewater and acclimatized activated sludge. 16 runs of experiments were conducted. The response of the experimental design was analyzed using ANOVA based on the p-value with 95% of confidence level. Later, experimental data was analyzed using Design Expert software to determine the most contributing factor and interaction between the factors.

**Table 1.** List of selected factors and level

| Factor | Level | |
|---|---|---|
| | -1 | +1 |
| WW/AS ratio | 1:1 | 3:1 |
| Type of system | Closed | Open |
| Temperature | 30 | 37 |
| Agitation (RPM) | 0 | 100 |
| HRT (hour) | 17 | 32 |

## 3 Results and Discussions

### 3.1 Screening of Factors Affecting on the Sulfide Removal (%)

The identification of the factors that to be effective on the sulfide removal was carried out through $2^4$ fractional design to determine the degree of the effect from the factors on the response. Independent factors such as WW/AS ratio, type of system, temperature, agitation and HRT have been taken into account to investigate their effect on sulfide removals (%) and the range of response was from 98.26 to 99.83%.

### 3.2. Main Effect and Interaction Effect between Factors on the Sulfide Removal (%)

ANOVA analysis of sulfide removal (%) using AS showed that p-values of agitation (C) and WW/AS ratio (E) were smaller than 0.05, which implies the significance of the independent factors. It is necessary to examine any interactions that are important in the experimental design analysis; the 2-way interactions show their significance at 95% confidence level. The regression equation (Equation (1)) represent the best description after the elimination of non-significant parameters ($p>0.01$) from the experimental results. The final empirical models in terms of actual parameters were determined as follows:

$$\text{Sulfide Removal} = +99.21 - 3.125E\text{-}3*A - 0.014*B - 0.12*C + 0.062*D - 0.076*E - 0.21*A*B + 0.29*B*E + 0.18*A*E - 0.18*B*C \quad (1)$$

where A is HRT, B is temperature, C is agitation, D is type of system and E is WW/AS ratio.

The Pareto Chart in Figure 1 showed the main effects and interaction effects of the factors for the sulfide removal. For main effect, it showed that agitation (factor C) gave the highest contribution for sulfide removal and followed by WW/AS ratio (factor E). For interaction effects, it showed that there were four interaction effects that contributed in removal of sulfide. There were two positive effects, interaction of temperature and WW/AS ratio (BE) and interaction of HRT and WW/AS ratio (AE). Positive effects mean that value of sulfide removal increased as the value of factors increased and vice versa for negative effects. The interaction of HRT and temperature (AB) and interaction of temperature and agitation (BC) showed negative effects to sulfide removal.

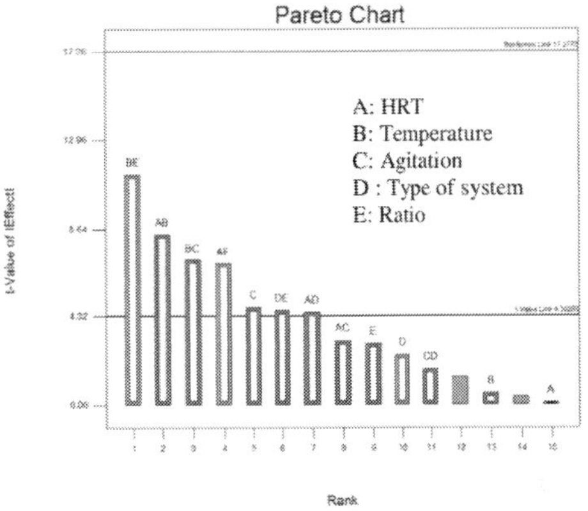


**Figure 1.** Contribution of independent and interaction effect to sulfide removal (%)

#### 3.2.1 Effect of Independent Processing Parameters on Sulfide Removal (%)

The effects of two independent variables on the sulfide removal are shown in Figure 2a and Figure 2b. Sulfide removal decreased with increasing of agitation rate as shown Figure 2a. From the figure, sulfide removal was lower at 100 rpm than at 0 rpm. However, there was no significant difference of sulfide removal between 0 rpm and 100 rpm of agitation where the sulfide removal achieved 99.3% and 99% at 0 rpm and 100 rpm, respectively. The difference of sulfide removal at both agitation rate maybe due to the sensitivity of the microbes to the agitation where the agitation reduced the performance of the microbes in AS and reduced the sulfide removal [9]. Figure 2b showed that sulfide removal was increased proportionally to WW/AS ratio where sulfide removal achieved 99.4% at 3:1 ratio and 99.2% at 1:1 ratio. Thus, it can be also concluded that there was still no significant difference of sulfide removal at both WW/AS ratios. This result is maybe affected by the application of AS for sulfide removal in this study. The acclimatized microbes in the wastewater can enhance degradation of organic pollutant compounds effectively [10].

DOI: 10.1051/matecconf/201711102001

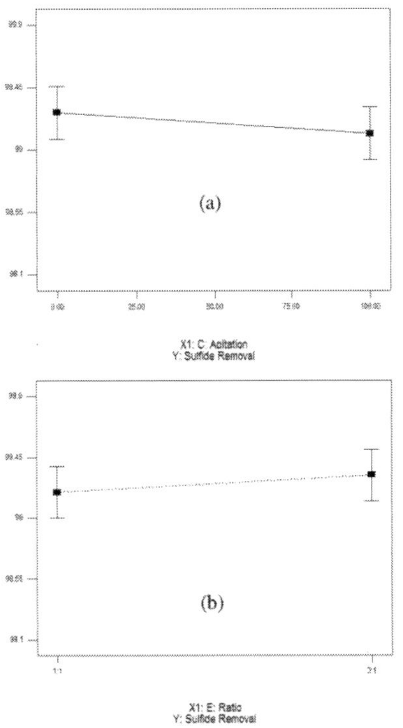


**Figure 2.** Most effective independent parameters in sulfide removal (%)

### *3.2.2 Interaction effects between factors on sulfide removal (%)*

Figure 3a showed the interaction effect between temperature and WW/AS ratios on sulfide removal at specific conditions which were at HRT of 32 hours, no agitation and at open system. From Figure 3a, it was observed that higher sulfide removal was achieved for 3:1 ratio than 1:1 ratio at higher temperature (37°C). At lower temperature (30°C), sulfide removal for 1:1 ratio was higher than 3:1 ratio. According to Siddique et al. [11], biological degradation is temperature dependent. At low temperature, it required more microbes to do the work while less microbes required at high temperature. Figure 3b showed the interaction effect between temperatures and HRT on sulfide removal. The best condition for other factors was set at: no agitation, open system and 3:1 for WW/AS ratio. The figure was also showed that the sulfide removal was observed to be increased with the HRT for both temperatures but it was higher at temperature 37°C than 30°C. According to Grady et al. [12], the maximum acceptable operating temperature for typical activated sludge is between 35°C to 40°C, which correspond to the mesophilic organisms. This range of temperature was the best for the microbes to remove wastewater constituents such as sulfide. Figure 3c showed the interaction effect between temperature and agitation on sulfide removal. The result shows the increasing of sulfide removal as the temperature increased when no agitation applied. At 100 rpm of agitation sulfide removal decreased when the temperature increased. The sulfide removal decreased with agitation because the microbes were sensitive to the presence of agitation and reduced the performance of the microbes in the wastewater [9]. Figure 3d showed the interaction effect between HRT and different WW/AS ratios on sulfide removal. From the result obtained in Figure 3d, sulfide removal increased proportionally with HRT values at 3:1 ratio and decreased when the HRT increased for 1:1 ratio. This observation is due to the low wastewater to AS volume ratio at 1:1 value where the wastewater represented as the food to the microbes. Thus, when the HRT increased, the food was limited to the microbes and will affect the sulfide removal process [11].

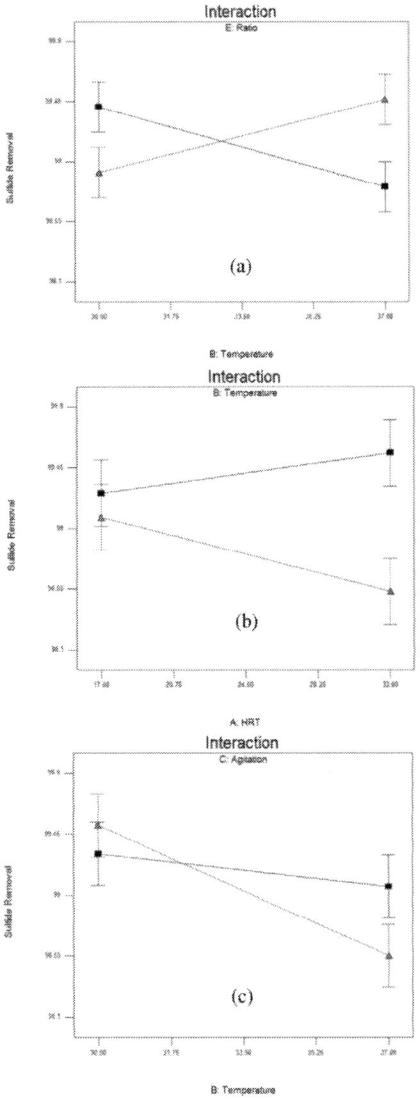

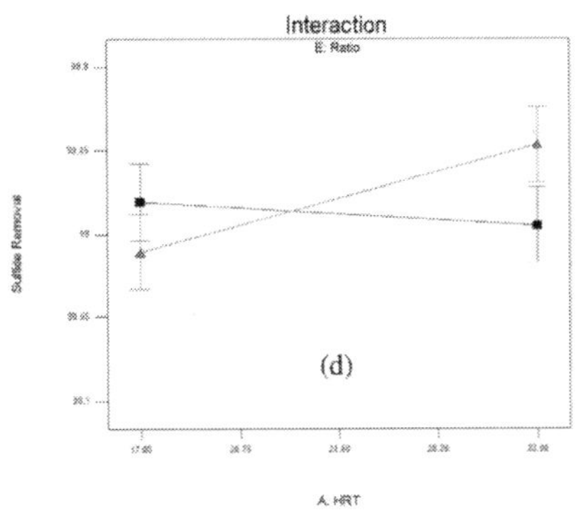


**Figure 3.** Interaction effect between factors on sulfide removal (%)

## 4. Conclusion

The sulfide removal was studied using $2^4$ fractional factorial design. The effect of five factors e.g. WW/AS ratio, type of system, temperature agitation and HRT which affecting the sulfide removal was investigated. The determination of significance effect was based on p-value < 0.05, among all independent factors studied only agitation has significant effect on the removals. The sequence of the contributing factors with respect to decreasing of % contribution to sulfide removal was found to be: agitation > WW/AS ratio > type of system > temperature > HRT. Amongst the operation factors in question, interaction factor of temperature*WW/AS ratio had the strongest effect on the sulfide removal (%). The results show that fractional factorial design is suitable in investigating the effect of large number of factors with a minimum number of experiments.

The authors would like to acknowledge the UMP Research Grant (Grant No. RDU 150396) from Universiti Malaysia Pahang, which has fully supported this research.

## References

1. Bhaskar, N, Georgi S.C. (2009). *Pollution Control Technologies* - Volume III. EOLSS Publications, **30** Sep 2009.
2. Amosa, M. K., Mohammed, I.A. and Yaro, S.A. (2010). Sulphide Scavengers in Oil and Gas Industry – A Review. *NAFTA* **61 (2)** 85-92
3. Gikunju, P., Njogu, P.M, Makhonge, P.X. (2016). Hydrogen sulphide concentrations a health risk to workers in geothermal power stations (A case study of Olkaria Geothermal Power Station in Kenya). *Prime Journal of Engineering and Technology Research (PJETR)*, Vol. **2(1)**, pp. 38-51.
4. Izadi, P., Rahimnejad, M. (2014). Simultaneous electricity generation and sulfide removal via a dual chamber microbial fuel cell. *Biofuel Research Journal*, **1**: 34-38
5. Dohare, D., Meshram, R. (2014). Biological Treatment of Edible Oil Refinery Wastewater using Activated Sludge Process and Sequencing Batch Reactors - A Review. *International Journal Of Engineering Sciences & Research Technology*, 251-260, 2014
6. Mkhize, S. P., Atkinson, B. W., Bux, F. (2000). *Assessment Of A Biological Nutrient Removal Process For Remediation Of Edible Oil Effluent*, Presented at WISA 2000
7. Bergquist, B. (2015). Analysis of an replicated $2^2$ factorial experiment performed in a continuous process.*Total Quality Management*, Vol **26**, No 10: 1083-1094
8. Saunders, I. W., Hoboken, N.J. (1992). Experimental design for continuous processes. *The Australian Journal of Statistics*, **34(1)** : 77-89
9. Silva, A. J., Domingues, M. R., Hirasawa, J. S., Varesche, M. B., Foresti, E. , Zaiat, M. (2011). Kinetic modeling and microbial assessment by fluorescent *in situ* hybridization in anaerobic sequencing batch biofilm reactors treating sulfate-rich wastewater. *Braz. J. Chem. Eng*. Vol.**28**:2.
10. Kumar, B. L., Sai Gopal, D. V. R. (2015). Effective role of indigenous microorganisms for sustainable environment. *3 Biotech*., **5(6)**: 867–876.
11. Siddique, T., Okeke, B.C., Arshad, M., Frankenberger, W.T.Jr. (2002). Temperature and pH effects on biodegradation of hexachlorocyclohexane isomers in water and a soil slurry. *J Agric Food Chem*., **50(18)**:5070-6.
12. Grady, C.P. Jr., Glen, T. D, Henry, C. L. (1999). *Biological Wastewater Treatment. 2nd ed.* New York: Marcel Dekker.

# Characteristics of PrekotAC as Formulated Filter Aids and Its Performance to Adsorb Volatile Organic Compound

*N. Masdiana*[1], *M. Rashid*[1]*, *S. Hajar*[1], *M.R. Ammar*[2]

[1]Air Resources Research Laboratory, Malaysian Japan International Institute Technology, 54100 UTM Kuala Lumpur
[2]AMR Environmental Sdn. Bhd., Taman Sri Pulai Perdana, 81110 Johor Bahru, Malaysia
*rashidyusof.kl@utm.my

**Abstract.** The characteristics of a formulated filter aids material known as PrekotAC which is a combination of an adsorbent, activated carbon with a pre-coating material was investigated in this study. PrekotAC, was formulated into two different weight ratios of 5:95 and 10:90 respectively. The relationship between the adsorption properties and the characteristics of the formulated materials in terms of its particle size distribution, particle density, bulk density, moisture content as well as its BET surface area was investigated. A specific amount of 100mg of formulated filter aids was placed in a fixed glass tube and subjected to a given concentration of toluene at the temperature of 25°C and 1.0 l pm gas flow rate. The adsorption properties of each filter aids material was tested under various time where the amount of toluene adsorbed by the filter aids was analysed using gas chromatography equipped with flame ionization detector (GC-FID). The ability of the formulated filter aids to adsorb toluene was presented and discussed in this paper.

## 1 Introduction

The environmental issue has become a major subject in the last few decades, affecting developed as well as those of developing nations. For an example, air pollution emission from waste incineration such as acid gases, particulate matter, heavy metals and volatile organic compounds (VOCs) are released into the atmosphere [1]. Organic compounds that include dioxin-furan is recognized to be one of the most potent pollutant resulted through post-combustion in incineration process.

Organic compounds are carbon-based compounds derived from many sources. It (author) was reported that 37% of total VOCs emissions comes from coal-fired power plants [2]. These emissions are of public concern especially VOCs considered as toxic chemicals causing severe health problems [3]. Prolonged exposure to VOCs has been shown to induce eye and throat irritation, damage to the liver and central nervous system as well as increase relative rates of leukemia and lymphoma [4]. Therefore, the removal of VOCs from a flue gas stream is of significant interest. There are five methods of removing VOCs from contaminated gas stream and these are adsorption, absorption, condensation, incineration, and biofiltration. However, adsorption seems to be the most commonly used in this case [5].

Activated carbon has been extensively used in variety of environmental applications as adsorbent in gas purification, solvent recovery and waste water treatment, due to its large surface area and high adsorption capacity [6]. Activated carbon has been used as an adsorbent in fabric filtration system which is commonly found in most waste incineration plant. In addition, filter aids is also applied to coat a layer of inert material onto each of the filter bag acting as a barrier for protection as well as to allow a uniform air flow passing through the filter media [7].

Thus, the aim of the present study is to investigate the characteristics of a formulated filter aids material which is a combination of activated carbon and precoating agent. In addition, the ability of the formulated filter aids to adsorb VOCs is presented and discussed in this paper.

## 2 Materials

Table 1 presents the specifications of the two main powder materials used in this study, which are activated carbon and PreKot™. The toluene solution of 99.99% purity (Merck) was used in the adsorption study.

**Table 1.** Specification of the materials used in the study.

| **Activated carbon** | **PreKot™** |
|---|---|
| Colour: black | Colour: snowy white |
| Origin: coal based | Origin: silica based |
| Surface area: 1094.96% | Surface area:2.52% |
| Moisture content:23.39% | Moisture content: 0.73% |

* Corresponding author: rashidyusof.kl@utm.my

DOI: 10.1051/matecconf/201711102002

## 3 Experimental methods

### 3.1 Formulation of the Filter Aids

PreKot™ and activated carbon were mixed into 5:95 and 10:90 of two different weight percentage respectively. Initially before formulation, the raw materials were dried in an oven (Memmert, Model UNB 200) for 24 hours at 110°C. Then, the properties of the formulated filter aids designated as PrekotAC were characterized in term of its surface area (BET), particle size distribution, particle density, bulk density and moisture content.

### 3.2 Material characterization

Particle size distribution was determined using a particulate sizer (Malvern Mastersizer 2000). The measurement was repeated two times and the average particle size distribution was taken by the software. The determination of the particle density of the material was measured using a gas pycnometer (Micromeritics AccuPyc II 1340), using nitrogen as a carrier gas. While the bulk density of the sample was measured based on the equation (1) where the sample was filled and compacted in a pre-weighed 10mL standard measuring cylinder up to the specific volume and re-weighed in order to determine the bulk density [8].

$$\text{Bulk density} = Wm / Vc \qquad (1)$$

Where;
$W_m$ = weight of the dried material (g)
$Vc$ = volume of the cylinder (10mL)

The specific surface area (SBET) was estimated by the BET equation (2). The porous structure of the sample was characterized by $N_2$ adsorption isotherms at 77 K using TriStar II 3020 V1.04. The specific surface area is an important characteristic in determining the chemical and physical interaction of the adsorbent and its surroundings. This is due to the fact that adsorption take place at the surface of particles.

$$(\upsilon/\upsilon m)=C(P/P_0)/\{[1-(P/P0)][1+(c-1)(P/P_0)]\} \qquad (2)$$

Where;
$\upsilon$ = volume of gas adsorbed per unit weight of material at pressure P
$\upsilon m$ = volume of gas adsorbed for monolayer coverage
$(P/P_0)$ = partial pressure of the gaseous adsorbate
C = constant

The moisture content of the formulated sample was determined based on a standard measurement (ASTM D2867-09) [9]. Approximately 2 g of the formulated sample was placed in a pre-weighed petri dish and exposed to the atmosphere for a few days. Then the sample was dried in an oven at 110°C for 24 hours to discard the moisture content in the sample. Later, the sample was removed from the oven and cooled to an ambient temperature in a desiccator before it was re-weighed. The experimental procedure was repeated twice and the average moisture content was taken. The moisture content of the sample was then calculated using equation (3) as written below.

$$\text{Moisture content (\%)} = [(S_i- S_a)/ (S_i- Wp)]\ x100 \qquad (3)$$

Where;
Wp = weight of the petri dish (g),
$S_i$ = initial weight of petri dish with sample (g)
$S_a$ = weight of petri dish with sample after drying(g)

### 3.3 Adsorption Study

#### *3.3.1 Sample preparation*

Figure 1 shows the schematic diagram of the adsorption experimental set up which consist of toluene solution, temperature sensor, adsorbent packed –glass (formulated filter aid), flowmeter and vacuum pump. Stopcock was used to control the toluene's flow during adsorbent exchange. Prior to start any experiment, all samples of formulated filter aids were dried for 24 hours in oven at 110°C to remove all adsorbed gases and moisture content [10]. A fixed glass with inside diameter of 4mm, outside diameter 8mm and total length 60mm was used for dynamic study. A 100mg of adsorbents were filled and compacted in a glass enclosed with glass wool and covered with a cap. After packaging, all prepared samples were stored in desiccator prior to use for sampling.

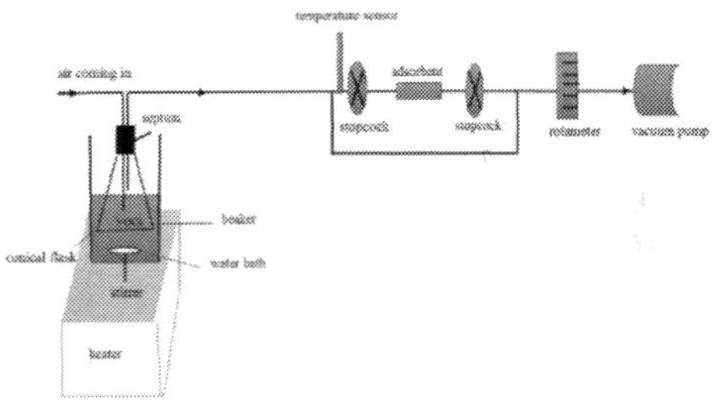


**Fig. 1.** Schematic diagram of experimental set-up

#### *3.3.2 Adsorption Test*

For each set of experiment, 100mL of toluene with 99.99% concentration was adsorbed by 100mg amount of total adsorbent dose at 25°C with 1lpm flowrate. Initially, 100mg adsorbent dose were packed in glass and placed in the adsorption system (Figure 1). The toluene's solution was bubbled and pumped through the system towards adsorbent. A water bath was used to keep the toluene solution in a desired constant temperature (50°C). Before the start of experimental run, toluene gas were freely pumped throughout the system for 10 minutes to ensure the gas distribute properly across the system. Sampling was done continuously and the effluent gas was collected at time intervals between 5 and 40 minutes and its concentration was determined by using gas chromatography equipped with flame ionization detector (GC-FID).

### 3.3.3 Analytical method

The sample was analysed using GC-FID with capillary column. FID detects analytes by measuring an electrical current generated by electrons from burning carbon particles in the sample. The adsorbed toluene was analysed using the NIOSH method [11]. The sample was desorbed in 2 mL of dichloromethane and allowed to stand with occasional shaking for 30 minutes [12]. A 1µL of sample extract was used for one injection. Each sample was injected twice and the average result was calculated. The oven temperature was set initially at 40°C for 3 minutes and then raised to 88°C per min. Both the auxiliary and detector temperatures were set at 250°C. The standard of toluene solution was used for quantification of the compounds.

## 4 Results and discussions

### 4.1 Material characterization

### 4.1.1 Characteristics in terms of particle size distribution

Figure 1 and 2 presents the particle size distribution of raw materials and formulated filter aids materials. Figure 1 showed that activated carbon has a finer particle size distribution compared to PreKot™. This results were in a good agreement with the previous study by Hajar *et al.* (2013) and Hajar *et al.* (2015) where it was reported that activated carbon has finer particle size distribution compared to PreKot™ [7,13]. By having a small particle material, it is good for adsorption purpose because as the particle size decreases, the degree of contact area between the particles increase and lead to a better adsorption properties.

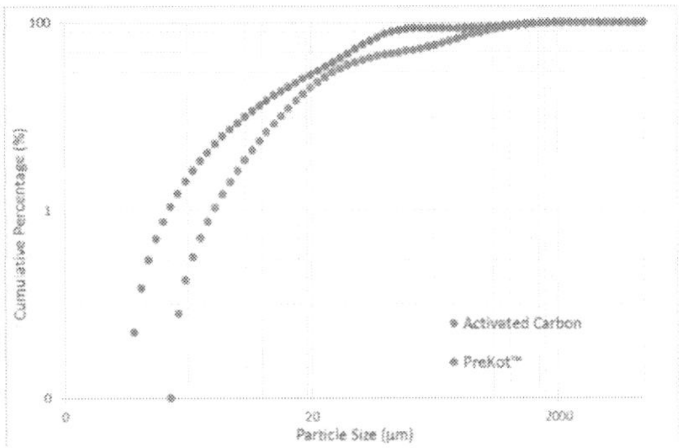


**Fig. 1.** Particle size distribution of raw materials

As shown in Figure 2, the formulated filter aids materials (PrekotAC 5:95 and PrekotAC 10:90) fitted perfectly with the original materials which proved that it has a very well distributed particle size distribution (Figure 2). It was reported that PrekotAC helps to improve the filtration performance by reducing the pressure drop across the filter cake as well as increases the filtration efficiency [7]. In addition, having a well distributed particle size distribution, PrekotAC could also be used as flue gas cleaning agent where it can be applied as a two in one filter aids material, pre-coating material as well as an adsorbent.

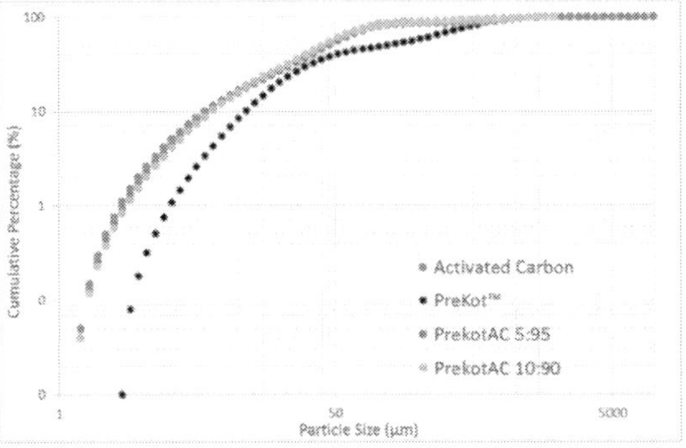


**Fig. 2.** Particle size distribution of raw materials and formulated samples

### 4.1.2 Characteristics in terms of bulk density and particle density

Figure 3 illustrates the results on the bulk density and particle density of the filter aids materials where it was found that activated carbon gives the highest while PreKot™ has the lowest density. PrekotAC laid nicely between their original materials with PrekotAC 5:95 showed slightly higher in bulk density (0.64 g/cm$^3$) and particle density (1.48 g/cm$^3$) compared to PrekotAC 10:90. There is not much different in bulk density and particle density for the formulated filter aids because of the close ratio of formulation.

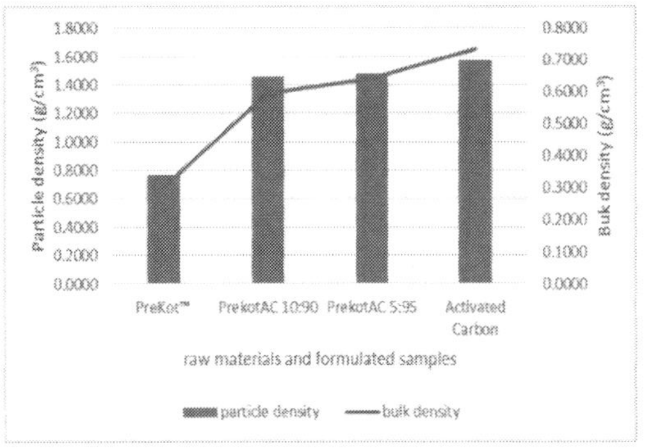


**Fig. 3.** Bulk density and particle density of raw materials and formulated samples

However, from the results, it showed that as the amount of PreKot™ in the formulation increases, the bulk and particle density decrease which explained the linear trend relationship of bulk density and particle density with the addition of PreKot™. It also showed that PreKot™ plays an important role in reducing bulk density and enhancing the porosity of the material for better filtration due to its particle characteristics [7].

By applying a filter aids material with a high porosity helps in reducing pressure drop across filter cake and also good for adsorption purpose. This is because, greater porosity materials lead to better roughness and larger specific surface area.

#### 4.1.3 Characteristics in terms of BET surface area

Specific surface area is a fundamental measurement in the field of fine particle characterization especially in adsorption process. Specific surface area is defined as the surface area per unit mass of sample. Figure 4 showed a BET surface area of raw materials and formulated samples which showed that activated carbon has the largest while PreKot™ has the smallest surface area with 1094.96 $m^2/g$ and 2.52 $m^2/g$, respectively.

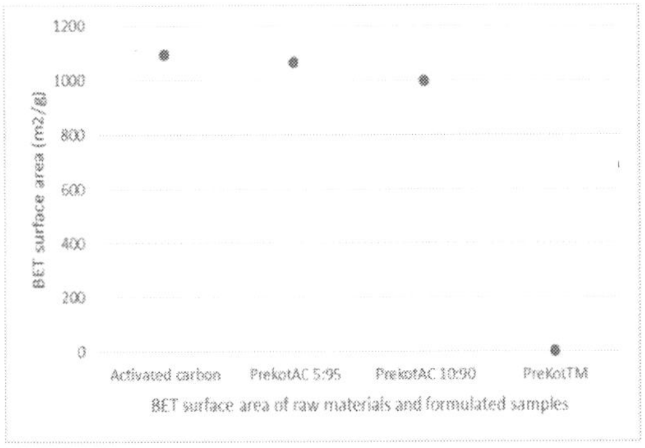


**Fig. 4.** BET surface area of raw materials and formulated samples

As shown in the figure, surface area consistently increases with the decreasing amount of PreKot™ with PrekotAC 5:95 gives higher surface area compared to PrekotAC 10:90. Both formulated filter aids have high surface area which is 1066.064 $m^2/g$ (PrekotAC 5:95) and 999.8710 $m^2/g$ (PrekotAC 10:90) where slightly lower than activated carbon alone. The large surface area accounts for the highly effective adsorptive characteristics.

#### 4.1.4 Characteristics in term of moisture content

Figure 5 presents the moisture content of the raw materials and formulated samples which showed that activated carbon has the highest while PreKot™ has the lowest moisture content with 23.39% and 0.73%, respectively. The formulated PrekotAC were consistent in term of their moisture content where PrekotAC 5:95 (22.60%) presents slightly higher moisture content compared to PrekotAC 10:90 (22.27%). The difference in moisture content due to the addition of the different amout of PreKot™. It was found that the addition of PreKot™ into the formulation helps to reduce the moisture content which is perquisite for adsorption purpose.

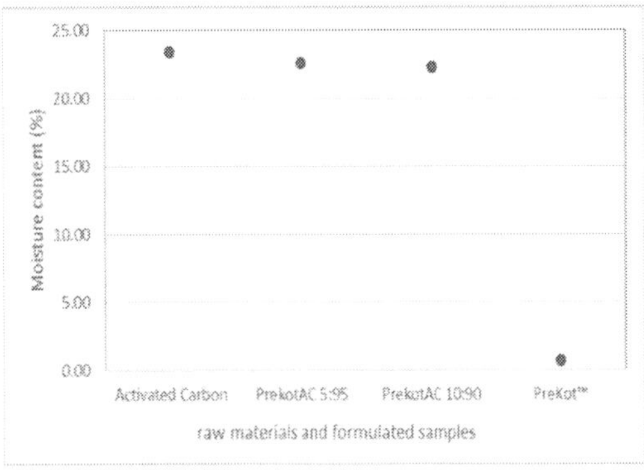


**Fig. 5.** Moisture content of raw materials and formulated samples

Moisture content is an important parameter since it directly influences the adsorption performance. According to Farzad *et al*, presence of water or moisture, trapped in activated carbon pores could lead to decrease in the amount of methane adsorption [14]. Therefore, low existence of moisture in the formulated material is better for adsorption process and indicates a good filter aids property [7]. However, the application of PrekotAC as a filter aids and flue gas cleaning agent may not be influence by moisture content since the operation occur in hot environment (~ 200°C) such as in incineration plant where high temperature will evaporate the moisture content exist in the material, thus maintained the performance as an adsorbent.

### 4.2 Toluene adsorption

#### 4.2.1 Comparison of adsorption performance between raw materials and formulated samples

Figure 6 shows the total toluene adsorbed in 100mg of filter aids materials under four different time duration of 5, 10, 30, and 40 minutes. Results showed that activated carbon gives the highest amount of toluene adsorbed (331µg) at 5 minutes compared to the other materials. It was also found that as time of adsorption increase, the amount of toluene adsorbed also increased. However, PreKot™ do not give a good result in adsorption where it was observed that the amount of toluene adsorbed decreased by the time of adsorption process. It showed that PreKot™ did not give much contribution on adsorption process compared to the other filter aids materials.

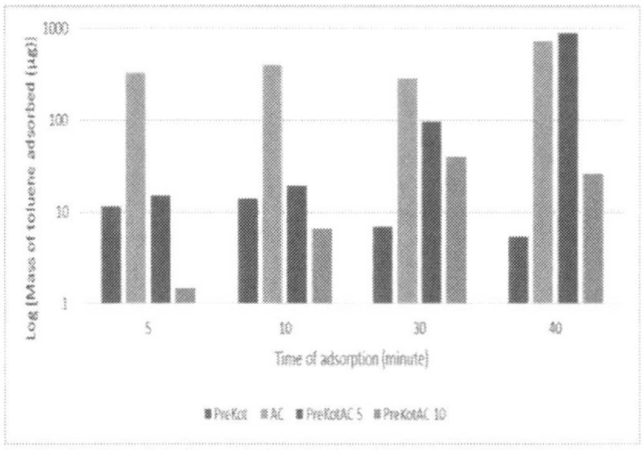


**Fig. 6.** Toluene's adsorption by raw materials and formulated samples at various time

As shown in the figure, PrekoatAC showed high adsorption of toluene, proving that the combination of activated carbon and PreKot™ still give a good adsorption properties. This can be explained by the fact that high number of surface area of formulated samples contribute to adsorb higher amount of VOC [14]. As the time increases, toluene loading volume increases until the saturation rate achieved. In this study, both formulated filter aids have a good potential as an effective adsorbent in fabric filter for flue gas cleaning purpose.

### 5 Conclusions

The formulated filter aids material PrekotAC showed a promising characteristic as an adsorbent for VOCs in fabric filter system. Having a small particle size with a large surface area will contribute to a better adsorption properties in a flue gas cleaning process. In addition, the improvement in terms of particle and bulk density would increase the porosity of material and maintain the adsorption performance of formulated material due to the higher surface area. It was also found that the combination of activated carbon and PreKot™ helps to reduce the moisture content in the formulated material. Results showed that PrekotAC has the capability to be applied as an adsorption for volatile organic compound (VOC) during flue gas cleaning process.

The authors gratefully acknowledge supervisor and Air Resource, MJIIT's team for their appreciable support.

## References

1. X. Su, L. Zhang, Y. Xiao, M. Sun, X. Gao, J. Su, J., evaluation of a flue gas cleaning system of a circulating fluidized bed incineration power plant by the analysis of pollutant emissions, Powder Technology, **286**, 9–15 (2015)
2. D. Prada-rodr, E. Fern, distribution of volatile organic compounds during the combustion process in coal-fired power stations, **35**, 5823–5831 (2001)
3. H. Zaitan, M. Manero, H. Valdés, L. Chimie De, D. Matière, C. Lcmc, M. Ben, application of high silica zeolite ZSM-5 in a hybrid treatment process based on sequential adsorption and ozonation for VOCs elimination, JES. 1–10 (2015)
4. P. Irigaray, J. A. Newby, R. R. Clapp, L. Hardell, V. Howard, L. Montagnier, D. Belpomme, lifestyle-related factors and environmental agents causing cancer : An overview. **61**, 640–658 (2007)
5. T. Cheng, Y. Jiang, Y. Zhang, S. Liu, prediction of breakthrough curves for adsorption on activated carbon fibers in a fixed bed, Carbon, **42**, 3081-3085 (2004)
6. K. P. Bhabendra, N. K. Sandle, effect of different oxidizing agent treatments on the surface properties of activated carbon, Carbon , **37**, 1323-1332 (1998)
7. S. Hajar, M. Rashid, A. Nurnadia, M. R. Ammar, formulation of a new pre-coating material for fabric filtration in air filtration system, Malaysia-Japan International Institute of Technology (MJIIT)- Japanese University Consortium (JUC) Joint International Symposium. 6th - 8th November. Hiratsuka, Japan (2013)
8. S. K. Theydan, M. J. Ahmed, optimization of preparation conditions for activated carbons from date stones using response surface methodology, **224**, 101-106 (2012)
9. ASTM D2867, standard test method for moisture content in activated carbon (2014)
10. F. Zeinali, A. A. Ghoreyshi, G. D. Najafpour, adsorption of dichloromethane from aqueous phase using granular activated carbon: Isotherm and breakthrough curve measurements, Middle-East Journal of Scientific Research, **5**, 191-198 (2010)
11. National Institute for Occupational Safety and Health (NIOSH), manual of analytical methods, fourth ed. hydrocarbons, aromatic (1994)
12. E. M. Martins, G. Arbilla, G. F. Bauerfeldt,M. de Paula, atmospheric levels of aldehydes and BTEX and their relationship with vehicular fleet changes in Rio de Janeiro urban area. Chemosphere, **67**, 2096-2103 (2007)
13. S. Hajar, M. Rashid, A. Nurnadia, M. R. Ammar, the characteristics of a formulated filter aids for fabric filters, Powder Technology, **283**, 315–320 (2015)
14. S. Farzad, V. Taghikhani, C. Ghotbi, B. Aminshahidi, L. E. Nemati, experimental and theoretical study of the effect of moisture on methane adsorption and desorption by activated carbon at 273.5 K, Journal of Natural Gas Chemistry, **16**, 22-30 (2007)

# Performance of PrekotAC Filter Aids on Pressure Drop across Two Different Filter Media

*S. Hajar*[1], *M. Rashid*[1*], *A. Nurnadia*[1], *N. Masdiana*[1], *M.R. Ammar*[2]

[1]Air Resources Research Laboratory, Malaysia-Japan International Institute of Technology, 54100 UTM Kuala Lumpur, Malaysia
[2]AMR Environmental Sdn. Bhd., Taman Sri Pulai Perdana, 81110 Johor Bahru, Malaysia
* Corresponding author: rashidyusof.kl@utm.my

**Abstract.** A study on the performance of formulated PrekotAC filter aids on pressure drop across two different types of filter media, i.e polytetrafluoroethylene (PTFE) and polyimide (P84) was carried out in a laboratory scale fabric filtration system. Filter aids is applied in order to prolong the lifespan of the fabric filter by reducing the pressure drop across a filter cake. PrekotAC which is a combination of 90%wt activated carbon and 10%wt PreKot™ was tested under various material loadings of 0.2, 0.4, and 0.6 mg/mm² at a constant filtration velocity of 5 m/min across its cake and filter media. The results showed that PrekotAC presents a lower pressure drop across the P84 compared to PTFE filter media under various material loadings. This is because of its different characteristics that effect the porosity of the filter media and allow a higher volumetric airflow passing through, resulting in lower pressure drop compared to PTFE filter media. In addition, the diversity in terms of particle size distribution of the formulated PrekotAC that increase its permeability property helps to reduce the pressure drop across the media and its filter cake. Thus, PrekotAC has a promising characteristic as a two in one filter aids, a pre-coating and adsorbent material for fabric filtration system.

## 1 Introduction

Filter aids is applied in fabric filtration system to reduce and prolong the lifespan of the fabric filter by reducing pressure drop across filter cake during filtration system. Pressure drop is one of the factors that need to be considered when evaluating a performance of a fabric filter system. Pressure drop is defined as the differential pressure across the system where it was reported that for a very well designed fabric filter, the maximum pressure is reached after a long period of filtration process before the fabric is being cleaned or replaced [1].

Cleaning frequency of a filter media is depending on the increment of the pressure drop across the filter cake during filtration process. As the frequency of cleaning increases, the maintenance cost also increases due to the higher probability of mechanical damages to the filter media during the process [2, 3].

Filter aids that applied in fabric filtration process works as a pre-coating material that coats the filter media during the process. In addition, filter aids helps to form a porous filter cake allowing a uniform airflow passing through it, resulting in low pressure drop [4, 5]. Filter aids has the capability to increase the porosity and reduce the compressibility of the accumulated filter cake [6]. It also improves the collection efficiency of particles and contaminants during filtration process [7].

The performance of fabric filter is characterized based on its pressure drop [8]. Thus, by using an appropriate filter media with a good filter aids material, helps to form a porous cake with low pressure drop across a filter cake.

In this study the performance of PrekotAC in terms of its pressure drop on two different filter media (PTFE and P84), under various material loadings of 0.2, 0.4 and 0.6 mg/mm² as well as a constant filtration velocity of 5 m/min was investigated. PrekotAC is a two in one filter aids material that consists of a combination of two different types of filter aids, i.e. activated carbon and PreKot™ that has the potential to form a porous filter cake due to its wide disparities in particle size distribution [9]. PrekotAC was tested under three different material loadings since these influence the pressure drop across the filtration process [10].

Hence, this study discussed on the effect of material loading and porosity of the filter aids on pressure drop across the filter cake. PrekotAC is a combination of two different types of material that will work as a two in one filter aids material (as an adsorbent and a pre-coating material) in fabric filtration system. By applying this material in the system, the all in one spray system can be introduced since these materials can work as a two in one filter aids material.

* Corresponding author: rashidyusof.kl@utm.my

## 2 Methodology

### 2.1 Filter Aids Material

PrekotAC filter aids was formulated by mixing two different types of filter aids material which is 90%wt of activated carbon with 10%wt of PreKot™ that works as an adsorbent and a pre-coating material, respectively. The properties of both materials used in this study are given elsewhere [9].

### 2.2 Experimental Procedures

The experiment was carried out in a fabricated laboratory scale filtration test system that consists of dust feeder, filter media holder, pressure manometer, particle counter, rotameter and a vacuum pump [7]. Pressure manometer (Extech Instrument Model HD755) were installed in the system in order to measure the pressure drop across filter media and cake during each run.

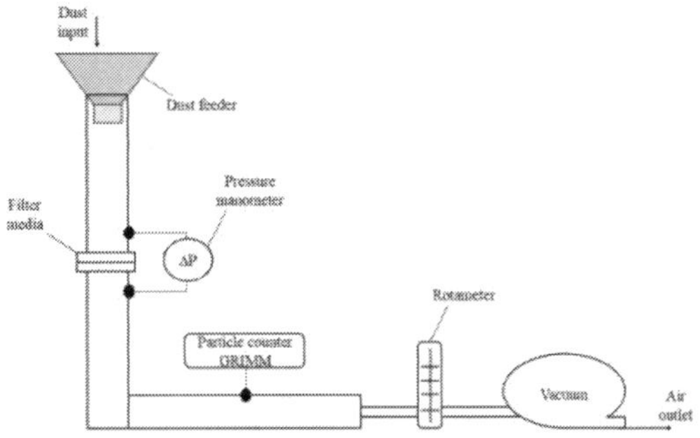


**Fig. 1.** Schematic diagram of the filtration test system

The performance of PrekotAC material on pressure drop across PTFE and P84 filter media under various material loading of 0.2, 0.4, and 0.6 mg/mm² and a constant filtration velocity of 5 m/min was observed in each run. Details on the specifications of both filter aids media are presented in Table 1.

**Table 1.** Specifications of PTFE and P84 filter media.

| Material | PTFE | P84 |
|---|---|---|
| Basic weight (g/m²) | 800 | 550 |
| Thickness (mm) | 1.3 | 2.4 |
| Working temperature (°C) | <200 | |

## 3 Results and Discussions

### 3.1 Effect of Different Filter Media on Pressure Drop

Figure 2 presents the results on pressure drop across two different filter media, PTFE and P84, under various material loading of 0.2, 0.4 and 0.6 mg/mm² and a constant filtration velocity of 5 m/min, which showed that P84 experienced a lower pressure drop compared to PTFE filter media in all conditions.

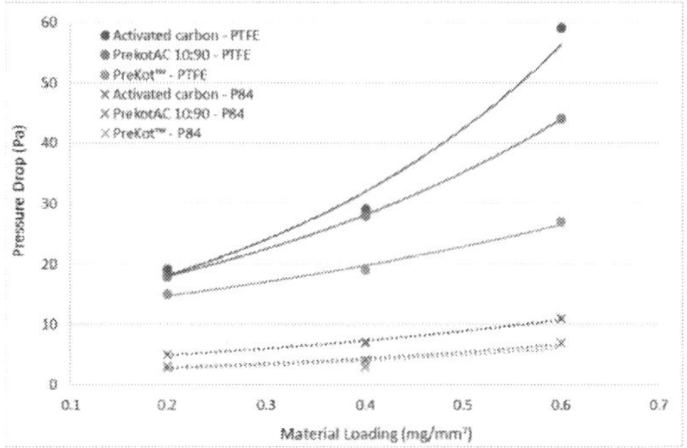


**Fig. 2.** Pressure drop across PTFE and P84 filter media.

This is due to the characteristics of the filter media where they have different structure and fiber diameter. As shown in Figure 3, it is found that P84 has a smaller fiber diameter of 15.3 μm compared to PTFE filter media (i.e. 75.0 μm). The average diameter of fiber in P84 is five times smaller than PTFE which gives a better airflow characteristic across the filter media.

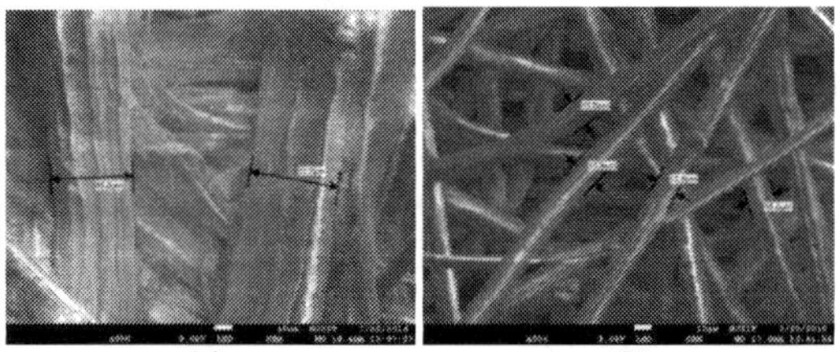

(a) (b)

**Fig. 3.** Structure and fiber diameter of, (a) PTFE and (b) P84 filter media.

Figure 4 presents the cross section microphotograph of the filter media which showed that P84 is characterized with loose structure compare to PTFE that explained the finding. Previous researcher reported that fibre arrangement in conventional PTFE filter media is dense and very thin that cause clogging of particles, resulting in high pressure drop across the filter media [11].

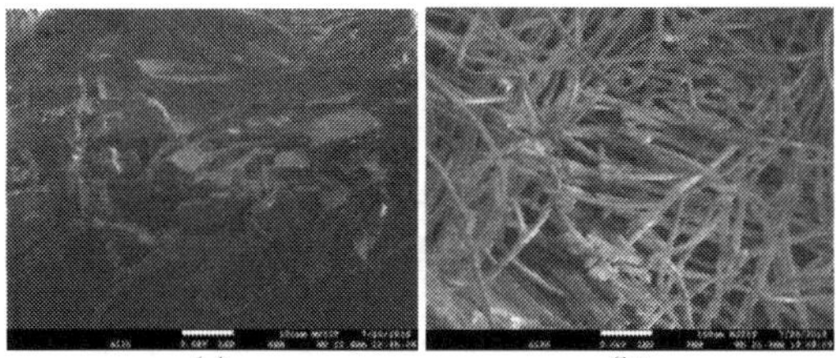

(a) (b)

**Fig. 4.** Cross section structure, (a) PTFE, (b) P84 filter media.

As in figure 2, it was also found that pressure drop across the filtration process increases with different amount of material loading of filter aids for both filter

DOI: 10.1051/matecconf/201711102003

media where higher material loading leads to a thicker filter cake that reduces the volumetric air flow that passing through, resulting in higher pressure drop across the filter cake [12-14]. This finding suggests that pressure drop across the filtration process depends on the characteristics of the filter media and filter aids [15].

It was reported that the root of pressure drop in fabric filtration process is the structure of the filter cake that denominated by the porosity [16]. The rise of material loading experiences a higher pressure drop which indicates the dependence of pressure drop on the porosity of the accumulated filter cake [17]. It was found that, as the dust cake thicken, it increases the resistance and reduce the porosity of the filter cake during the filtration process [2].

Thus, it can be concluded that pressure drop is highly dependent on the characteristics of the filter media and accumulated filter cake. Results showed that pressure drop across P84 is lower due to its structure where it has loose and porous filter media compared to PTFE. However, both filter media marked the highest pressure drop across the filter media at the highest amount material loading of 0.6 mg/mm² since higher material loading leads to thicker filter cake resulting in higher pressure drop.

### 3.2. PrekotAC as a Filter Aids

As in figure 2, the formulated filter aids material, PrekotAC, exhibits a lower pressure drop compared to the original material, activated carbon, even at the highest material loading of 0.6 mg/mm². The characteristics of the filter aids having a wide disparities particle size distribution that form a porous filter cake which allowed easy passage of air flow leads to low pressure drop across filter cake [18].

Particle size distribution of filter aids plays an important role in its performance as a filter aids which effect the performance in terms of its pressure drop in filtration system. Activated carbon has a lot finer particle size (80% of particles ≤75 µm) compared to PreKot™ (25% of particles ≤75 µm) which make it an ideal formulation for a good filter aids [19].

It was reported that pressure drop entailed by aerosol filtration is smaller for higher particle size. This is likely due to the fact that larger particles have a smaller specific area entailing a smaller pressure drop across the filter cake [20]. It was also found that pressure drop increased with a decreasing average particle diameter. The permeability that also referring to its porosity increases with the particle size since larger particles give rise to larger pores and channels for better air flow [21].

The newly formulated filter aids material PrekotAC has the capability to reduce its pressure drop across the filter cake due to its non-uniform particle size fractions which help to form a porous filter cake and allow a uniform airflow passing through the filter media. As pressure drop is one of the most important criteria that influences the operating cost a filtration system, it appears that a considerably amount of savings could be achieved with the use of PrekotAC.

## 4 Conclusions

Performance of PrekotAC filter aids on pressure drop across two different filter media had been investigated and reported in this paper. Results showed that P84 exhibits lower pressure drop compared to PTFE filter media due to their different fiber structure. The findings suggests that PrekotAC is a potential filter aids material for filtration system for its ability to reduce pressure drop across the filter cake even under high material loading.

## References

1. A.K. Choudhary, and A. Mukhopadhyay, Particulates: Media selection for pulse jet fabric filtration, *Filtration and Separation*. **49**, 34-35 (2013a)
2. K.J. Jeon, and Y.W. Jung, A Simulation Study on the Compression Behaviour of Dust Cakes, *Powder Technology*, **141**, 1-11 (2004)
3. A.K. Choudhary, and A. Mukhopadhyay, Particulates: Selection of Cleaning Pulse (a) Pressure for Pulse Jet Fabric Filtration, *Filtration and Separation*, **50**, 28-30 (2013b)
4. D. Purchas, 1st edition, Handbook of Filter Media, *Elsevier Advanced Technology* (1996)
5. M. Saleem and G. Krammer, Effect of Filtration Velocity and Dust Concentration on Cake Formation and Filter Operation in a Pilot Scale Jet Pulsed Bag Filter, *Journal of Hazardous Materials*, **144**, 677-681 (2007)
6. T. Barnett, Improving the Performance of Fabric Filter Dry Dust Collection Equipment, *Filtration and Separation*, **37**, 28-32 (2000)
7. C. Boittelle, C. Poupot, V. Milisic, and M.M. Peucheot, Advances in the Precoat Filtration Process, *Separation Science and Technology*, **43**,1701-1712 (2008)
8. A. Nurnadia, M. Rashid, S. Hajar, and M.R. Ammar, Penetration of Particle across a Newly Formulated Filter Aids Material, *Proceeding of 3rd International Science Postgraduate Conference* (2015)
9. A. Nurnadia, M. Rashid, S. Hajar and M.R. Ammar, Effect of Filter Aids on Pressure Drop and Permeability of Two Different Filter Media, *Advanced Materials Research*, **1125**, 245-249 (2015)
10. D. Thomas, P. Contal, V. Renaudin, P. Penicot, D. Leclerc, and J. Vendel, Modelling Pressure Drop in HEPA Filters during Dynamic Filtration, *Journal of Aerosol Science*, **30**, 235-246 (1999)
11. L. Bao, K. Seki, H. Niinuma, Y. Otani, R.O.T. Balgis, L. Gradon, and K. Okuyama, Verification of slip flow in nanofiber filter media through pressure drop measurement at low pressure conditions, *Separation and Purification Technology*, **159**, 100-107 (2016)

12. T.G. Chuah, and J.P.K. Seville, Numerical Simulation of Dust Cake Build up and Detachment on Rigid Ceramic Filters for High Temperature Gas Cleaning, *Journal the Institution of Engineers,* Malaysia, **65**, 1-2 (2004)
13. S. Hajar, M. Rashid, A. Nurnadia, and M.R. Ammar, The Effects of Newly Formulated Filter Aids Material Loading on Pressure Drop and Particle Penetration, *Advances in Engineering Research* (2016)
14. K.M. Lee, Y.M. Jo, J.H. Lee, and J.A. Raper, Assessment of Surface and Depth Filters by Filter Quality, *Powder Technology,* **185**, 187-194 (2008)
15. H. Leibold, and J.G. Wilhelm, Investigations into the Penetration and Pressure Drop of HEPA Filter Media during Loading with Submicron Particle Aerosols at High Concentrations, *Journals of Aerosol Science,* **22**, S773-S776 (1991)
16. J.H Choi, S.J. Ha, and Y.O. Park, The Effect of Particle Shape on the Pressure Drop across the Dust Cake, *Korean Journal Chemical Engineering,* **19**, 711–717 (2002)
17. C. Kanaoka, and M. Amornkitbamrung, Effect of Filter Permeability on the Release of Captured Dust from a Rigid Ceramic Filter Surface, *Powder Technology,* **118**, 113-120 (2001)
18. S. Hajar, M. Rashid, A. Nurnadia, and M.R. Ammar, The Characteristics of a Formulated Filter Aids for Fabric Filters, *Powder Technology,* **283**, 315-320 (2015)
19. S. Hajar, M. Rashid, A. Nurnadia, H. Norelyza, and M. Ammar, PrekotAC as Filter Aids for Efficient Dust Separation in a Fabric Filter, *Jurnal Teknologi,* **67**, 29-31 (2014)
20. C.B. Song, H.S. Park, and K.W. Lee, Experimental Study of Filter Clogging with Monodisperse PSL Particles, *Powder Technology,* **163**, 152-159 (2006).
21. M. Mayerhofer, J. Govaerts, N. Parmentier, H. Jeanmart and L. Helsen, Experimental Investigation of Pressure Drop in Packed Beds of Irregular Shaped Wood Particles, *Powder Technology*, **205**, 30-35 (2011)

# Effect of membrane on carbonation and carbon dioxide uptake of *Chlorella* sp.

*Emma* Suali*, *Rosalam* Sarbatly, *SM* Anisuzzaman, *Farhana* Abd. Lahin, *Mohd Asyraf* Asidin, *Tiffanyca* Jusnukin

Faculty of Engineering, Universiti Malaysia Sabah, 88400 Kota Kinabalu, Malaysia

**Abstract** Recent studies showed that as low as 5% $CO_2$ increased microalgae growth. However, common bioreactor operation resulted in low carbonation due to poor $CO_2$ mass transfer and this inhibited $CO_2$ uptake of microalgae. Although bubbling increases mass transfer of $CO_2$-$O_2$ exchange, preserving high dissolved $CO_2$ remains the most challenging of microalgae cultivation in bioreactor. In order to increase high dissolved $CO_2$ and $CO_2$-$O_2$ exchange, this study employed two types of membrane; hollow-fibre membrane for carbonation and hydrophobic membrane for deoxygenation. It was found that membrane increased carbonation from 20 % to 75 % when operated at control $CO_2$ concentration. The hollow-fibre membrane capable of creating as small as 2 mm bubble which effective for high carbonation. At the same time, it increased $CO_2$ uptake up to 85% in bioreactor. The hydrophobic membrane removed 43% $O_2$ from the bioreactor. Both membranes increased mass transfer of $CO_2$-$O_2$ exchange in bioreactor which stimulated microalgae growth.

Keywords: Membrane; microalgae; carbonation; deoxygenation; *Chlorella* sp.

## 1 Introduction

Carbonation of microalgae in bioreactor has reported since 1990s [1,2]. However, study of hollow fibre membrane rarely discussed although it increases $CO_2$ uptake [3-5]. Microalgae has potential to produce sufficient biomass if cultivated under $CO_2$ with a mean to remove photosynthetic $O_2$ [6]. Microalgae growth associated with $CO_2$ uptake, which increased under carbonation.

In general, bubbling provides aeration to microalgae and this resulted in low $CO_2$ mass transfer because of bubble formation [7-10]. This causes most $CO_2$ released to the bioreactor headspace and lessen its potential for microalgae growth. Microalgae lack in vascular tissue and an internal means to capture $CO_2$ outside watery media [11,12]. Thus, carbonation by membrane ensures $CO_2$ entrapped in the watery media. In addition, membrane capable to scatter $CO_2$ in the media and this increases potential use of $CO_2$ [13-15].

Another challenge of microalgae cultivation in bioreactor is high $O_2$ concentration. Microalgae produced $O_2$ as a side product of photosynthesis, which occurring simultaneously with $CO_2$ uptake. Oxygen reduces efficiency of continuous photosynthesis and thus interfere the $CO_2$ uptake in bioreactor [16,17]. Common cultivation uses aeration to remove $O_2$. However, aeration removes $CO_2$ and decreases carbonation efficiency due to unstable bubbling created during aeration. An optimum $CO_2$ – $O_2$ exchange required at least two membranes each acts as $CO_2$ carrier and $O_2$ removal. Therefore, this study discussed hydrophobic hollow fibre membranes as $O_2$ removal in bioreactor and its effect on microalgae growth.

Carbonation affects media pH and microalgae tolerance on pH level varies among species [18,19]. This study has employed *Chlorella* sp. which known has high tolerance on pH level. The tolerance of *Chlorella* sp. was emphasize further in this study.

## 2 Materials and Methods

### 2.1 *Chlorella* sp and cultural condition

This study employed *Chlorella* sp from Borneo (found at 6 °N and 116 °E). In this study, the *Chlorella* was cultivated under lessened carbon source in an adjusted standard Jaworski medium. Thus, the medium solution has no contents of $NaHCO_3$, Cyanocobalamin, Thiamine HCl and Biotin. About 0.0401 kg/m$^3$ (40 ppm) of *Chlorella* presents in the medium and used as a basis analysis on the *Chlorella* growth. The $CO_2$ gas enters Membrane 1 at controlled flowrates and microalgal media enter Membrane 2 for $O_2$ removal.

### 2.2 Membrane bioreactor set up

* Corresponding author: emma.suali@gmail.com

So far, no membrane can add and remove two different gases simultaneously. Thus, this study employed a hydrophobic-membrane with a model number UFS220 for carbonation and labelled as Membrane 1. The other membrane, 2 X 6 Radial Flow type superphobic membrane supplied by Liqui-Cel for $O_2$ removal and labelled as Membrane 2. Table 1 shows feature of Membrane 1 and Membrane 2 whereas others feature was published in previous work [10].

Figure 1 shows membranes position in the experimental setup. Membrane 1 connected to bioreactor 1 (BR1) and Membrane 2 connected to both bioreactors for $O_2$ removal. Two fluorescence white cool lamps lit up the bioreactors at 296 μE/m$^2$s and a Lux meter with a model number LX-101 used to measure the light intensity.

**Table 1.** Characteristic of membranes

| | **Membrane 1** | **Membrane 2** |
|---|---|---|
| Number of fibres | 2400 | 3550 |
| Fibre pore size (μm) | 0.01 to 0.1 | 0.03 to 0.06 |
| Length (mm) | 495 | 146 |
| Total surface area (m$^2$) | 0.80 | 0.50 |
| Shell diametre (mm) | 60 | 55.60 |
| Capacity (m$^3$/s) x 10$^{-5}$ | 2.7 | 0.16 to 1.6 |
| Fibre thickness (mm) | 0.65 | N/A |
| Shell side volume (m$^3$) x 10$^{-3}$ | 0.43 | 0.08 |
| Volume occupies by hollow fibre (m$^3$) x 10$^{-3}$ | 2.5 | N/A |
| Housing material | Polyvinyl chloride | Polyethylene |
| Membrane material | Polysulfone | Polypropylene |

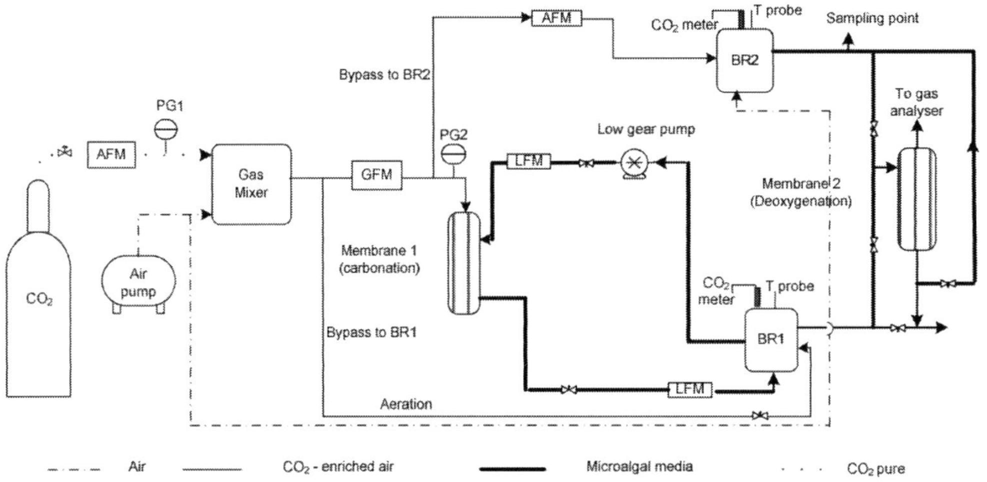


**Fig.1.** Membrane bioreactor set up

### 2.3 Carbonation and bubble measurement

The $CO_2$ feed was controlled from 0.49 x 10$^{-3}$ kg/m$^3$ to 3.11 x 10$^{-3}$ kg/m$^3$ for each experiment. The concentration of $CO_2$ in media and at the headspace of bioreactor was measured according to previous work [10]. The carbonic acid in media was estimated based on the total fraction of hydrogen ion to hydroxyl ions, which represented as pH reading [20].

The membrane created small bubbles in bioreactor, which was estimated according to Rodrigues and Rubio [21] method. The tubular bioreactor was labelled with numerical estimation from 0 cm to 100 cm as aid to the bubble estimation.

### 2.4 Oxygen removal by Membrane 2

Membrane 2 acted as $O_2$ removal prior to $CO_2$ depletion in the bioreactor. Membrane 2 is a superphobic membrane, which have high resistance to water and microalgae attachment at the membrane walls.

### 2.5 $CO_2$ uptake and microalgae growth

The $CO_2$ uptake by microalgae is a measured of (a) dissolved $CO_2$, (b) carbonic acid in media, (c) $CO_2$ at the bioreactor headspace and (d) $CO_2$ in the membrane. The $CO_2$ meter measures $CO_2$ concentration from 1.49 x 10$^{-3}$ kg/m$^3$ to 1490 x 10$^{-3}$ kg/m$^3$ (1 ppm to 1490 ppm) in media and 0.1 % to 100 % in the bioreactor headspace. At the same time, a reading of microalgae growth, pH, and dissolved $O_2$ was recorded daily for three weeks. A UV-Vis spectrophotometer with UV-Vis bandwidth of 2.0 nm was used to measure growth of microalgae. The absorbance was recorded at a wavelength of 657 nm.

## 3 Results and Discussion

### 3.1 Effect of membrane integration

Figure 2a shows dissolved $CO_2$ in a bioreactor with and without membrane. At 0.49 x $10^{-3}$ kg/m$^3$, the dissolved $CO_2$ was 0.1 x $10^{-3}$ kg/m$^3$ in bioreactor without membrane. This equivalent to 20% of $CO_2$ feed. In comparison, the dissolved $CO_2$ in membrane integrated-bioreactor was 0.4 x $10^{-3}$ kg/m$^3$. This equivalent to 82% of $CO_2$ feed and 4 times higher than in bioreactor without membrane. At 3.11 x $10^{-3}$ kg/m$^3$, about 42 % and 29 % of $CO_2$ was dissolved in media for integrated and non-integrated bioreactor, respectively. Thus, further increase in $CO_2$ feed was considered as ineffective beyond 3.11 x $10^{-3}$ kg/m$^3$.

The experimental result suggests that up to 82 % of $CO_2$ feed was delivered by membrane to the bioreactor and entrapped within the microalgae culture. Nearly 18 % was released into the headspace of the bioreactor, piled up within the membrane and converted to carbonic acid. From the perspective of Henry's law, the highest $CO_2$ could dissolve in water was approximated about 83 % of total supply (dissolved $CO_2$ = 0.8317 x $CO_2$ inlet) [22]. This means that 83 % of $CO_2$ was dissolved in water at atmospheric pressure and room temperature (28 °C). However, the equilibrium of $CO_2$ dissolves with $CO_2$ feed which is the solubility of $CO_2$ at natural conditions takes longer period to achieve compared with membrane-integrated bioreactor. This was a reason for drawback in microalgae cultivation with added $CO_2$. During carbonation, pH level was controlled at tolerable conditions for microalgae growth.

The acidic level of media in bioreactor was indicated by pH reading. From Figure 2b, pH level decreased as dissolved $CO_2$ increased. The natural conversion of $CO_2$ into carbonic acid in the presences of water caused the final reading of pH in the media as low as 4.82 [23]. It is desired to have high dissolved $CO_2$ compared to carbonic acid for microalgae cultivation. Microalgae ability that can use dissolved $CO_2$ without depending on carbon concentrating mechanism (CCM) increased $CO_2$ uptake in bioreactor. In comparison, carbonic acid conversion to useable carbon such as carbonate required CCM activity. Some microalgae have low CCM activity thus incapable to convert carbonic acid to useable carbon for growth. Microalgae can survive if less than 5% of carbonic acid in media [24,25]. Therefore, a small fraction of carbonic acid resulted from the reaction with water is sufficient for growth. Cultivation with more than 5% carbonic acid can cause acidic to the media.
Figure 2b shows that allowed $CO_2$ feed is less than 0.5 x $10^{-3}$ kg/m$^3$ $CO_2$ to prevent extreme acidic condition. In overall, Figure 2 shows that carbonic acid increased with $CO_2$ feed. It was found that $CO_2$ conversion to carbonic acid agreed with Henry's law statement whereas the equilibrium constant of carbonic acid in water is approximated at 1.7 x $10^{-3}$ at room temperature [26]. Most of the $CO_2$ gas remains as $CO_2$ in water. The remaining $CO_2$ was scattered to another part of the membrane bioreactor. Thus, further study was conducted to evaluate the $CO_2$ distribution.

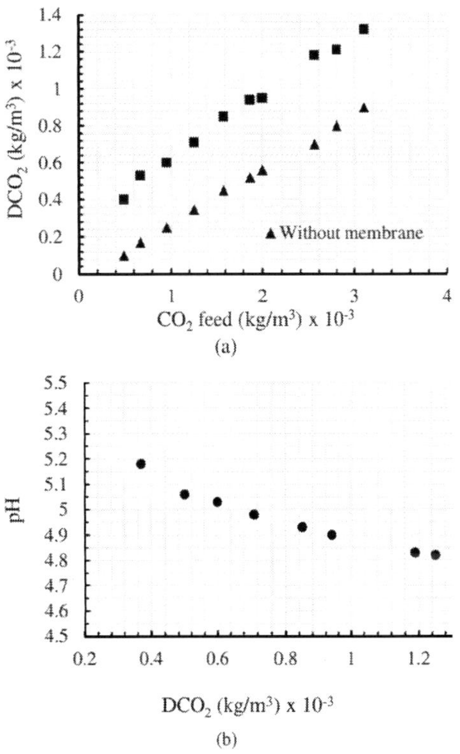


**Fig.2.** (a) Comparison of dissolved $CO_2$ in bioreactor; (b) pH

### 3.2 Distribution of $CO_2$ in bioreactor during carbonation

Carbon dioxide distributed in bioreactor mainly as dissolved gas in media, gas at the bioreactor headspace, gas piled up within the membrane and chemically converted to carbonic acid. Figure 3a, Figure 3b, Figure 3c and Figure 3d show the fraction of this $CO_2$, respectively. The membrane increased $CO_2$ delivery to media from 40 % to 80 % of $CO_2$ feed and thus reduced $CO_2$ at the bioreactor headspace up to 18 %. About 14 % amassed in the membrane and the smallest fraction, less than 1 % converted as carbonic acid in media.

In Figure 3a, dissolved $CO_2$ increased as the $CO_2$ feed increased. However, low increment of dissolved $CO_2$ was found when fed with $CO_2$ above 2.5 x $10^{-3}$ kg/m$^3$. This suggests that $CO_2$ should not feed beyond 2.5 x $10^{-3}$ kg/m$^3$. There are three main reasons for the loss which were presented in Figure 3b, Figure 3c and Figure 3d.

In Figure 3b, the accumulated $CO_2$ within the membrane was 0.02 x $10^{-3}$ kg/m$^3$ to 0.42 x $10^{-3}$ kg/m$^3$ compared to the $CO_2$ feed which was 0.49 x $10^{-3}$ to 3.11 x $10^{-3}$ kg/m$^3$. This equivalent to 4 % to 14 % of $CO_2$ feed. This shows that the higher the $CO_2$ feed, the higher possibility of the $CO_2$ accumulated within the membrane. This considered as a loss to the cultivation of microalgae. However, a simple flushing by air at the final stage of the carbonation lessens the gas build-up in the membrane.

The remaining $CO_2$ was stored as gas at the bioreactor headspace. In Figure 3c, up to 1.42 x $10^{-3}$ kg/m$^3$ of $CO_2$ feed was found in the bioreactor headspace. This equivalent to 0.46 % of $CO_2$ feed. The stability of bubble formation during carbonation has affected gas concentration at the bioreactor headspace although it lower than $CO_2$ piled-up in the membrane. The $CO_2$ at the headspace lessen the possibility of $CO_2$ uptake by *Chlorella*. It is possible to use $CO_2$ outside the watery media but takes longer period to achieve equilibrium between the gas and liquid phase. This affects microalgae growth. Thus, $CO_2$ dissolved into media have more benefit for microalgae growth compared to $CO_2$ outside the media.

The smallest fraction of the $CO_2$ feed was found as carbonic acid. The carbonic formation in microalgal-media has a slight difference compared with water. The converted $CO_2$ was in the range of 0.01 x $10^{-3}$ kg/m$^3$ to 0.03 x $10^{-3}$ kg/m$^3$ as shown in Figure 3d. This equivalent to 0.006 % to 0.02 % of the $CO_2$ feed. However, this amount caused acidic in microalgal media. Some microalgae species have high tolerance to the acidic environment. In this study *Chlorella* sp. showed no decrease in microalgae growth at low pH which suggested that *Chlorella* sp. might has high tolerance on low pH. Adding alkaline such as NaOH probably can reduce the effect of acidic condition in media. However, the effect of alkaline on pH was not evaluated in this study. In overall, the use of membrane to aid carbonation lower $CO_2$ amount at the headspace and within the membrane. It increased carbonation and ease $CO_2$ addition into the *Chlorella* sp. culture. The key to high increase in carbonation with the aid of membrane is the ability of membrane to control $CO_2$ feed and thus control over the bubble formation.

The bubbling is the apparent reason for low carbonation in bioreactor which aerated without membrane. Some study has reported that carbonation of microalgae culture better achieved by bubbling column than common aeration [3,27]. However, bubbling promotes low carbonation in bioreactor. The use of bubbling column increases microalgae growth up to 0.63 per day [28]. Thus, this study presents discussion on bubbling by membrane.

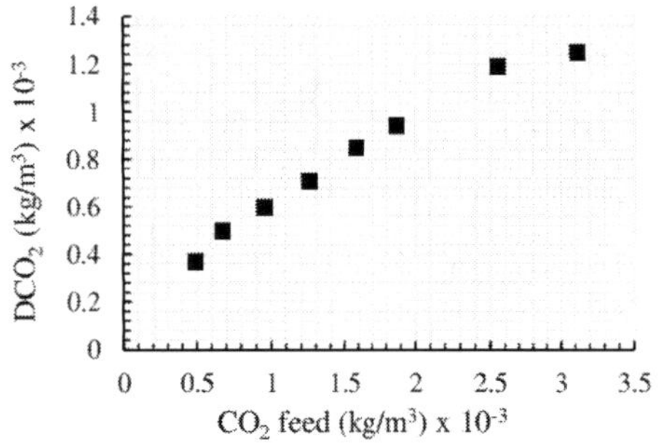


(a)

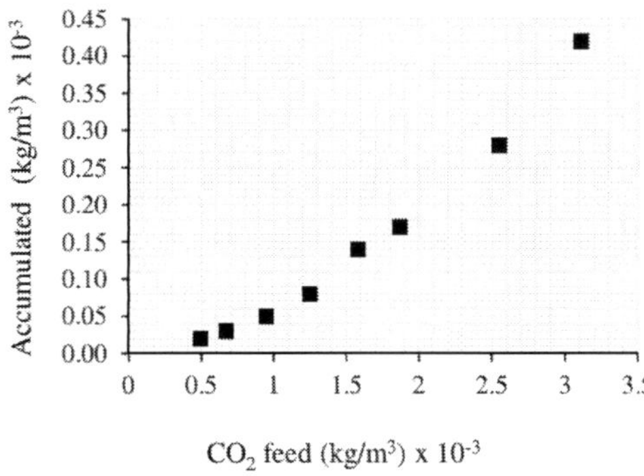


(b)

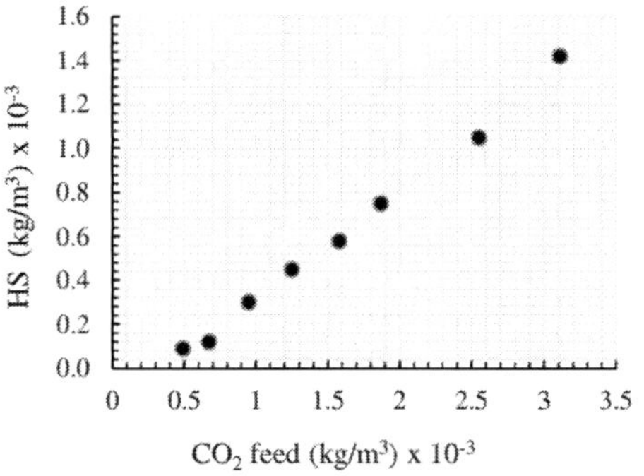


(c)

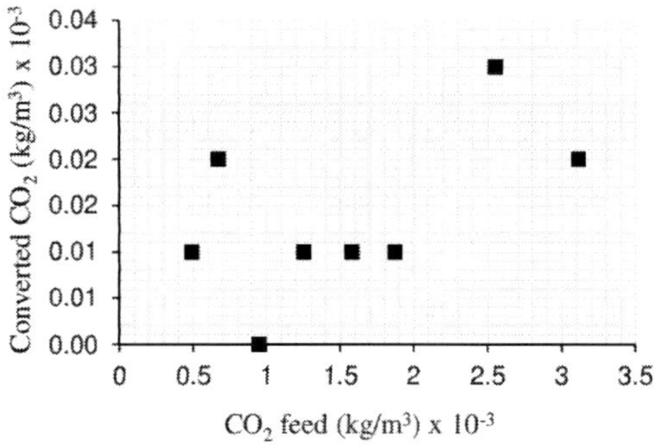


(d)

**Fig.3.** (a) Dissolved $CO_2$; (b) Amount of $CO_2$ in bioreactor headspace; (c) Accumulated $CO_2$ within the membrane; (d) Converted to carbonic acid

### 3.3 Relationship of carbonation and bubbling

Membrane distributes $CO_2$ in bioreactor and the key for achieving high dissolved $CO_2$ depends on membrane ability to control the size of the bubble. The bubble in bioreactor without membrane and aerated with standard air composition was found in the range of 5 mm to 8 mm in diameter. The bubble was in spherical shape. At constant flow rate, the bubble size was remained constant. The use of membrane resulted in small bubble ranges from 1 mm to 2 mm. Figure 4 shows bubble size compared with dissolved $CO_2$. Figure 4 shows that the bigger the bubble size, the lower amount of $CO_2$ dissolved. Flow rate also was found affected the bubble formation. Less than 2 mm bubbles observed at 0.3 x $10^{-5}$ $m^3/s$. The dissolved $CO_2$ was found 70 % of the $CO_2$ feed. The bubble size increased with the gas feed. From 0.3 x $10^{-5}$ $m^3/s$ to 0.7 x $10^{-5}$ $m^3/s$ the bubble size was found from 1 mm to 2 mm and the dissolved $CO_2$ decreased from 70 % to 60 %.

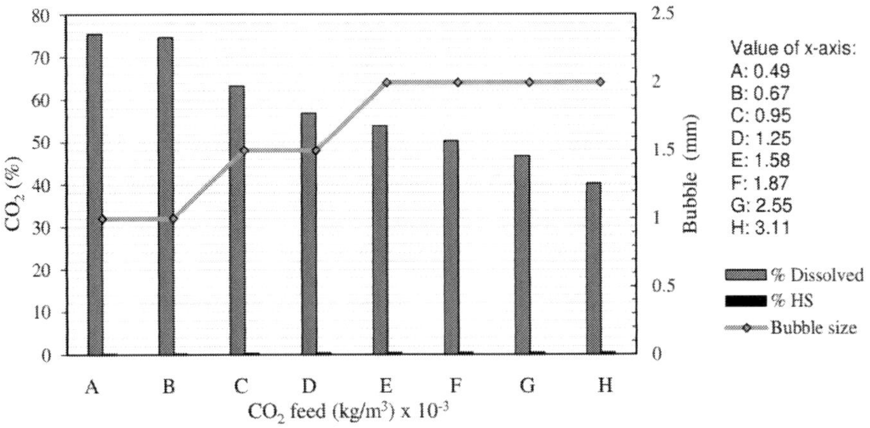


**Fig.4.** Effect of bubble on $CO_2$ concentration in media and bioreactor headspace

In continuous carbonation and cultivation, the bubble size decreased as the inlet media flow rate increased. At constant gas flow rate, which is 1.3 x $10^{-5}$ $m^3/s$, the bubble size decreased from 2 mm to barely seen by eyes as a side effect of increasing liquid flowrate. The effect of liquid and gas flow rate in bioreactor was discussed in previous work [10].

As conclusion, the bubble size affected $CO_2$ in media and bubble created between 1 mm to 2 mm resulted in 70 % dissolved $CO_2$. When bubbled at 5 mm and above, the dissolved $CO_2$ was found less than 30 %. The large bubble created without membrane indicates that membrane plays a major effect on the bubbling formation. In addition, the inlet flow rate affected size of bubble. This result agreed with findings in bubbling column [29-31]. The instability of inlet flow rate and high Reynolds number creates large bubbles. The large bubble fetched up the $CO_2$ gas and escaped into the bioreactor headspace. Logically, microalgae only uses $CO_2$ that is in the water because of its cell phycology that does not have vascular tissue [32]. Thus, the $CO_2$ released into the headspace lessened the potential of $CO_2$ uptake by microalgae. The relation of carbonation and $CO_2$ uptake was measured further in this study.

### 3.4 Membrane carbonation and $CO_2$ uptake

The purpose of this study is to examine the range of tolerable carbonation or dissolved $CO_2$ by *Chlorella* sp. This work also aims to find a way to increase $CO_2$ uptake of *Chlorella* sp. Preliminary study shows that $CO_2$ uptake was in the range of 20 % to 85 % regardless of $CO_2$ feed. The highest $CO_2$ uptake was recorded at 2 mm bubble, followed by 1 mm. The lowest $CO_2$ uptake was when the cultivation was bubbled at 1 mm. It shows that at 1 mm, *Chlorella* sp. used only 20 % of the total $CO_2$ feed. At the same time, the specific growth was recorded as 0.17 per day, which is the lowest among bubble size. However, this value higher compared to fast-growing grass which in the range of 0.09 $d^{-1}$ to 0.59 $d^{-1}$ [33].

The uneven distribution of $CO_2$ in media was identified as a main reason for low $CO_2$ uptake. The distribution of $CO_2$ bubbled at 1 mm was evaluated with the pH measure of the media. The pH was measured at the bottom, middle and surface of the media. Less than 0.2 % of $CO_2$ feed at the bioreactor headspace shows that most $CO_2$ entraps in the media. The *Chlorella* sp. growth which is 0.17 per day was due to insufficient $CO_2$ in the bioreactor. In overall, the $CO_2$ uptake by microalgae is best achieved when carbonized with bubble size approximately 2 mm where the $CO_2$ uptake was up to 85 %. This resulted in *Chlorella* sp. specific growth of 0.21 $d^{-1}$. This show that 2 mm was effective for both biomass and $CO_2$ uptake.

The $CO_2$ uptake in bioreactor associated with $O_2$ produced in the bioreactor. The total $O_2$ produced in bioreactor was 30 % and the dissolved $O_2$ was 18 %. The higher the growth rate, the higher the $O_2$ produced in the bioreactor. The $O_2$ generation also indicates the efficiency of $CO_2$ uptake. In ideal photosynthetic theory, each molecule of $CO_2$ produces one molecule of $O_2$. The $O_2$ in bioreactor headspace is in balance with the dissolved $O_2$. The presence of dissolved $O_2$ in the media causes photorespiration, thus preventing the photosynthetic enzyme to drive $CO_2$ uptake by microalgae. As low as 20 % of $O_2$ in media enough to hinder the photosynthesis cycle [14]. The $O_2$ reacts with RubisCO and produces phosphoglycolate, which inhibits the enzymes of photosynthesis.

The main factor of affecting $CO_2$ uptake was the bubbling within the bioreactor and the preliminary microalgae density in media. The density of microalgae affects light penetration inside the bioreactor. In extremely low microalgae density, light was easy to reach to each microalgae cell. However, the $CO_2$ uptake was found less compared to denser *Chlorella* sp. concentration. In addition, at extremely dense microalgae cell, more $O_2$ was produced. The produced $O_2$ inhibits the continuous photosynthesis cycle. Large bubbles in media capable of removing the dissolved $O_2$. Thus, the $O_2$ generated does not prevent the photosynthesis of *Chlorella* sp. as most of the dissolved $O_2$ released into the bioreactor headspace. The average ratio of dissolved $O_2$ to $O_2$ in bioreactor headspace was 4:1. Other means of removing $O_2$ without creating large bubble is applying hydrophobic membrane instead of aeration. Thus, preliminary effect of secondary membrane was discussed in this study.

### 3.5 Effect of secondary membrane to $O_2$ removal and *Chlorella* sp.growth

The Membrane 2 lessened $O_2$ in media from 8.5 x $10^{-3}$ kg/m$^3$ to 4.9 x $10^{-3}$ kg/m$^3$ at third week of cultivation. Another set of experiment which ran simultaneous also shows similar result. The BR1 which integrated with Membrane 1 was bubbled at 1 mm to 2 mm and BR2 without Membrane 1 ran at 5 mm to 10 mm. Both bioreactor were equipped with Membrane 2. About 41 x $10^{-3}$ kg/m$^3$ of $CO_2$ was depleted in BR1. This equivalent to 89 % of the $CO_2$ feed. The overall dissolved $O_2$ and $O_2$ at the bioreactor headspace was about 24 % and 12 %, respectively. In comparison, BR2 shows 5 % and 21 % depletion of dissolved $O_2$ and $O_2$, respectively. The large bubble size in BR2 has aided the release of $O_2$ from the media. However, this resulted in low growth of *Chlorella* sp. which was only up to 0.19 per day (specific growth rate).

As a conclusion, the secondary membrane increased $CO_2$ uptake of *Chlorella* sp. The secondary membrane has no direct effect on the $CO_2$ uptake. However, both membranes have improved $CO_2$ uptake and $O_2$ removal from the bioreactor, thus, increased the *Chlorella* sp. growth.

## 4. Conclusion

This study presents an evaluation of membranes to increase carbonation of microalgae media. About 82 % of $CO_2$ feed was successfully delivered to the media with the aid of Membrane 1 and this resulted in low pH of media. The membrane also distributed $CO_2$ as dissolved $CO_2$ in media, piled up in membrane, accumulated at headspace and converted to carbonic acid. This study suggested that 2 mm bubble size is optimum for high carbonation. Finally, the secondary membrane has indirectly aided in increasing the $CO_2$ uptake by removing $O_2$ from the media.

## Acknowledgements

This study was supported by FRGS grant no. FRG0416-TK-1/2015 and UMS grant with grant no. GUG0052-TK-2/2016 and GUG0050-TK-2/2016.

## References

1. Y. K. Lee, H.K Hing, Appl. Microbiol. Biotechnol. **31**, 298-301 (1981)
2. S. Abinandan, S. Shanthakumar, 3 Biotech. **6**, 1-9 (2016)
3. R. Putt, M. Singh, S. Chinnasamy, K.C. Das, Bioresour. Technol. **102**, 3240-3245 (2011)
4. R. Sarbatly, E. Suali, F.A. Lahin, C.K. Chiam, In: *Advances in Bioprocess Technology*, Springer International Publishing, 371-386 (2015)
5. F.V. Winck, D.O.P. Melo, D.M. Riano-Pachon, M.C.M. Martins, C. Caldana, A. F. G. Barrios, Front. Plant Sci. **7**, 43 (2016)
6. S. P. Singh, P. Singh, Renew. Sust. Energ. Rev. **38**, 172-179 (2014)
7. H. S. Fogler, V.K.Verma, Chem. Eng. Sci. **26**, 1391-1400 (1971)
8. A. P. Carvalho, F.X. Malcata, Biotechnol. Prog. **17**, 265-272 (2001)
9. C. Vejrazka, M. Janssen, G. Benvenuti, M. Streefland, R.H Wijffels, Appl. Microbiol. Biotechnol. **97**, 1523–1532 (2013)
10. R. Sarbatly, E. Suali, Algal. Res. **5**, 274-282 (2014)
11. L.Tomaselli, In: *Handbook of microalgal culture*, Biotechnol. Appl. Phycol., Blackwell Science Ltd, pp. 565 (2014)
12. J.A. Raven, J. Beardall, J. Exp. Bot. **67**, 1-13 (2016)
13. A. Contreras, F. Garcia, E. Molina, J.C. Merchuk, Biotechnol. Bioeng. **60**, 317-325 (1998)
14. F.C. Rubio, F. G. Fernandez, J.A. Perez, F.G. Camacho, E.M. Grima, Biotechnol.Bioeng. **62**, 71-86 (1999)
15. E. Molina, J. Fernandez, F.G. Acien, Y. Chisti, J. Biotechnol. **92**, 113-131 (2001)
16. C. Brindley, C., F.G. Acien, J.M. Fernandez-Sevilla, Biotechnol. Bioeng. **106**, 228-237 (2010)
17. M.Ota, Y. Kato, M. Watanabe, Y. Sato, R.L.Smith, S. Rosello-Sastre, C. Posten, H. Inomata, Bioresour. Technol. **102**, 3286–3292 (2011)
18. J.C. Goldman, Y. Azov, C.B. Riley, M.R. Dennett, J. Exp. Mar. Biol. Ecol. **57**, 1-13 (1982)
19. E. Touloupakis, B. Cicchi, A.M.S Benavides, G.Torzillo, Appl. Microbiol. Biotechnol. **100**, 1333-1341 (2016)
20. C.Y.Chen, E.G. Durbin, Mar. Ecol. Prog. Ser. **109**, 84-93 (1994)
21. R.T. Rodrigues, J. Rubio, Miner. Eng. **16**, 757-765 (2003)
22. M.L. Davis, S.J. Masten, *Principles of environmental engineering and science*, McGraw-Hill, New York (2004)
23. J. Thielmann, N.E. Tolbert, A. Goyal, H.Senger, Plant. Physiol. **92**, 622-629 (1990)

24. S.A. Schoenberg, R. Benner, A. Armstrong, P. Sobecky, R.E Hodson, Appl. Environ. Microbiol. **56**, 237-244 (1990)
25. R.C.D.Goswami, N. Kalita, M.C. Kalita, Ann. Biol. Res. **3**, 499-510 (2012)
26. J.M.Allen, W.X. Balcavage, B.R.Ramachandran, A.L. Shrout, Environ. Toxicol. Chem. **17**, 1216-1221 (1998)
27. A.S.Miron,C.F. Garcia, A.C. Gomez, E.M. Grima, Y. Chisti, *AIChE Journal*, **46**, 1872-1887 (2000)
28. G.Najafpour, N. Bakuei, G. Amini, M. Jahanshahi, Indian J. Chem. Technol. **22**, 20-25 (2016)
29. S.V.Gnyloskurenko, T. Nakamura, Mater. Trans. **44**, 2298-2302 (2003)
30. S.M. Walke, V.S. Sathe, Int. J. Chem. Eng. Appl. **3**, 25-30 (2012)
31. S.A. Sulaiman, N.Z.Z. Kamarudin, J. Appl. Sci. **12**, 2464-2468 (2012)
32. V.B.Rastogi, *Modern Biology*. Pitambar Publishing (1997)
33. E. Suali, R. Sarbatly, S.R.M. Shaleh, *International Conference on Applied Energy*, Jul 5-8, Suzhou China (2012)

# Policosanol extraction from beeswax and improvement of the purity

*Anakhaorn* Srisaipet*, *Sudarat* Phromchan, and *Thanyarat* Jaipaeng

Department of Chemistry, Maejo University, Chiangmai, Thailand

**Abstract.** Policosanol is a mixture of high molecular weight aliphatic long chain alcohols (20-36 carbon atoms). It has been use in pharmaceutical composition and food supplements. This research aimed to isolate and improve the purity of policosanol extracted from beeswax. Triglycerides and other impurities were eliminated from beeswax by refluxing with hexane followed by isopropanol. The purified beeswax was hydrolyzed by refluxing with 1 M ethanolic NaOH for 2 hours. Purification of policosanol was performed by extracting the hydrolyzed product with acetone at 50-60 °C for 3 hours and it was stored at 4 °C for precipitation. The precipitate was refluxed with heptanes followed by washing with hot water. The heptanes layer was kept for policosanol precipitation at 4 °C. The purity of policosanol was confirmed by TLC and high performance liquid chromatography (HPLC). The yield of purified policosanol was 13.23-13.89 %.

## 1 Introduction

Policosanols (PC) are a mixture of high molecular weight of primary aliphatic long chain alcohol (20-36 carbon atoms) containing mainly docosanol, tetracosanol, hexacosanol, octacosanol and triacontanol [1]. Policosanol have effect potential to reducing LDL peroxidation [2-5], plated aggregation, prevention and treatment of cardiovascular disease [2-8] and proliferation of smooth muscle cell. Nowadays, it is used to nutraceuticals and food supplementary in the world market.

Policosanol was originally isolated from plant wax and animal wax. Beeswax is mainly complex mixture of fatty acids and aliphatic long chain alcohols. The monoester of carbon atom 24 to 34 atoms alcohols are about 40% of the composition of beeswax [9]. Thereby, Miguel and Fernando, 2013 [10] reported the carbon length 18 to 34 atoms of aliphatic alcohol which extracted from Portuguese beeswax. Irmak et al., 2006 [1] reported the main component in policosanols of beeswax from USA showing more than 40% is triacontanol (C30-OH). However, the content of beeswax composition including to another substance are various with origin, extraction and purification technique [1, 10-12].

Policosanol can be extracted by hydrolysis or saponification reaction with NaOH or KOH base in various conditions [1, 13-19]. The researchers have investigated on the use of enzyme catalyst methanolysis in supercritical carbon dioxide [20], using High ultrasound (HIU) as catalyst [21] and using pressurised solvent extraction [22]. The purification step of policosanol has to use of organic solvents for purifying [1, 4, 11, 13-19, 22] and following by crystallization and using a silica column to purify [23]. The most commonly technique for policosanol analyzation and characterization can be done by Thin layer chromatography (TLC), Gas chromatography (GC) and High performance liquid chromatography (HPLC) or combining these technique [4, 16-17, 24-26].

The suitable solvent selection is the most important step in policosanol purification for optimising recovery of aspiring to the main components from a complex mixture. Solvent–solute interactions relate to dispersion and and/or multi-polar interactions. The dielectric constant identifies the polarity of a solvent and a key parameter to determine solute–solvent interactions. Moreover, a decrease in dielectric constant of a solvent with increasing temperature, accordingly lowering the polarity of the solvent [27]. Thus, temperature can be used for guide line to match the solvent properties to a recovery of the solute of the interest compound.

In this study, the policosanol was extracted from beeswax by hydrolysis reaction and improvement on the purity of extracted policosanol was done by considering the effect of solvent and temperature collective on purity and recovery of policosanol. The purity of policosanol was confirmed by TLC and HPL.

## 2 Materials and Methods

### 2.1 Materials and regents

Beeswax as policosanol source was obtained from bees' farm at a local farm located in Chiang Mai Province, Thailand. Standard aliphatic long chain alcohols and Stearic acid were supplied by Sigma Aldrich, Thailand. Chemical reagents in analytical grade were provided

* Corresponding author: anakhaorn@mju.ac.th

DOI: 10.1051/matecconf/201711102005

from LabScan (Bangkok, Thailand). Commercial rice bran oil (as triglyceride) in food grade was purchased from local supermarket in Chiangmai, Thailand.

### 2.2 High Purity Beeswax Preparation

The compositions of beeswax were studied by TLC. The spot on TLC plate were compost of beeswax, rice bran oil (as triglyceride; TG), standard free fatty acid (steric acid; C18:0) and standard ester (steric acid methyl ester derivative).

High Purity Beeswax were prepared by dissolving crude beeswax in hexane (30 g wax/300 ml) and it was refluxed at 65 °C for 3 hrs with continuous stirring. The refluxed beeswax was washed with hot hexane in twice times. The washed beeswax was refluxed again with isopropanol at 80 °C for 3 hrs and washed again with isopropanol. The purity of beeswax was checked by TLC.

The developing solvent for TLC was a mixture of hexane, ethyl acetate and acetic acid (90: 9: 1, v/v/v) and bands developed by iodine resublimed.

### 2.3 Extraction of Policosanol by Hydrolyzation

The high purity beeswax was hydrolyzed with 1 M ethanolic NaOH by refluxing with continuous stirring at 80-85 °C for 2 hrs. The sample mixtures of complete hydrolyzed beeswax was tasted by TLC with chloroform: hexane: acetic acid (70: 30: 1, v/v/v) as developing solvents.

### 2.4 Purification of Policosanol

Approximately 15 g of complete hydrolyzed beeswax and 250 ml of acetone were refluxed for 6 hrs at 50-60 °C and it was stored at 4 °C for precipitation. The precipitate was refluxed with 250 ml of heptanes at 50-60 °C followed by washing with 150 ml of hot water in three times. After that the heptane layer was separated from the solution and it was kept for policosanol precipitation at 4 °C (applied from Gemble et al. (2003)) [23]. The policosanol precipitated were filtered and dried at room temperature. The purification of policosanol test was done by the checking of free fatty acid (FFA) by TLC for primary test. Moreover, the quantification of FFA were studied by HPLC via free fatty acid as external standard. The FFA analysis was carried out by HPLC using acetronitrile/methanol (4:1) solvent system as a mobile phase and C18 (Hewlett Packard) HPLC column (125 mm x 4.0 mm i.d.).

## 3 Results and Discussion

### 3.1. High Purity Beeswax Preparation

The compositions of beeswax material are show in Figure 1. It is compost of wax ester, triglyceride and free fatty acid which is closely to the report of Gemble et al. 2003 [23] and Tolloch, 1980, 1971 [24-25]. The triglyceride and free fatty acid were in case of rancidity via lipid oxidation and soap forming via hydrolysis with strong base (saponification). Thus, the removal of the impurities in beeswax is necessary process of wax ester hydrolyzation.

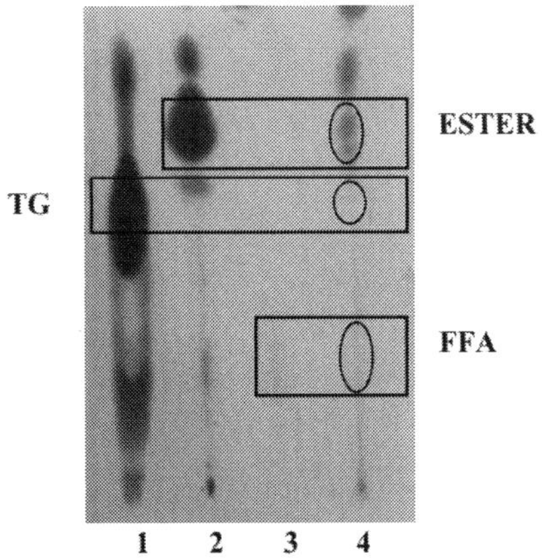


**Fig. 1.** Thin layer chromatography show crude beeswax composition. (1, rice bran oil (TG); 2, std. fatty acid methyl ester (FAME); 3, std. stearic acid (FFA); 4. Crude beeswax)

The elimination of triglycerides and free fatty acids were made by dissolving beeswax in hexane and reflux for 30 minutes following by washing with hot hexane. TLC plate of the impurities elimination show in Figure 2. Triglycerides and free fatty acids cannot be removed in the first step (refluxing and washing with hexane). However, the remains of the impurities can be removed in second step of refluxing and washing again with isopropanol. We had been success in elimination of the triglyceride and free fatty acids as a resulting in Figure 3. The high purity of wax ester was used as material for policosanol extraction.

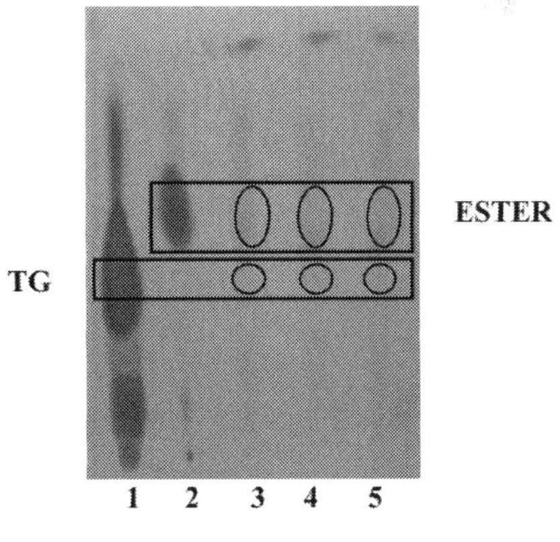


**Fig. 2.** Thin layer chromatography show purified beeswax composition using hexane in purification. (1, rice bran oil (TG); 2, std. fatty acid methyl ester (FAME); 3, purified beeswax (refluxed with hexane); 4, purified beeswax (refluxed and washed with hexane 1st time); 5, purified beeswax (refluxed and washed with hexane 2nd time)

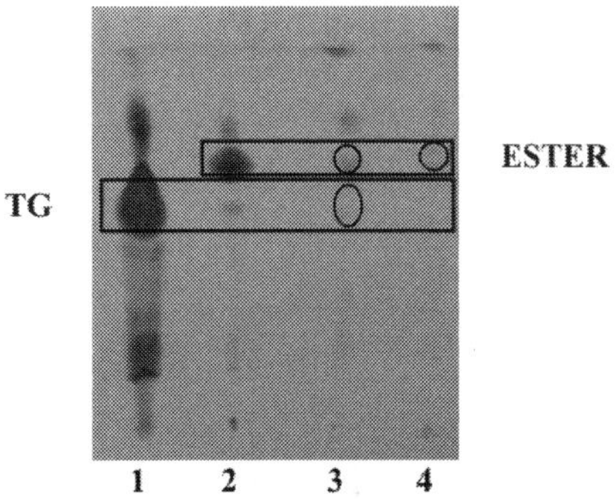


**Fig. 3.** Thin layer chromatography show purified beeswax composition using isopropanol in purification. (1, rice bran oil (TG); 2, std. fatty acid methyl ester (FAME); 3, purified beeswax (refluxed with isopropanol); 4, purified beeswax (refluxed and washed with isopropanol)

### 3.2 Extraction of Policosanol by Hydrolyzation

The extraction of beeswax by base catalyzed hydrolysis can be done by refluxing with 1 M ethanolic NaOH at 80-85 °C for 2 hrs as shown in Figure 4. The TLC plate had been present to complete hydrolyzation of the high purity wax ester. Thereby the wax ester did not appear on the TLC plate due to the fact that hydrolyzation of the wax to long chain fatty acids and long chain aliphatic alcohols. The researchers have been using NaOH or KOH base prepared in organic solvent for catalyst the hydrolysis reaction [1, 13-19]. However, the advantage of NaOH solution in ethanol is to protection the possible emulsion formation of the reaction.

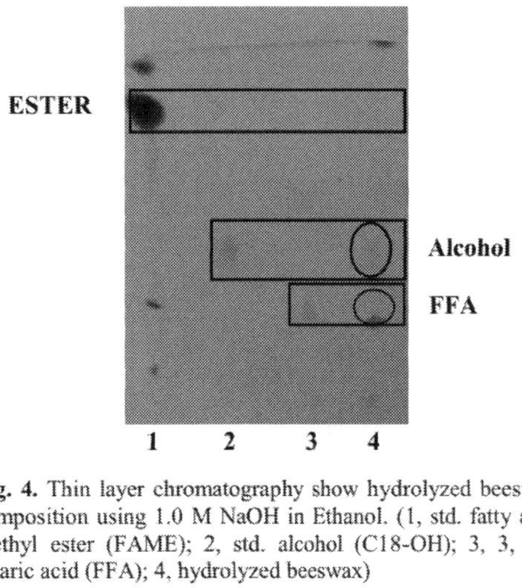


**Fig. 4.** Thin layer chromatography show hydrolyzed beeswax composition using 1.0 M NaOH in Ethanol. (1, std. fatty acid methyl ester (FAME); 2, std. alcohol (C18-OH); 3, 3, std. stearic acid (FFA); 4, hydrolyzed beeswax)

### 3.3 Purification of Policosanol

Beeswax is a complex mixture of compounds with a large range of long chain alcohol molecular and long chain fatty acid structures so the solubility of these molecules would be different in various organic solvent types. The aim of purification the hydrolyzed wax ester is to get rid of long chain fatty acid. The difference properties of solvent were used in two purification step. The result of policosacol purity was tasted by TLC showing in Figure 5. The high dielectric constant solvent (acetone) using in the first step cannot use to purify the policosanol due to sodium salt of long chain fatty acid can be dissolved in acetone too. However, the policosanol was repeated extraction via low dielectric constant solvents (hepane) in the second step of this study. It is capable of efficiently extracting policosanol thereby the residual sodium salt of long chain fatty acid can be eliminated by washing with hot water. The result of policosanol purity was confirmed by HPLC displaying in Figure 6. Qualification of the free fatty acids were carried out using of stearic acid as an external standard (Figure 6 (A)). The high purity of policosanol extracted was indicated in Figure 6 (B). It was not found free fatty acids residual to show in the chromatograms. The suitable solvent selection is the most important step in policosanol purification for optimizing recovery of aspiring to the main components from a complex mixture. The dielectric constant is a measure of the polarity of a solvent and a key parameter in determining solute–solvent interactions [27]. Moreover, a store of a complex mixture of the extracted substance with solvent at low temperature (4 °C) exhibit in increasing of the extracted substance precipitation. It is similarly to previous report of Gemble et al. 2003 [23] which use cooling technique at 2 to -10℃ for 18 hours to purify the policosanol. The selection of solvent used and precipitation at low temperature is an important role in the complete of policosanol purification process. We had been success in purification of policosanol extracted from beeswax thereby the total yield of high purity policosanol was 13.23-13.89 %.

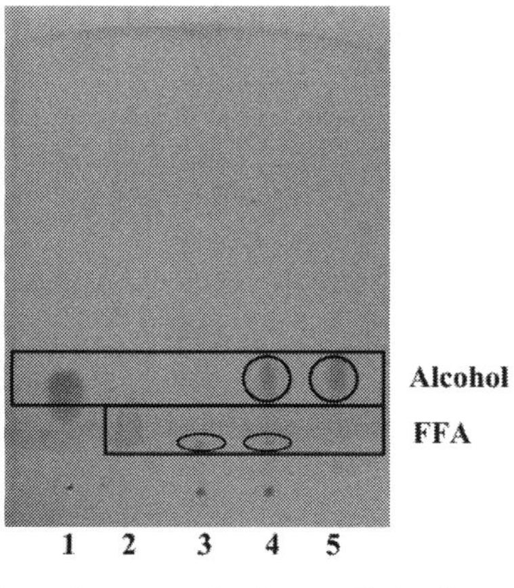


**Fig. 5.** Thin layer chromatography show purified policosanol. (1, std. alcohol (C18-OH); 2, std. stearic acid (FFA); 3, hot water layer; 4, purified policosanol (refluxed with acetone); 5, purified policosanol (refluxed with heptane and washed with hot water)

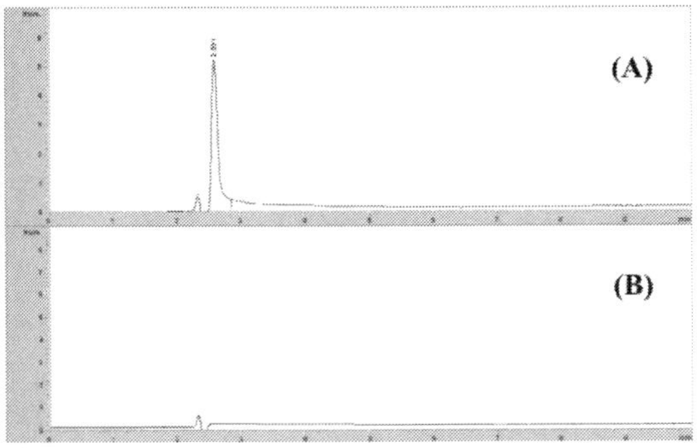


**Fig. 6.** Chromatogram shows high purity policosanol. (A; Std. stearic acid (FFA); B, purified policosanol)

## 4 Conclusions

The mixture of long chain alcohols key compounds within the beeswax have been extracted, purified, identified and quantified. All the technique in the process of policosanol extraction had significant effects on purity, recovery yields and policosanol composition. Moreover, the solvent selectivity and temperature collective on exhibit in improving efficiency of purification process. The high purity of policoanol extracted from beeswax have many potential in applications.

The authors wish to thank National Research Council of Thailand and Department of Chemistry, Maejo University, Chiang Mai, Thailand for financial support.

## References

1. S. Irmak, N. T. Dunford, J. Milligan, Food Chemistry, **95** (2006)
2. V. Fraga, R. Menendez, A.M. Amor, RM. Gonzalez, S. Jimenez, R. Mas, Archives of Medical Research, **28**, 3 (1997)
3. S. Kato, K.I. Karino, S. Hasegawa, J. Nagasawa, A, Nagasaki, M. Eguchi, T. Ichinose, K. Tago, H. Okumori, British Journal of Nutrition (London), **73** (1995)
4. I. Gouni-Berthold, H.K. Berthold, American Heart Journal, **143** (2002)
5. S. Irmak, N. T. Dunford, Journal of Agricultural and Food Chemistry, **53** (2005)
6. S. Valdes, M.L. Arruzazabala, E. Fernandez, International Journal of Clinical Pharmacology Research, **16** (1996)
7. F. Francini-Pesenti, F. Brocadello, D. Beltramolli, M. Nardi, L. Caregaro, American Journal of Health-System Pharmacy, **60**, 11 (2003)
8. S. Harrabi, S. Boukhchina., P. M. Mayer, H. Kallel, Food Chemistry, **115** (2009)
9. Jackson, M. A. and Eller, F. J., The journal of Supercritical Fluids, **37** (2006)
10. M. Miguel, M.N. Fernando, Food Chemistry, **136** (2013)
11. B. Ilaria, P.C. Maria, Journal of Chromatography A, **1028** (2004)
12. J.J. Jimenez, J.L. Bernal, S. Aumente, M.J.D. Nozal, M.T. Martin, J.Jr. Bernal, Journal of Chromatography A, **1024** (2004)
13. N. T. Dunford, E. Jeff, Bioresource Technology, **101** (2010)
14. O.C., Aicha, B.M, Mahamad, K, Belhassen, B. Sadok, P. Claude, K. Habib, Journal of Agricultural and Food Chemistry, **58** (2010)
15. S. Irmak, R.S. Jonnala, F. MacRitchie, Journal of Cereal Science, **48** (2008)
16. P. Adhikari, K.T. Hwang, J.N. Park, C.K. Kom, Journal of Agricultural and Food Chemistry, **54** (2006)
17. K.T. Hwang, C.L. Weller, S.L. Cuppett, M.A. Hanna, Cereal Chemistry, **81**, 3 (2004)
18. N. Gladys, B. Paul, G.D. Micheline, P. Patrick, H.L. Marie, Phytochemistry, **61** (2002)
19. M.F. Wang, H.L. Lian, L. Mao, J.P. Zhou, H.J. Gong, B.Y. Qian, Y. Fang, J. Li, Journal of Agricultural and Food Chemistry, **55** (2007)
20. A.J. Michael, J.E. Fred, Journal of Supercritical Fluid, **37** (2006)
21. G. Cravotto, A. Binello, G. Merizzi, M. Avogadro, European Journal of Lipid Science and Technology, **106** (2004)
22. N. T. Dunford, S. Irmak, R. Jonnala, Food Chemistry, **119** (2010)
23. W.R. Gemble, Z. Liu, D.T. Bailey, P.P. Perez, D.P. Stull, S.L. Richheimer, R.L. Nochols, R, Lenobel, United States Patent (2003)
24. A.P. Tulloch, Bee World, **61**, 2 (1980)
25. A.P. Tulloch, Chemistry and Physics of Lipids, **6**, 3 (1971)
26. Y. Lin, R. Mike, P.J.W. Reggy, A.T. Elke, M. Gerald, S. Aafje, W.M. Gert, Metabolism, **53**, 10 (2004)
27. Abboud, J.-L. M., Notario, R., Pure and Applied Chemistry, **71**, 4 (1999)

# Surfactant-less Sol-Gel Technique for the Synthesis of Mg-ZnO Nanoparticle

*Wafaa* A. Kadhim[1,2], *Mohd Hasbi* Bin Ab. Rahim[1]

[1]Faculty of Industrial Sciences & Technology, Universiti Malaysia Pahang, Gambang, 26300 Pahang, Malaysia
[2]EDP Sciences, Production Department, 91944 Les Ulis Cedex A, France

**Abstract.** In this study, sol–gel technique has been used to prepare nanoparticle Mg doped ZnO with 2 wt. % at low pH without using any capping or surfactant agent. The Mg-ZnO nanostructure, surface area and porosity, surface morphology and element analysis was analyzed by X-Ray Diffraction (XRD), Fourier Transforms Infra-Red (FTIR), N2 Physisorption and Field Emission Scanning Electron Microscopy with energy dispersive X-ray spectroscopy (FESEM-EDX). The characterizations confirmed that the surfactant is not necessary for sol-gel synthesis technique, whereby highly crystalline material with smaller crystallite size (30nm) and high surface area (21.7) was obtained. Besides, the synthesis approach is useful for accurately immobilized the required amount of Mg as doping element on ZnO material with the accuracy up to 99.5% confirmed through EDX analysis.

## 1 Introduction

The important properties of nanostructured materials have started motivation among scientists to explore the possibilities of using them in various technological applications. In particular, the electronic and optical properties of nanostructured materials have been of interest because of their potential applications in the fabrication of microelectronic and optoelectronic devices [1-3]. Materials in nanoscale in size will typically having higher surface area compared to their bulk counterpart of similar mass. One benefit of greater surface area can dramatically improve the reactivity of surface dependent chemical processes especially in catalysis application. Nano engineered catalysts have benefited the oil and chemical production industries. Besides, energy technology sectors such as batteries, fuel cells, etc can potentially enhanced by incorporating nanomaterials. Indeed, the physico-chemicals characteristics of nanoscale materials are size dependent and this offers the possibility for researchers either in academic or industry to fine-tune a material property of interest to suite their niche application[4].

ZnO for example, is considered to be one of suitable metal oxides[5] that can be used at a nanoscale level. ZnO itself has normally a hexagonal or wurtzite structure and it is well-known as an n-type II–VI semiconductor with a wide direct band-gap of about 3.37 eV and a large exciting binding energy of 60 meV [6]. From this point of view nanostructured ZnO powders display an excellent potential in many applications such as gas sensors [7], solar cells [8], and photocatalyst with high chemical activity [9]. On the other hand, the ZnO characteristics can be enhanced by incorporating other element on its. For instance, addition of magnesium ($Mg^{2+}$) with ionic radius of 0.66 Å, which is very close to ionic radius of $Zn^{2+}$ (0.74 Å). Therefore, substitution of Zn by Mg does not give rise to significant changes in lattice constants. It is possible to obtain a wide band gap ZnO-Mg NPs by doping ZnO with a suitable concentration of Mg as reported by several research groups [10-12]

Recently, several chemical techniques, such as co-precipitation, sol–gel[13], hydrothermal [14], wet chemical [15], precipitation [1] and combustion method [16] have been used to prepare nanocrystalline materials. Among them, sol-gel technique gives us excellent thermal stability with controllable material purity. However, most of them utilised surfactant in the synthesis route [17]

Thus, this study aims to synthesize ZnO and Mg-ZnO through modification of existing sol-gel techniques. The simplicity with less starting material will open up the possibility in the production of targeted material in large scale.

## 2 Materials and method

ZnO and Mg-ZnO materials were prepared by using zinc acetate dehydrate $Zn(CH_3COO)_2{\cdot}2H_2O$ (99.5%), Magnesium (II) nitrate hexahydrate $Mg(NO_3)_2{\cdot}6H_2O$ (99.5%) and oxalic acid $C_2H_2O_4{\cdot}2H_2O$ (99.5%) from (R&M chemicals), ethanol (99.99% ) from (Fisher chemical) and acetone $CH_3OCH_3$ (99.5%) from (Friendemann Schmidt chemical). As calculated percentages of Mg and Zn precursor was prepared by dissolved it in ethanol. Then, oxalic acid solution was added slowly with continuous stirring until the gel is formed. Next, the gel are filtered and washed with acetone prior to drying step for 4 h in 100 °C. As dried powder then subjected to calcination process for 2 h at 400 °C under static air environment.

X-ray diffractometer (RIGAKU, MINIFLEX II), FTIR spectrometry (Perkin-Elmer, Spectrum 100), BET (MICROM ERETICS, ASAP 2020) and FESEM (JEOLEVO-50, Japan) are used in powder characterizations.

## 3 Results and discussion

The XRD spectrum of the 2% Mg on ZnO nanoparticle is shown in Figure 1. However, only diffraction peaks correspond to ZnO were observed. This finding might due to high dispersion of Mg into the lattice of the ZnO crystal or due to the small amount of Mg used in doping.

The peak positions of each sample exhibit the hexagonal wurtzite (a=3.24 Å, c=5.188 Å) structure of ZnO which were confirmed from the ZnO (ICDD - 10800075). It is important to note that the availability of Mg was confirmed through EDX analysis.

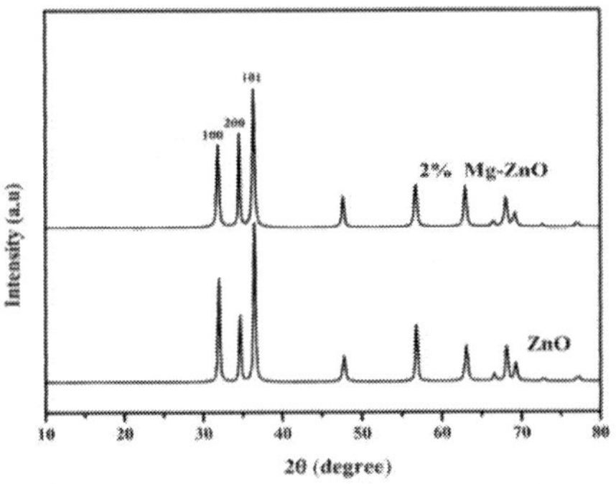


**Fig. 1.** XRD diffractogram of ZnO and 2wt%.Mg-ZnO.

The crystallite size calculated using Scherrer's equation (equation 1) [18] demonstrated around 36nm for ZnO and 30nm for 2% Mg-ZnO (Table 1). The reduction in the crystalline size was due to the cations segregation at the boundaries of the grain and that will prevent the growth of the grain by restricting the grain direct contact.

$$D = \frac{K\lambda}{\beta \cos\theta} \quad (1)$$

λ is the x-ray wavelength,
K is shape factor (0.89),
β is the peak broadening at half maximum for Cu kα
θ is the Bragg angle.

In addition to XRD analysis, FTIR spectroscopy was used to study the structure of Mg-ZnO sample and the KBr technique has been used to prepare the sample.

**Table 1.** Crystal properties of ZnO and Mg-ZnO.

| Catalyst | Diffraction angle (2θ) | Lattice parameter | | | *Crystallite size (nm) |
|---|---|---|---|---|---|
| | | a (Å) | c (Å) | c/a | |
| ZnO | 36.41 | 3.240 | 5.188 | 1.601 | 36 |
| 2% Mg-ZnO | 36.46 | 3.227 | 5.232 | 1.621 | 30 |

Figure 2 shows the FTIR spectrum for ZnO and Mg-ZnO where the band in range (400- 600) $cm^{-1}$ can be attributed to the Zn-O stretching mode. The bands in range ( 1400-1650) $cm^{-1}$ correspond to stretching of C–O and C=O at while the stretching vibration at the range (3200 to 3600) $cm^{-1}$ corresponds to the stretching of OH bond the water molecule bound on the surface of sample.

In case of Mg-ZnO doping there are changes in the positions and sizes of FTIR peaks, which is a clear indication of Mg incorporation with in ZnO host lattice. The shifted peak from 593$cm^{-1}$ to 598$cm^{-1}$ for (Zn–O) also favors the amalgamation of Mg doping in the lattice.

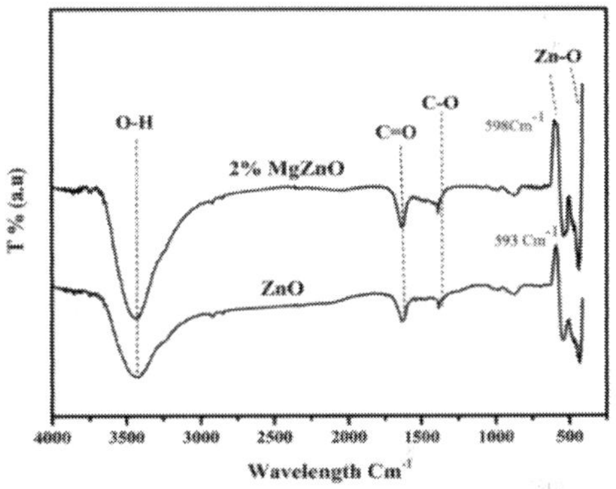


**Fig. 2.** FTIR spectra of 2wt. %Mg-ZnO

The surface area and porosity of prepared powders were analysed with $N_2$-physisorption technique. The surface area of Mg-ZnO sample was found to be bigger than ZnO (Table 2). The adsorption-desorption isotherms (Figure 3) of the Mg-ZnO to be of type IV in nature and exhibited a H1 hysteresis loop, which is typical of mesopores solids [18, 19].

**Table 2.** BET surface area and pore volume data of ZnO and Mg-ZnO.

| Catalyst | BET Surface area ($m^2$/g) | Pore Volume ($cm^3$/g) |
|---|---|---|
| ZnO | 17.5 | 0.079 |
| 2%Mg-ZnO | 21.7 | 0.098 |

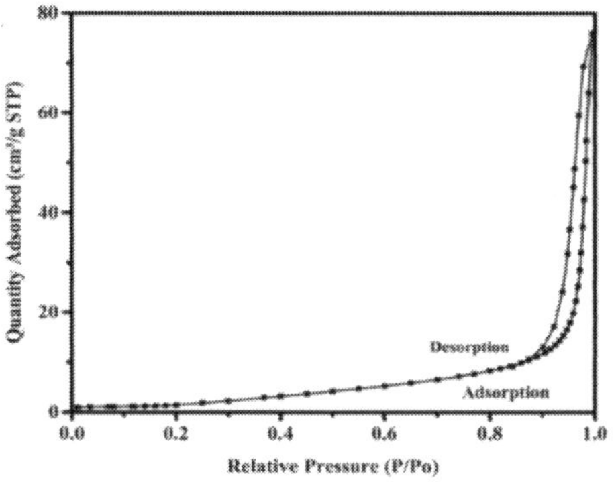


**Fig. 3.** Nitrogen adsorption-desorption isotherms of 2wt% Mg-ZnO

Surface morphology of Mg-ZnO with 2 % loading is illustrated in Figure 4. It can be seen from the micrographs that Mg-ZnO show sphere in-shape, the sphere particle for doped samples is closely connected as straight line-structure. The average crystallite size calculated from micrographs is slightly higher than XRD. The sizes observed from FESEM were ~ 40nm. However, it is important to note that the modified sol-gel technique used in this study able to accurately use to immobilize Mg with the accuracy up to 99.5% with the calculated and required Mg doping.

**Fig. 4.** FESEM and EDX of 2wt% Mg/ZnO

## 4 Conclusions

In concise, Mg-ZnO material with controlled size, uniform distribution and good structural have been successfully synthesized using simple, inexpensive sol-gel method without using capping agent. The synthesis approach produced highly crystalline material with high purity and structurally developed in ordered morphology.

## Acknowledgements

The authors are grateful for the research funding (RDU130317) from Universiti Malaysia Pahang, Kuantan, Malaysia.

## References

1 R. Viswanatha, T. Venkatesh, C. Vidyasagar, Y. A. Nayaka, and Y. Arch, Archives of Applied Science Research, **4**, 480-486, (2012)
2 C. Yan, L. Nikolova, A. Dadvand, C. Harnagea, A. Sarkissian, D. F. Perepichka, D. Xue, and F. Rosei, Adv. Mater., **22**, 1741-1745, (2010)
3 C. Yan, A. Dadvand, F. Rosei, and D. F. Perepichka, J. Am. Chem. Soc., **132**, 8868-8869, (2010)
4 S. Suwanboon, P. Amornpitoksuk, and A. Sukolrat, Ceram. Int., **37**, 1359-1365, (2011)
5 N. Kumar, A. Dorfman, and J.-i. Hahm, Nanotechnology, **17**, 2875, (2006)
6 J. Xu, Y. Cui, Y. Han, M. Hao, and X. Zhang, RSC Advances, **6**, 96778-96784, (2016)
7 J. Xu, J. Han, Y. Zhang, Y. a. Sun, and B. Xie, Sensors and Actuators B: Chemical, **132**, 334-339, (2008)
8 J. Huang, Z. Yin, and Q. Zheng, Energy & Environmental Science, **4**, 3861-3877, (2011)
9 T. Rattana, S. Suwanboon, P. Amornpitoksuk, A. Haidoux, and P. Limsuwan, J. Alloys Compd., **480**, 603-607, (2009)
10 H. Li, Y. Zhang, X. Pan, H. Zhang, T. Wang, and E. Xie, J. Nanopart. Res., **11**, 917-921, (2009)
11 Y.-J. Lin, P.-H. Wu, C.-L. Tsai, C.-J. Liu, Z.-R. Lin, H.-C. Chang, and C.-T. Lee, Journal of Applied Physics, **103**, 11370-9, (2008)
12 R. Ding, C. Xu, B. Gu, Z. Shi, H. Wang, L. Ba, and Z. Xiao, Journal of Materials Science & Technology, **26**, 601-604, (2010)
13 S. Kumara, A. Kumara, F. Singhb, and A. Kapoora, Stress (GPa), **3**, 1.712, (2013)
14 B.-C. Kim, S.-M. Kim, J.-H. Lee, and J.-J. Kim, J. Am. Ceram. Soc., **85**, 2083-2088, (2002)
15 R. Zamiri, B. Singh, I. Bdikin, A. Rebelo, M. Scott Belsley, and J. M. F. Ferreira, Solid State Commun., **195**, 74-79, (2014)
16 S. Sundar Manoharan and S. Arora, Materials Science and Engineering: B, **162**, 68-73, (2009)
17 Y. L. Zhang, Y. Yang, J. H. Zhao, R. Q. Tan, P. Cui, and W. J. Song, J. Sol-Gel Sci. Technol., **51**, 198-203, (2009)

DOI: 10.1051/matecconf/201711102006

18 V. Etacheri, R. Roshan, and V. Kumar, ACS Applied Materials & Interfaces, **4**, 2717-2725, (2012)
19 S. P. Meshram, P. V. Adhyapak, D. P. Amalnerkar, and I. S. Mulla, Ceram. Int., **42**, 7482-7489, (2016)

# Pre-dispersing Process of Montmorillonite by Water and Toluene: Influence on Thermal Properties of Ethylene Vinyl Acetate / Montmorillonite Nanocomposites

*Azlin Fazlina* Osman[1*], *Tuty Fareyhynn* Mohammed Fitri[1], *Yusrina* Mat Daud[1], *Md* Rakibuddin[2] and *Supri* A. Ghani[3]

[1] Center of Excellence Geopolymer and Green Technology (CEGeoGTech), School of Materials Engineering, Universiti Malaysia Perlis, Arau, Perlis 02600, Malaysia.
[2] Department of Chemistry, Indian Institute of Technology Kharagpur, India.
[3] Faculty of Engineering Technology, Universiti Malaysia Perlis, Kampus UniCITI, S. Chuchuh, Padang Besar, Perlis, Malaysia.

**Abstract.** Organanically modified montmorillonite (organo-MMT) was incorporated into ethylene vinyl acetate (EVA) copolymer to improve its thermal stability. A processing procedure, the so called 'pre-dispersing' process of the organo-MMT was introduced prior to melt compounding process of both constituents in order to facilitate the nanofiller dispersion in the EVA matrix. Water and toluene were used as pre-dispersing medium, while magnetic stirring and ultrasonication were utilized as pre-dispersing method. The effects of pre-dispersing medium and method on thermal behaviour of neat EVA and EVA nanocomposites were analysed. It was anticipated that improvement in organo-MMT dispersion would enhance the matrix-nanofiller interactions, thereby the thermal properties of the resultant EVA/MMT nanocomposite. Based on thermal studies by thermogravimetric analysis (TGA), the organo-MMT nanofiller pre-dispersed by ultrasonication in water medium for 2 minutes (MMT(W)2m_u) resulted in most significant thermal stabilizing effect to the EVA copolymer. This was due to the significant improvement in the organo-MMT dispersion when the above mentioned pre-dispersing parameters were employed. Apparently, the result indicates that the high temperature behaviour of the nanocomposite can be affected by the strength of interphase interactions between the matrix and nanofiller which is also influenced by the dispersion of the organo-MMT.

## 1 Introduction

Ethylene vinyl acetate (EVA) copolymer is a polymeric material that exhibits softness and flexibility similar to elastomeric materials, yet it can be processed like other thermoplastic [1]. EVA is recyclable, offers excellent optical clarity, barrier properties, low-temperature toughness and stress-crack resistance [1, 2]. However, to be used as advanced material, further improvement in EVA thermal performance is needed, due to its tendency to thermally degrade under harsh environment and elevated temperatures. The enhancement in EVA thermal performance can be achieved by adding small amount of nanoclay as reinforcing filler [3, 4]. The resultant 'EVA nanocomposite' can perform greater thermal stability as compared to the neat EVA due to the dispersion of more thermally stable layered silicate material in the host copolymer structure. However, the production of fully exfoliated polymer/nanoclay nanocomposites is crucial to obtain an actual benefit of nanoscale reinforcement in the polymer matrix [5]. Dispersion of nanoclay filler must be optimized to effectively shield the polymer chains during thermal degradation process. Previous studies showed that poor dispersion of MMT clay inhibits significant thermal stabilizing effect of this nanofiller to the host EVA copolymer [3, 4]. In this work, pre-dispersing process of the organically modified montmorillonite (organo-MMT) was done prior to melt compounding process with the EVA copolymer. The purpose was to obtain EVA/MMT nanocomposite with well exfoliated and dispersed nanofiller and subsequently improve its thermal stability. Magnetic stirring and ultrasonication were used to pre-disperse the organo-MMT under water or toluene medium. The effect of both pre-dispersing medium and method on thermal properties of neat EVA and EVA nanocomposites was observed and compared.

## 2 Materials and Method

### 2.1 Materials

The EVA copolymer used as matrix material was commercially known as COSMOTHENE EVA H2181. The resin was supplied by the Polyolefin Company (Singapore) Pte. Ltd. The EVA composed of 18 wt% of vinyl acetate (VA) and 82 wt% ethylene. The organo modified montmorillonite (organo-MMT), which contains 35-45% of dimethyl dialkyl (C14-C18) amine

* Corresponding author: azlin@unimap.edu.my

DOI: 10.1051/matecconf/201711102007

as an organic surfactant was used as nanofiller. This organo-MMT is commercially known as Nanomer (types 1.44P) by Nanocor, Inc. Distilled water and toluene were used as pre-dispersing medium to allow swelling of the organo-MMT. Toluene (ACS, ISO, Reag.PhEur grade) was supplied by Merck Pty Ltd.

### 2.2 Samples Preparation

The organo-MMT powder was pre-dispersed in two types of liquid medium which were distilled water and toluene in order to allow swelling of nanoplatelets prior to melt compounding with the EVA matrix, The organoclay was prepared as 20% solution in toluene and distilled water respectively and the suspension was stirred at room temperature by using two methods which were ultrasonication and magnetic stirring. After that, the suspension (in water) was placed in the air circulated oven for 48 hours. The suspension of MMT/toluene was left unattend under fume hood for several hours before being placed in the oven for 24 hours at temperature of 50°C, to ensure the solvent was fully evaporated. Lastly, the dried organo-MMT powders were grinded and sieved to reduce agglomeration. The EVA nanocomposites were prepared according to the formulation summarized in Table 1. The nanocomposite samples were prepared by melt compounding process of EVA copolymer with constant amount of organo-MMT nanofiller (5 wt%) at processing temperature of 160℃ using internal mixer (Brabender plasticoder) machine. Then, by using compression moulding machine, the compound was compressed into 1 mm thick sheets.

Table 1. Formulation of EVA/MMT nanocomposites

| **Nanofiller (Organo-MMT)** | | **Nanocomposite (EVA+ 5% MMT)** |
|---|---|---|
| **Pre-dispersing parameter** | **Acronym** | **Acronym** |
| None | MMT | EVA+MMT |
| 20% Organo-MMT suspension in toluene pre-dispersed by magnetic stirrer | MMT(TL)_m | EVA+MMT(TL)_m |
| 20% Organo-MMT suspension in water pre-dispersed by magnetic stirrer | MMT(W)_m | EVA+MMT(W)_m |
| 20% Organo-MMT suspension in water pre-dispersed by ultrasonic probe for 2 minutes | MMT(W)2m_u | EVA+ MMT(W)2m_u |

### 2.3 Thermogravimetric analysis (TGA)

TGA was carried out by using TGA Pyris Diamond Perkin Elmer equipment, to study on thermal stability of neat EVA and EVA nanocomposite samples. The weight loss of specimen against temperature was measured from 30°C to 700°C and with a constant heating rate of 10°C/min, under nitrogen atmosphere.

### 2.4 Transmission electron microscopy (TEM)

TEM analysis was carried out using JEOL JEM2010 electron microscope (Japan) operating at 200 kV. Prior to viewing, the samples were cut approximately 300 nm thickness using glass cutter on a Leica Utracut Ultramicrotome (UCT) instrument at Tg temperature = -80 °C maintained by using liquid nitrogen.

## 3 Results and Discussion

TGA analysis was conducted to study and compare the thermal stability of neat EVA and EVA nanocomposites containing both pre-dispersed and non-pre-dispersed organo-MMT when heated up to their decomposition temperature. The thermal stability was assessed through the determination of the maximum temperature for degradation ($T_{dmax}$) of neat EVA and EVA nanocomposites and the rate of mass loss. Table 2 shows the thermal degradation peak temperature of all materials during the first and second step of mass losses, while Figure 1 illustrated TGA curves of the neat EVA and EVA nanocomposites respectively.

**Table 2**. Thermal degradation peak temperature ($T_{dmax}$) of EVA and EVA nanocomposites.

| Material | 1st step mass loss | 2nd step mass loss |
|---|---|---|
| | $T_{dmax}$ (°C) | $T_{dmax}$ (°C) |
| EVA | 347 | 497 |
| EVA+MMT | 346 | 492 |
| EVA+MMT(TL)_m | 351 | 523 |
| EVA+MMT (W)_m | 349 | 506 |
| EVA+MMT (W)2m_u | 355 | 530 |

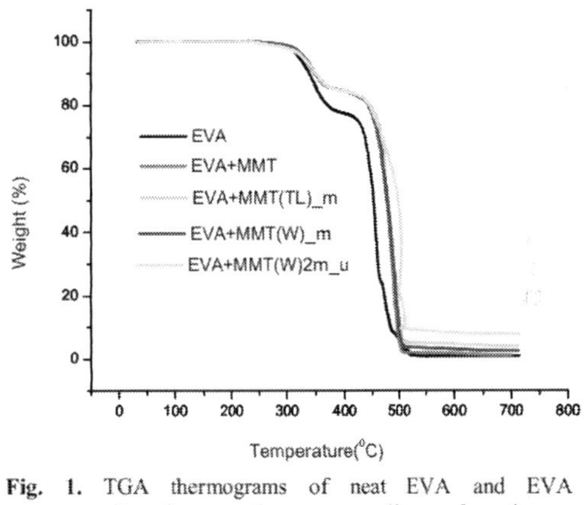


**Fig. 1.** TGA thermograms of neat EVA and EVA nanocomposites incorporating non-pre-dispersed and pre-dispersed organo-MMTs by various pre-dispersing parameters.

Basically, when subjected to high temperatures, the polymer–filler, filler–filler and filler-aggregate interactions can bring more pronounce effect on the decomposition behaviour of the host polymer, thereby its thermal stability. This is due to the fact that there are larger deviations between the underlying thermodynamics of these distinct materials [3]. High temperature behaviour of the EVA nanocomposite can be much more affected by the strength of interphase interactions between all the components presence in the material. According to Figure 1 and Table 2, the decomposition process of EVA is generally occurred in two steps (referred to as first step mass loss and second step mass loss). The first step mass loss between 300 and 400℃ was due to the deacetylation reaction, in which implicates the releases of gaseous acetic acid and formation of C=C along the polymer backbone [3-6]. The second step mass loss in the range between 400 and 600℃ was attributed to the oxidation and volatilization of hydrocarbons produces from the decomposition of the EVA copolymer backbone [3-6]. The $T_{dmax}$ of the first step mass loss of nanocomposites containing EVA+MMT (non-pre-dispersed) occurred at lower temperature as compared to neat EVA. This would possibly due to the catalytic effect from the degradation of the organic surfactant [3-5]. The degradation of dimethyl dialkyl (C14-18) amine, that used to surface modified the MMT accelerated the degradation of polyvinyl acetate (PVA) to acetic acid. When compared with the ones containing the pre-dispersed nanofillers, apparently, this nanocomposite shows lower $T_{dmax}$ of the first step mass loss. Poorer EVA-nanofiller interactions may lead to more severe catalytic effect on the deacetylation of EVA during the decomposition process. This is because, weaker interactions between MMT nanoplatetets and the EVA molecular chains provides less shielding for the degradation process of the organic surfactant. As the organic surfactant degrades faster, it will bring stronger catalytic effect to the deacetylation of the EVA. In contrary, the $T_{dmax}$ of the first step mass loss for all nanocomposites containing the pre-dispersed organo-MMTs shows higher in values compared to neat EVA. This was due to enhanced MMT dispersion (Figure 2) causing greater nanofiller-PVA chains interactions and more restricted polymer chain mobility. Consequently, higher temperature was needed to break the bonds.

The $T_{dmax}$ values of second step mass loss revealed more significant effect of the pre-dispersing procedure to the thermal stability of the resulting EVA nanocomposites. The $T_{dmax}$ of all nanocomposites containing the pre-dispersed nanofillers occur at much higher temperature as compared to neat EVA and EVA+MMT nanocomposite. This shows that the capability of the well dispersed and interacted nanofiller to enhance the thermal stability of the host polymer is more obvious at high temperature decomposition process. In agreement with the statement of Ray & Okamoto [6], the nanoclays can act as a heat barrier, which can well assist in the formation of char during thermal decomposition of the polymer. Based on Figure 2, the better dispersed and exfoliated nanofiller (EVA+MMT(W)2m_u) shows the greatest thermal stability among all the materials as it exhibits the highest $T_{dmax}$ for both first and second step mass losses. In addition, it produced the highest content of char residue (~4wt %) as compared to EVA+MMT (non-pre-dispersed) and other nanocomposites. Without the nanofiller addition, the neat EVA only exhibit less than 1wt% residue ash when decomposed. Finally, the findings also revealed that the application of sonic waves in water medium can more efficiently disperse the organo-MMT nanoplatelets than using magnetic stirring in toluene medium. During ultrasonication, the hydrodynamic effect can produce greater implosion energy of the cavitation bubbles which can penetrate in between the organo-MMT interlayers, thereby facilitate the nanoplatelet/nanolayer exfoliation. Consequently, the obtained pre-dispersed nanofiller can more efficiently disperse in the host EVA copolymer and act as thermal stabilizing agent.

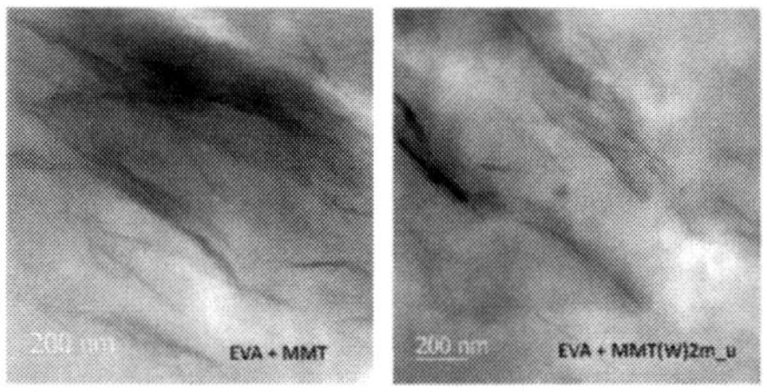


**Fig. 2:** TEM images of EVA+MMT and EVA+MMT(W)2m_u nanocomposites which indicate that the nanocomposite containing the pre-dispersed MMT has greater nanofiller dispersion and smaller tactoid size.

## 4 Conclusion

Based on the TGA results, the MMT(W)2m_u is the best nanofiller to enhance the thermal stability of the EVA copolymer. The improved exfoliation and dispersion of the organo-MMT upon the pre-dispersing procedure resulted in enhanced interactions between this nanofiller with the EVA copolymer chains, thereby producing a more thermally stable EVA nanocomposite system.

## Acknowledgement

The authors thank Ministry of Higher Education, Malaysia for funding the project through Fundamental Research Grant Scheme (FRGS) (No: 9003-00473).

## References

1. J. K. Fink, *Ethylene Vinyl Acetate Copolymers*, in *Handbook of Engineering and Specialty Thermoplastics: Polyolefins and Styrenics*, John Wiley & Sons, Inc., Hoboken, NJ, USA, 187-209 (2010).
2. A.F. Osman, A.M. Alakrach, H. Kalo, W.N.W Azmi, F. Hashim, RSC Adv., **5**, 31485 (2015).
3. A.F. Osman, H. Kalo, M.S. Hassan, T.W. Hong, F. Azmi, , J. of Appl. Polym. Sci., **133**, 43204 (2016).
4. A.F. Osman, A.R. Abdul Hamid, Md. Rakibuddin, G.K. Weng, R. Ananthakrishnan, S.A. Ghani, Z. Mustafa, J. of Appl. Polym. Sci., **134**, 44640 (2017)
5. A.F. Osman, T.F. Mohammed Fitri, Md. Rakibuddin, F. Hashim, S.A.T.T. Johari, R. Ananthakrishnan, R. Ramli, Materials Science and Engineering C, **74**, 194 (2017).
6. S.S Ray, & M. Okamoto. Polymer/layered, Prog. Polym. Sci., **28** 1539 (2003).

# The effect of particle size and solvent type on the gallic acid yield obtained from *Labisia pumila* by ultrasonic extraction

*Afiqah* Yeop[1,3], *Jessinta* Sandanasamy[1,3], *Sook Fun* Pang[1,3], *Sureena* Abdullah[1,3], *Mashitah* M. Yusoff[2], and *Jolius* Gimbun[1,3*]

[1]Centre of Excellence for Advanced Research in Fluid Flow (CARIFF), Universiti Malaysia Pahang, Gambang 26300, Pahang, Malaysia
[2]Faculty of Industrial Science and Technology, Universiti Malaysia Pahang, Gambang 26300, Pahang, Malaysia
[3]Faculty of Chemical and Natural Resources Engineering, Universiti Malaysia Pahang, Gambang 26300, Pahang, Malaysia

**Abstract.** This paper presents the effect of particle size and solvent type on the gallic acid (GA) extraction yield obtained from *Labisia pumila*. The GA extraction from a grinded dried plant material (66.66 µm to 1055.18 µm) was performed using the sonicator probe. The GA quantification and identification was performed using a liquid chromatography with photodiode array detector. Exact match between the spectra obtained from the plant extract and external standard was found indicating a presence of GA. In addition, matching of residence time between the standard and plant extract are also observed. It was found that particle size significantly affects the yield of GA and the highest yield (0.150 mg GA/g DW) was obtained using the particle size around 125 µm. Solvent type does also plays an important role in GA extraction. The result shows that extraction using water as solvent yielded about 29% higher GA content (0.126 mg GA/g DW) than that of 10% ethanol (0.098 mg GA/g DW). The findings in this work may serve as a useful guide to select the particle size and solvent used to maximise GA extraction from *L. pumila*.

## 1 Introduction

*Labisia pumila*, (vernacular name: *'kacip fatimah'*) had been recognised in the Malaysian National Key Economic Areas (NKEA) to be developed as an herb for commercial purposes. It is an extensively consumed herb in Southest Asia regions. *L. pumila* traditionally used to induce and facilitate childbirth, as health supplement, as treatment for rheumatism, intestine infection, prostate cancer, dysentery, dysmenorrheal, flatulence and gonorrhea [1-3]. Previous scientific studies revealed that the extract of *L. pumila* enclosed many medically worthy bioactive compounds such as phenolics, flavonoids, vitamins and proteins that poses numerous biological effects (antibacterial, antioxidant, anticancer, antiobesity, cytotoxic) [4-8]. Many authors had recognized gallic acid as the core compound of *L. pumila* [5, 9-10]. Gallic acid is an organic phenolic acid that is commonly found in foods such as apples, tea leaves, blueberries, and walnuts. Recently, gallic acid derivatives considerably improve the assessment of mechanisms of action in human and murine cell lineages with promising antitumor activity [11]. Ishihara and Sakagami [12] had reported the cytotoxic activity of gallic acid against human leukemia (HK-63) cell lines whereas Saleem et al. [13] reported that gallic acid exhibits cytotoxic potential on HOS-1 cell lines. However, limited study is available concerning the effect of particle size and solvent on the gallic acid extraction from *L. pumila*.

Extraction of gallic acid from *L. pumila* is an important step before it can be developed further as supplement or functional foods. Many conventional methods (maceration, soxhlet and decoction) had been applied intensively by several researchers to extract the plant polyphenols [14-17]. Recent studies also focus on the modern method such as ultrasound assisted extraction (UAE). UAE was shown to be the most effective extraction method based on high yield, high selectivity and reduction in extraction time and solvent consumption. The UAE technique reduced the inner and external mass transfer limitation and hence increases the yield of extraction [18]. In addition, ultrasonic waves can break the cell membrane which enhances the inner mass transport. Thus UAE was employed in this work to extract gallic acid from *L. pumila*.

Besides the extraction technique, the particle size of plant also affects the extraction yield. The smaller the particle size of the material, the shorter the path of the solvent has to travel; which indirectly decreases the time for maximum phytochemical extraction [19]. Smaller particles also have a much larger surface area and hence providing a greater mass transfer rate. According to Fonseca et al. [20], grinding the material indirectly damages the plant cellular which leads to better extraction. For instance, Baldosano et al. [21] reported that finer particle produced higher yield of tannin, similar finding are also reported by Roriguez-Miranda et al. [22].

Water and other organic solvents (ethanol, acetonitrile, acetone, hexane and diethyl ether) are usually used for extraction of bioactive compounds [23]. The difference in polarities of the solvents may influence

* Corresponding author: jolius@ump.edu.my

the solubility of the chemical constituents and their extraction yield. According to Pang et al. [24], the yield of methoxylated and hydroxylated polyphenols is greatly affected by the solvent type used. They found that aqueous alcohol is the best solvent to simultaneously extract the methoxylated and hydroxylated polyphenols [24, 25]. Recently, Sandanasamy et al. [26] reported that 10% ethanol yielded the highest yield of total phenolic and flavonoid extraction from *L. pumila*, however the effect of solvent on the yield specific polyphenol such as gallic acid is not available. Therefore, this work aims to elucidate the effect of particle size and solvent type on the gallic acid extraction from *L. pumila* using UAE method.

## 2 Materials and methods

### 2.1 Chemicals and plant material

Standard GA was obtained from Acros Organics, Fisher Scientific (Leicestershire, UK) whereas analysis grade ethanol was obtained from Merck (Darmstadt, Germany). HPLC grade trifluoroacetic acid (TFA) and acetonitrile was obtained from Fisher Scientific (Leicestershire, UK) and Merck (Darmstadt, Germany), respectively. *L. pumila* var. *pumila* was procured from a trader in Kuantan, Malaysia. The plant materials were washed with deionized water to remove impurities and were dried under open air. Subsequently, the material is oven-dried at the temperature of 37°C until no significant change in weight. The dried plant was grounded into powder and were sieved with analytical sieve shaker (Retsch AS 200, Germany) into five groups of mesh (≤125, >125 - ≤250, >250 - ≤500, >500 - ≤800 and >800 μm). The moisture content was determined using a moisture analyzer (AND MS-70, Japan). The powderized plant was preserved in an air-tight plastic at room temperature prior to extraction.

### 2.2 Ultrasonic assisted extraction (UAE)

Extracts were prepared according to different mesh using Qsonica Q700 (Newtown, USA) equipped with a standard probe. A volume of 100 mL of solvent (water and 10% ethanol) were added into weighted powder (dry weight) at the solid-to-solvent ratio of 0.05. The mixture was then immediately sonicated at the amplitude of 90% for the extraction time of 10 min. The supernatant was then separated by centrifugation (Eppendorf 5810 R, Hamburg, Germany) at 10 000 rpm for 10 min to obtain a clear solution. The extracts were then stored at -80°C prior to GA quantification.

### 2.3 Particle size distribution

The particle size distribution of plant powder was measured using laser diffraction particle size analyzer (Malvern 2000 mastersizer, Malvern Instruments Co., Worcestershire, UK) equipped with an automated dry powder dispersion unit (Scirocco 2000). The distribution was characterized by the volume weighted mean. Prior to analysis of the distribution, the refractive index of the powder were determined by calibration curve at different concentration (5, 10, 20, 30, 40 and 50%) of powder in water suspension. The curve was extrapolated to 100% to obtain the refractive index value of the dry powder.

### 2.3 Gallic acid quantification

The quantification of GA was performed on Waters Acquity UPLC H-Class (Milford, USA) fitted with Purospher® STAR RP-18 Endcapped Hibar UHPLC column (2μm, 100 x 2.1 mm) (Merck, Darmstadt, Germany). The UPLC system is equipped with photodiode array detector operated via Waters Empower 3 software. The operating conditions were at room temperature, with injection volume of 3 μL and flowrate at 0.17 mL/min. The mobile phase consists of 0.1% TFA (solvent A) and 0.1% TFA in acetonitrile (solvent B). The performed elution scheme were: 0-2.0 min, 90% A; 2-2.4 min, 90-88% A; 2.4-3.1 min, 88% A; 3.1-4.3 min, 88-85% A; 4.3-5.5 min, 85% A; 5.5-6.7 min, 85-80% A; 6.7-7.5 min, 80% A; 7.5-8.8 min, 80-75% A; 8.8-9.7 min, 75% A; 9.7-10.5 min, 75-70% A; 10.5-11.5 min, 70% A; 11.5-12 min, 70-5% A; and finally washing the column with isocratic 5% A; and reconditioning with 90% A for 3 min. The extract and standard were filtered with 0.22 μm nylon membrane prior to injection. The peak of GA in comparison with the standard were detected at 280 nm.

### 2.4 Statistical analysis

The influence of the different particle size was assessed via an analysis of mean and standard deviation using analysis tools in Microsoft Excel 2013.

## 3 Results and discussion

### 3.1 Quantification of gallic acid in *L. pumila*

The GA was identified by means of the retention time and UV spectra of the standard. The quantitative analysis performed by determining the concentration of the GA peak area in comparison with the results of a calibration series using standards obtained from Fisher Scientific. The seven point calibration curve in the concentration range of 0.001-0.5 mg/mL showed good linearity ($R^2$ = 0.9985) (Fig. 1). UV spectra of GA from *L. pumila* extracts were found to be in a good match with the standard compound (Fig. 2). Thus, this confirmed its presence in the extracts. Fig. 3 shows the UPLC overlay chromatogram profile of *L. pumila* extract and standard. The developed UPLC separation requires a run time of 23 min and GA were spotted at 2.255 min. Therefore, this method is capable for an accurate qualitative and quantitative analysis of gallic acid from *L. pumila* extract. Thus similar method was used to quantify gallic acid from the extracts throughout this work.

 DOI: 10.1051/matecconf/201711102008

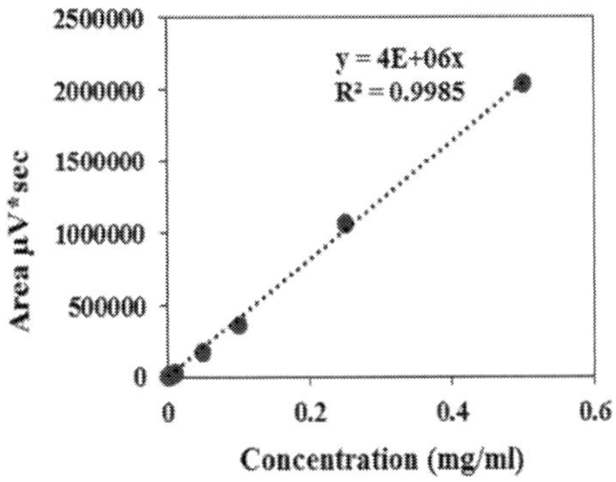


**Fig. 1.** Calibration curve of standard GA

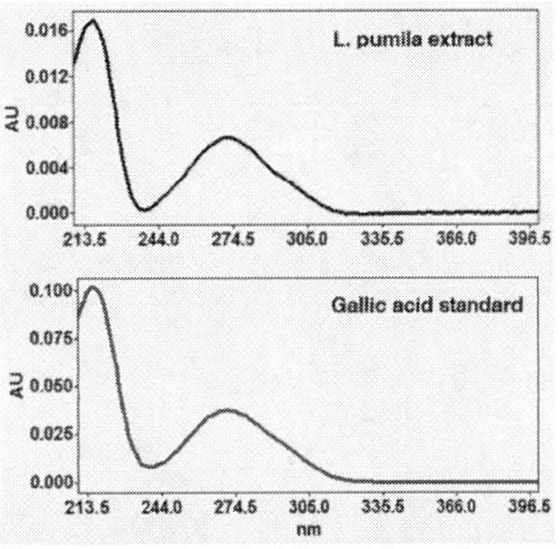


**Fig. 2.** Spectra of standard GA and *L. pumila* extract.

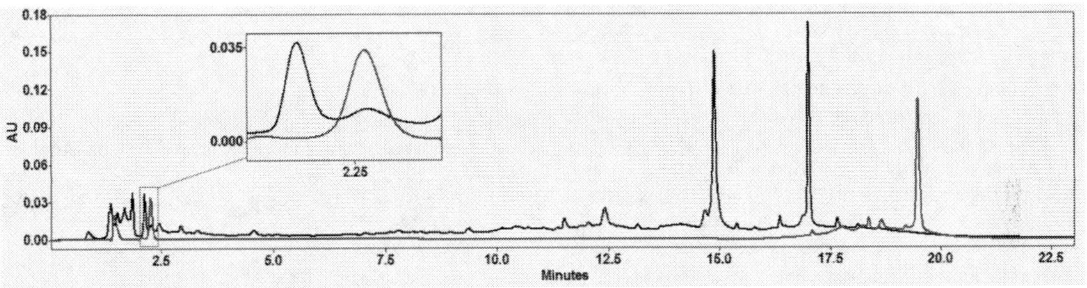


**Fig. 3.** UPLC overlay chromatogram of GA standard and *L. pumila* extract recorded at 280 nm

## 3.2 Effect of particle size on gallic acid recovery

Grinding and classification according to the particle size were made before the extraction of GA was performed to increase surface area and hence the mass transfer efficiency. Furthermore, grinding breaks the plant cell wall, thus facilitating active compounds release to the extraction solvent and enhancing the yield.

The effect of particle size on the extraction yield during UAE was examined under conditions of similar extraction volume (100 mL), solid-to-solvent ratio (1:20), amplitude (90%) and time (10 min). The *L. pumila* powders consisted of particles with the smallest diameter of 1.4 µm, while the largest size detected was 2000 µm (Fig. 4). The particle size distribution curves (mesh size of >125 - ≤250, >250 - ≤500, >500 - ≤800 and >800 µm) is a monomodal with a log normal type distribution. The particle obtained from mesh size of ≤125 µm found to be slightly bimodal due to the fact that all particles smaller than 125 µm falls under this size. It is possible that the grinded particles still have many finer particle classes, but it cannot be separated due to the limitation of the equipment. Nevertheless, fine particle <125 µm is not good for extraction as they tend to float. Hence, it is not important do further separate the particle smaller than 125 µm. The powders had average particle size of 66.66 µm to 1055.18 µm and surface weighted mean ranging from 31.24 µm to 642.77 µm (Table 1).

It was found that the recovery of GA from *L. pumila* is significantly affected by the particle size. The highest extraction yield of GA (0.15 mg GA/g DW) was obtained using particle size of >125 - ≤250 µm (Fig. 5). Particle size of ≤125 µm is smaller than the one in the range of >125 - ≤250 µm, in which supposed to give better yield of GA due to its higher surface area. However, the very small particles are more prone to float in the extraction solvent and therefore, the contact with the solvent is limited [27]. Therefore, the GA yield for particle <125 µm is lesser than that of >125 - ≤250 µm.

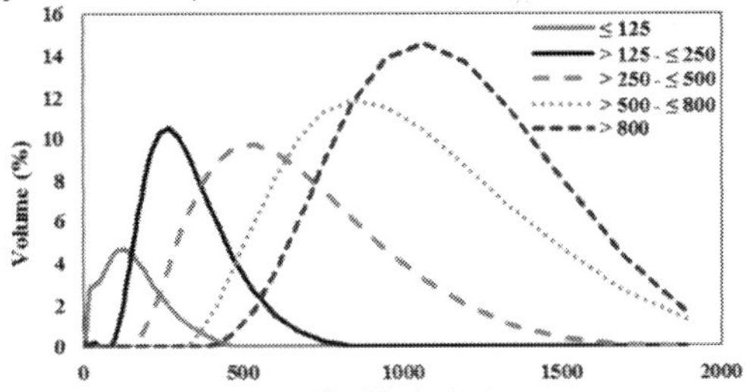


**Fig. 4.** Particle size distribution of *L. pumila* powder

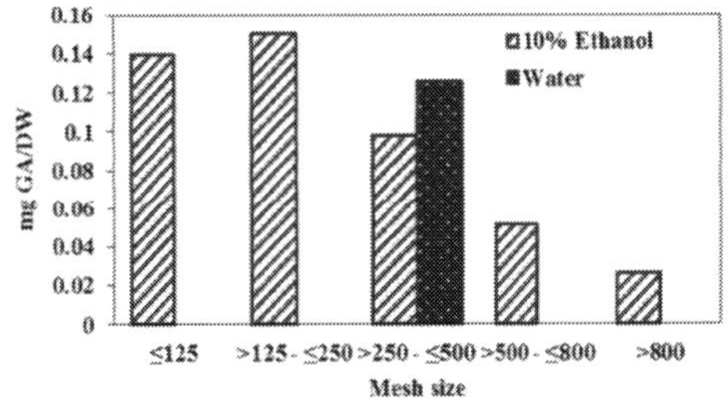


**Fig. 5.** Quantification of GA in different mesh size of *L. pumila*

**Table 1.** Average particle size and surface weighted mean of *L. pumila* powders

| Mesh size range (μm) | d(0.5) (μm) | D[3,2] (μm) |
|---|---|---|
| ≤125 | 66.66 ± 1.99 | 31.24 ± 0.36 |
| >125 - ≤250 | 252.15 ± 5.57 | 140.97 ± 2.11 |
| >250 - ≤500 | 517.10 ± 0.25 | 453.46 ± 20.48 |
| >500 - ≤800 | 858.15 ± 10.29 | 804.90 ± 9.34 |
| >800 | 1055.18 ± 34.18 | 642.77 ± 469.40 |

### 3.3 Effect of type of solvent on gallic acid recovery

Generally ethanol and water are mostly used in the extraction of bioactive compounds as they have been classified as GRAS (Generally Recognized as Safe) solvents [28]. The results showed that the extraction using water as solvent yielded about 29% higher GA content (0.126 mg GA/g DW) than that of 10% ethanol (0.098 mg GA/g DW) for particle size ranged from >250 - ≤500 μm. Solvent of different polarities have effect on the efficiency of extraction of polyphenolic from plants. In terms of polarity, water has higher polarity than that of 10% ethanol. In this case, GA which is classified as a polar phenolic compound [29] is more soluble in water i.e. solvent of higher polarity. Thus explaining why the GA yield is higher when water is used instead of aqueous ethanol. Paini et al. [30] explained that the higher vapour pressure and viscosity of ethanol with respect to water can decrease the force of the implosion of the cavitation bubbles. Consequently, the implosion cause less effective solid material disruption which is vital for gallic acid release from the plant material. Earlier, Paini et al. [30] also obtained lower total phenolic content from agri-food waste with high percentage of ethanol via UAE. Similarly, Fecka and Turek [31] also reported a higher yield of polyphenol with solvent containing higher water content from maceration of thyme and marjoram.

## 4 Conclusion

The UPLC separation method developed in this work for *L. pumila* extracts had successfully identified and confirmed the presence of GA by the means of spectral and residence time match. The recovery of GA generally increase as the particle size decreases, however the highest yield (0.15 mg GA/g DW) was obtained from particle size distribution in the range of >125 - ≤250 μm. Water provide about 29% higher yield of GA than that of aqueous 10% ethanol via UAE extraction. The findings in this work may serve as a useful guide to maximise GA extraction from *L. pumila*.

We acknowledge funding from Ministry of Higher Education Malaysia FRGS/1/2016/TK02/UMP/02/4 (RDU160124) and Universiti Malaysia Pahang (PGRS 160301).

## References

1. M.H. Ibrahim, H.Z.E. Jaafar, E. Karimi, A. Ghasemzadeh, Sci. World J. **2014**, 1-13 (2014)
2. N. Jamaludin, R. Hasham, I. Ware, M.R. Sarmidi, R. Abd Aziz , J. Technol. **77**, 61–66 (2015)
3. N. A. Kamarudin, M. Markom, J. Latip, Indian J. Sci. Technol. **9**, 3–7 (2016)
4. Z. Ali, I. Khan, Phytochemistry **72**, 2075–2080 (2011)
5. E. Karimi, H.Z.E. Jaafar, Molecules **16**, 6791–6805 (2011)
6. L.S. Chua, N.A. Latiff, S.Y. Lee, C.T. Lee, M.R. Sarmidi, R.A. Aziz, Food Chem. **127**, 1186–1192 (2011)
7. N.A. Al-Mekhlafi, K. Shaari, F. Abas, R. Kneer, E.J. Jeyaraj, J. Stanslas, N.H. Lajis, Phytochemistry **80**, 42–49 (2012)
8. A. Pandey, S. Bani, P.L. Sangwan, Adv. Res. **2**, 556–570 (2014)
9. A. Mohd Azrie, A. Luqman Chuah, K.Y. Pin, H.P. Ta. J. Chem. and Pharma. Res. **6**, 172–332 (2014)
10. Mohd Abdah, Siti Nadiah, Sarmidi, Mohammad Roji, Yaakob, Harisun and I. Ware, J. Technol. **69**, 65–68 (2014)
11. C. Locatelli, F.B. Filippin-Monteiro, T.B. Creczynski-Pasa, European J. Med. Chem. **60**, 233–239 (2013)
12. M. Ishihara, H. Sakagami, Anticancer Res. **23**, 2549-2552 (2003)
13. A. Saleem, M. Husheem, P. Harkonen, K. Pihlaja, Ethnopharmacol. J. **81**, 327-336 (2002)
14. Y. Pan, K.H. Tiong, B.A. Abd-Rashid, Z. Ismail, R. Ismail, J.W. Mak, C.E. Ong J. of Ethnopharmacol. **143**, 586–591 (2012)
15. H.K. Choi, D.H. Kim, J.W. Kim, S. Ngadiran, M.R. Sarmidi, C.S. Park, Journal of Biosci. Bioeng. **109**, 291–296 (2010)
16. A. Pandey, S. Bani, P. Sangwan, S. Koul. J. Ethnopharmacol. **132**, 309–315 (2010)
17. A.H. Al-Adhroey, Z.M. Nor, H.M. Al-Mekhlafi, R. Mahmud J. Ethnopharmacol. **132**, 362–364 (2010)
18. S.F. Pang, M. Yusoff, A. Luqman, C. J. Gimbun, Sci. Technol. Prog. **1**, 10–13 (2015).
19. J. Shi, H. Nawaz, J. Pohorly, G. Mittal, Y. Kakuda, Y. Jiang, Food Rev. Int. **21**, 139–166 (2005)
20. J.M. Fonseca, J.W. Rushing, R.L. Thomas, M.B. Riley, N.C. Rajapakse, J. Herbs Spices Med. Plants **12**, 139-152 (2006)
21. H.Y. Baldosano, M.B.M.G. Castillo, C.D.H. Elloran, F.T. Bacani, DLSU Res. Cong. **3**, 4–9 (2015)
22. J. Rodríguez-Miranda, B. Hernández-Santos, E. Herman-Lara, C.A. Gómez-Aldapa, H.S. Garcia, C.E. Martínez-Sánchez, CyTA J. Food **12**, 9–15 (2014)
23. A. Dailey, Q.V. Vuong, Cogent Food Agri. **7**, 1–10 (2015)
24. S. F. Pang, M. Z. Lau, M. M. Yusoff, J. Gimbun, Materials Science Forum, **890**, 155-158 (2017)
25. S. F. Pang, M. M. Yusoff, J. Gimbun, Food Hydrocolloids, **890**, 155-158 (2017).
26. J. Sandanasamy, M. M. Yusoff, J. Gimbun, Materials Science Forum, **890**, 167-170 (2017).
27. C.L. Paiva, W.P. Evangelista, V.A.V. Queiroz, M.B.A. Gloria, Food Chem. **173**, 224-230 (2015)

28. Y.M. Monroy, R.A.F. Rodrigues, A. Sartoratto, F.A. Cabral, J. Supercrit. Fluids **116,** 10-19 (2016)
29. D. Boskou, M. Tsimidou, G. Blekas, Polar Chemical Compounds in Olive Oil, Chemistry and Technology, (Champaign, Illinois: AOCS Press, 2006)
30. M. Paini, A.A. Casazza, B. Aliakbarian, P. Perego, A. Binello, G. Cravotto, Int. J. Food. Sci. Technol. **51,** 349-358 (2016)
31. I. Fecka, S. Turek, Food Chem. **108,** 1039-1053 (2008)

# The effect of auxiliary energy on rhein, kaempferol and astragalin extraction from *Cassia alata*

*Yi Ling* Yeong[1], *Sook Fun* Pang[1,2] *and Jolius* Gimbun[1,2*]

[1]Faculty of Chemical & Natural Resources Engineering, Universiti Malaysia Pahang, 26300 Gambang, Pahang, Malaysia
[2]Centre of Excellence for Advanced Research in Fluid Flow, Universiti Malaysia Pahang, 26300 Gambang, Pahang, Malaysia

**Abstract.** This paper presents the effect of auxiliary energy on rhein, kaempferol and astragalin extraction from *Cassia alata*. The effect of auxiliary energy was examined by performing extraction using either the ultrasonic assisted extraction (UAE) or microwave assisted extraction (MAE). A dried plant material with size ranging from 125 μm to 800 μm was used throughout this work. The rhein, kaempferol and astragalin quantification and identification was performed using ultra-performance liquid chromatography coupled with quadrupole time-of-flight mass spectrometer. The presence of rhein, kaempferol and astragalin were identified and compared with external standards.The residence time, neutral mass and mass fragmentation pattern between the standard and plant extract are also observed. It was found that the extraction yield of rhein, kaempferol and astragalin were higher with UAE method compared to MAE method due to degradation of active component occur. The findings in this work may serve as a useful guide to select the extraction method used to maximise the yield of rhein, kaempferol and astragalin extraction from *C. alata*.

## 1 Introduction

*Cassia alata* (Family: Leguminosae) is a beautiful flowering shrub that grows about 2 to 3 m in height. The leaves are pinnate and 40 to 60 centimetres long. *C. alata* produces beautiful yellow flowers in a column that look like yellow candlesticks. Due to its beauty, it is often cultivated as an ornamental plant. It is well known as ringworm shrub or ringworm *Cassia* in many countries due to the application of its fresh leaves in the treatment of skin diseases such as ringworm, itching, ulcers, eczema, scabies and pruritus. It is widely distributed in many tropical countries such as Malaysia, Indonesia, India, America, Fiji, Brazil and Africa. [1-3][4]. This herbal medicine has been recognized for centuries as a traditional medicine owing to its remarkable medical properties [2]. In Nigeria, the leaves of *C. alata* are used in the skin infection treatment and the juice expressed from the young leaves can be applied topically to the affected parts [1]. In South America, it is used for skin diseases, asthma, snake bites, fever, stomach problems and venereal disease. In Africa, the boiled leaves are used to treat hypertension, whereas in Thailand, the leaves are boiled and drunk to hasten delivery. Moreover, *C. alata* also used as a cure for poisonous bites and for venereal eruptions in India [4]. The leaf extract not only act as antibacterial and antifungal but also perform anti-inflammatory and antioxidant functions. The extracts of *C. alata* have been used in cosmetic product as a formulation for dermatological skin care products [3].

The major bioactive components of *C. alata* are anthraquinones and polyphenols. The leaves of *C. alata* have been qualitatively analysed for the presence of pharmacologically active anthraquinones which are rhein, aloe-emodin, chrysophanol, emodin, and physcion. These anthraquinones perform various functions which are antioxidant, antitumor, antimicrobial, antifungal, cytotoxic, and hypoglycaemic activities. Besides that, *C. alata* also contain flavonoid kaempferol which has been reported to have anticancer properties in several studies [2].

The extraction method is crucial as it will influence the yield, recovery and quality of the bioactive components in the final extract [4-7]. Previously, researchers had applied soxhlet extraction and maceration to extract bioactive components in *C. alata* [8-11]. The aforementioned technique belongs to a conventional method which needs higher temperatures and longer time in order to obtain the extract. Apart from that, higher temperature that is often used in conventional extraction process may cause a thermal degradation of the bioactive components. An alternative extraction method, Ultrasonic Assisted Extraction (UAE) was carried out by Sousa et al. [12] as this method can reduce the mass transfer limitation for both internal and external transport and with the aid of the ultrasonic wave, cell membrane of the plant may break hence facilitating release of bioactive components to the bulk liquid. However, to our knowledge, there is no study in the literature on the extraction of bioactive components from *C. alata* by using Microwave Assisted Extraction (MAE). MAE has shown to be more efficient for

* Corresponding author: jolius@ump.edu.my

extraction of bioactive components than soxhlet, maceration and UAE because of short operating times [4]. MAE enhances the efficiency of extraction as compared to conventional extraction since microwaves interact with the polar molecules in the extraction media, heat is generated and the internal pressure of the solid material is increased [5, 13]. Yet, other bioactive components such as rhein, kaempferol and astragalin are affected by microwave irradiation or high temperature, resulting in degradation [4]. This leads to the necessity of developing this method for extraction in *C. alata* hence this is the aim of this work.

# 2 Materials and methods

## 2.1 Chemicals and plant material

Standard rhein, kaempferol and astragalin were obtained from Sigma-Aldrich whereas analysis grade ethanol was obtained from Merck (Darmstadt, Germany). HPLC grade trifluoroacetic acid (TFA) and acetonitrile was obtained from Fisher Scientific (Leicestershire, UK) and Merck (Darmstadt, Germany), respectively. *C. cassia* was procured from a trader in Sungai Lembing, Malaysia. The dried plant was grounded into powder and the moisture content was determined using a moisture analyzer (AND MS-70, Japan). The powdered plant was preserved in an airtight plastic at room temperature to prevent moisture absorption prior to extraction.

## 2.2 Ultrasonic assisted extraction (UAE)

Extracts were prepared using Qsonica Q700 (Newtown, USA) equipped with a standard probe. A 100 mL of solvent (water and 50% ethanol) was added into weighted powder (dry weight) at the solid-to-solvent ratio of 0.05g/100 ml. It was understood from Pang et al. [14] that 50% aqueous ethanol solvent provides an excellent extraction of both the methoxylated and hydroxylated phenolic compound. Thus the same solvent is used in this work. The mixture was then immediately sonicated at the amplitude of 90% for the extraction time of 9 min. The supernatant was then separated by centrifugation (Eppendorf 5810 R, Hamburg, Germany) at 10 000 rpm for 15 min to obtain a clear solution. The extracts were then stored at -80°C freezer to prevent the degradation of bioactive components in the extract.

## 2.3 Microwave assisted extraction (MAE)

Extracts were prepared using SP-Microwave (CEM) which is controlled by Synergy$^{TM}$ software using a method adapted from our earlier work [14]. A 5 mL of solvent (water and 50% ethanol) was added into weighted powder (dry weight) at the solid-to-solvent ratio of 0.05g/ml. The mixture was then immediately put in the microwave at the power of 200W and extracted for 5 min. The supernatant was then separated by centrifugation (Eppendorf 5810 R, Hamburg, Germany) at 10 000 rpm for 15 min to obtain a clear solution. The extracts were then stored at -80°C freezer to prevent the degradation of bioactive components in the extract.

## 2.4 Quantification of rhein, kaempferol and astragalin

The quantification of kaempferol, rhein, and astragalin was performed on Waters UPLC-TUV-QTOF Xevo G2S. Waters Acquity UPLC I-Class (Waters Corporation, Milford, USA) fitted with Acquity UPLC HSS T3 column (2.1x75mm, 1.8μm) and a Acquity UPLC HSS T3 VanGuard column guard (2.1x5mm, 1.8μm) was used for analysis. The column temperature is 30°C with injection volume of 1 μL and flowrate at 600 μL /min. The mobile phase consists of water in 0.1% formic acid (solvent A) and acetonitrile in 0.1% formic acid (solvent B). The performed elution scheme were: 0-2.0 min, 85% A; 2-6 min, 50% A; 6-10 min, 5% A; 10-11 min, 85% A; and the MS system was performed on Waters Xevo G2S QTOF with ionization ESI negative mode, capillary voltage of 1.5 kV. The collision energy is ramp from 10 to 40 eV and acquisition mass range from 100 to 1500 m/z.

# 3 Results and discussion

## 3.1 Comparison of ultraviolet and mass spectrometry of MAE extract

The extracts obtained in *C. alata* by using UAE and MAE were analysed by using UPLC-TUV–QTOF MS. This is an analytical technique in which the separation performed by liquid chromatography and analyses with UV detector and a mass spectrometer. This technique is high in selectivity and sensitivity, hence often used for plant sample characterization. Fig. 1. Show the UV chromatogram and MS chromatogram of MAE extract. More compound peaks are detectable under MS compared to UV. UPLC-QTOF MS has an advantage over the other method such as GC-MS because it can be used to separate the plant extracts containing compounds that are highly polar and less volatile [9]. Thus, separation of phenolic compounds in the *C. alata* extracts was conducted via LC-MS and LC-UV. It is known that the sensitivity of LC-MS is 10 times more than that of LC-UV. Therefore, LC-MS is preponderance for microanalysis. Apart from that, LC-MS can identify the component by retention time (t(R)) and m/z, whereby it has high selectivity and exclusion for the determination of bioactive component. Meanwhile, the LC-UV is simpler, lower cost and has a better separation of components. Therefore, it is preferred for identification of abundant component, which is easier to separate [15]. Owing to the nature of the sample in this work, the UPLC-QTOF-MS was used to compare the active compounds obtained from MAE and UAE.

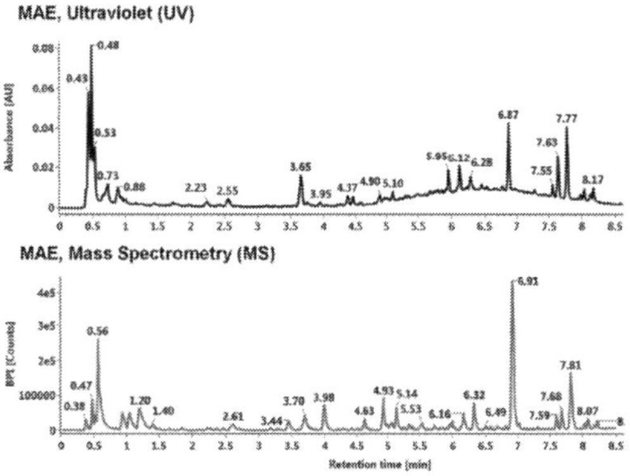


**Fig. 1.** Microwave assisted extraction (MAE) Comparison of Ultraviolet (UV) and Mass Spectrometry (MS).

### 3.2 Mass spectrometry comparison between the MAE and UAE extract

The sonication in UAE produced the mechanical waves that will generate micro-cavitation in the liquid surrounding the plant material, disrupting the plant cell wall and releasing the extract. The kinetic energy is introduced to the whole bulk liquid volume by the cavitation bubbles collapse at the interface, hence increasing the mass transfer across the sold-liquid interface. This will induce penetration of the solvent into cell membrane walls and facilitating the release of cells contents and improving the mass transfer [16].

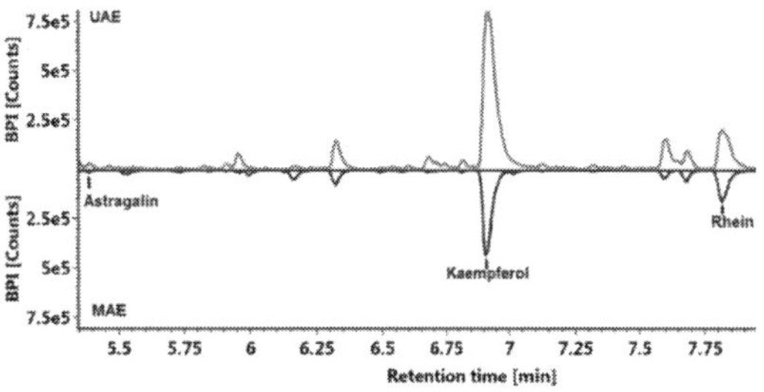


**Fig. 2.** Comparison of UAE and MAE in liquid chromatography mass spectrometry (LCMS) base peak ion chromatogram. UAE (red line), MAE (blue line).

The base peak ion chromatogram of MAE and UAE extract were compared in Fig. 2. It can be observed that the chemical profile of the MAE extracts is slightly different from that of UAE extracts. The three major compounds, namely, kaempferol, rhein and astragalin were identified by comparing the retention times, UV spectra and MS data of the separated compounds with their respective standards. Kaempferol elutes at 6.92 minutes, rhein elutes at 7.82 minutes and astragalin elutes at 5.39 minutes. The MAE is an efficient method for extraction of bioactive components from plant materials because of its shorter operational times. However, rhein, kaempferol and astragalin may be affected by the microwave irradiation or high temperature, resulting in degradation which adversely affecting the yield [4]. Therefore, the yield of extraction by MAE is not always higher than that of UAE.

### 3.3 Comparison of rhein, kaempferol and astragalin yield

The yield of bioactive components such as kaempferol, rhein, and astragalin by using UAE is higher than that of MAE as shown in Fig. 3. The yield of kaempferol and astragalin by UAE is doubled to that of MAE. For instance, UAE yielded 0.32 mg/g of kaempferol whereas only 0.14 mg/g was obtained from MAE. Similarly for astragalin the yield from UAE (0.02 mg/g) is doubled to that of MAE (0.01 mg/g). The different between UAE (0.09 mg/g) and MAE (0.06 mg/g) on the yield of rhein is not a much as that of kaempferol and astragalin, although it is clear that UAE is a better method for extracting rhein than MAE. As mentioned earlier this is due to cell disruption by sonication in UAE which enhanced the release of active compound from the plant material.

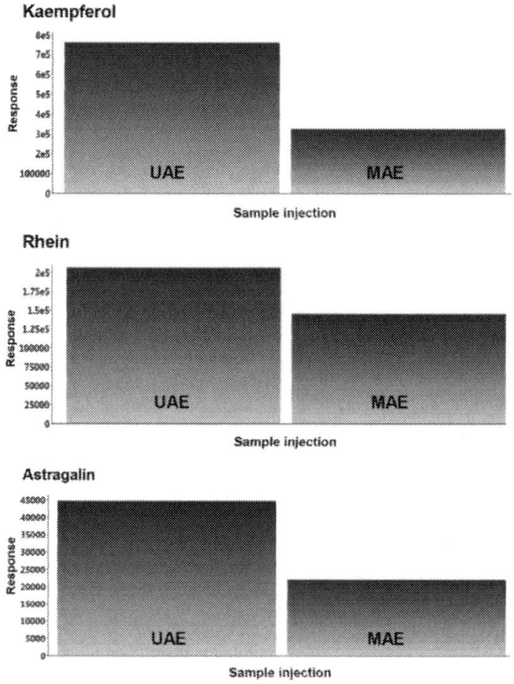


**Fig. 3.** Comparison of the yield of bioactive components by using UAE and MAE extraction method.

In the microwave extraction, the atomic kinetic energy is introduced and the heat propagates in the liquid phase, hence increasing the diffusion rate, while sonication introduced the kinetic energy to the whole volume as the cavitation bubbles collapse at the interface hence improving the mass transfer across the solid-liquid interface. Compared to MAE, UAE has the main advantage of working closer to ambient temperatures, thus avoiding the thermal overexposure [17]. In this case, rhein kaempferol and astragalin may suffer from thermal degradation induced by the excessive heat from the microwave irradiation in MAE.

## 4 Conclusion

The liquid chromatography with mass spectrometry detector method developed in this work for *C. alata* extracts had successfully identified and confirmed the presence of rhein, kaempferol and astragalin in the extracts by matching the mass spectra and residence time obtained from the plant extracts and external standard. It was found that UAE provide a higher extraction yield of rhein, kaempferol and astragalin from *C. alata* leaves compared to MAE. In this work the sonication increased the yield of bioactive component extraction from *C. alata* by at least 30% compared to that of MAE.

## References

1. P. L. Rajagopal, R. Majetha, K .R. Sreejith, S. S. Kiron, S. Aneeshia, K. Premaletha, J. Int. Aca. Res. Multidiscip. **2**, 280-307 (2014)
2. V. E. Fernand, D. T. Dinh, S. J. Washington, S. O. Fakayode, J. N. Losso, R. O. V. Ravenswaay, I. M. Warner, Talanta. **74**, 896-902 (2008)
3. D. E. Okwu and F. U., Nnamdi, Pelaga. Res. Lib. **2(2)**, 247-254 (2011)
4. C. A. L. Escobar, F. P. Capote, M. D. L. de Castro, Talanta. **144**, 522-528 (2015)
5. S. F. Pang, M. Z. Lau, M. M. Yusoff, J. Gimbun, Mater. Sci. Forum (**890**), 155-158 (2017)
6. J Sandanasamy, M. M. Yusoff, J. Gimbun, Mater. Sci. Forum (**890**), 167-170 (2017)
7. S. L. Nguang, Y. L. Yeong, S. F. Pang, J. Gimbun, Ind. J. Sci. Technol. (**10**)**2**, (2017)
8. W. F. Sule, I. O. Okonko, S. Omo-Ogun, J. C. Nwamze, M. O. Ojezele, O. J. Ojezele, J. A. Alli, E. T. Soyemi, T. O. Olaonipekun, J. Med. Plants Res. **5(2)**, 176-183 (2011)
9. W. Gritsanapan and P. Mangmeesri, J. Health. Res. **23(2)**, 59-64 (2009)
10. C. A. Alalor., C. I. Igwilo., E. Jeroh, Asian. J. Biol. Sci. (**5**), 120-125 (2012)
11. O. Oladeji., F. Adelowo., K. Odelade., S. Aremu., M. Adisa, CIBTech. J. Pharma. Sci. **5(3)**, 40-51 (2016)
12. A. D. Sousa, A. I. V. Maia, T. H. S. Rodrigues, K.M. Canuto, P. R. V. Ribeiro, R. D. C. A. Pereira, R. F. Vieira, E. S. de Brito, Ind. Crops Prod. (**79**), 91-103 (2016)
13. A. E. Ince, S. Sahin, G. Sumnu, J. Food. Sci. Technol. **51(10)**, 2776-2782
14. S. F. Pang, M. M. Yusoff, J. Gimbun, Food Hydrocoll. (**37**), 159-165, (2014)
15. R. H. Liu, A L. Jiang, L. L. Bian, F. Shao, Z. L. Ma, B. Y. Yu, US Nat. Lib. Med. **30(3)**, 285-289 (2007)
16. A. Alupului, I. Calinescu, V. Lavric, AIDIC. Conference. Series. (**9**), 1-7 (2009)
17. V. Mandal, Y. Mohan, S. Hemalatha, Pharma Rev. **1**, 7-18 (2007)

# Sustainability Management Program for Industries- A Case Study

*Alwin* Long Su Weng [1], *Samson* Mekbib Atnaw [2,*], *Roziah* Binti Zailan [2] and *Muhamad* Muhtazam Noor Din [3]

[1]Universiti Kebangsaan Malaysia LESTARI Institute for Environment & Development
[2]Universiti Malaysia Pahang, Faculty of Engineering Technology, Gambang, Pahang, 26300, Malaysia
[3]Malaysian Green Technology Corporation, Green-Living-Energy Services

**Abstract.** This research studied the effectiveness of Sustainability Management Program in improving production efficiency of the manufacturing site with verified result using the regression analysis. For this study, a dairy manufacturing industry located in Malaysia was selected and major energy consuming equipment in the industry plant were identified. Sustainability Management Program (SMP) was carried out for three years and energy consumption and product has improved regression coefficients of 0.625 in 2013, 0.826 in 2014, and 0.878 in 2015 as the manufacturing site becomes more energy efficient. This suggests that the energy management should be carried out in a continuous manner with energy management team responsible for energy saving practices.

## 1 Introduction

Global statistics revealed that industrial energy consumption accounts for about 40% of generated electricity [1, 2]. Increasing technological development over the years has therefore attracted complex designs for energy consuming equipment. Miniaturization, micro-fabrication of devices and systems are somewhat as a result of a rise in the use of energy saving techniques. However, efficiency improvement technique as a means of generating more economic output with less energy as input is key to achieving a sustainable development [3]. Utilization index of energy supplied to a system cannot be 100%, thereby obeying the third law of thermodynamics. Energy consumption by industrial systems is rather a form of conversion from its initial form to another [4]. An efficient management of how this transformation takes place in an industrial system is therefore of necessity and it must have continuity in order to be sustainable. Organisational structural planning of energy saving policy remains principal to its growth [5]. Regardless of a cost-benefit analysis and value addition of energy, increasing energy cost in industrial systems remains detrimental to process output cost and competitiveness [6, 7].

Over the years, research has been narrowed to design and development of low-energy consuming and conserving devices [8, 13]. Recent microprocessors built in industrial systems devices allow for power management through dynamic voltage scaling (DVS) thereby reducing voltage consumption and increasing power savings [13]. Limiting the energy dissipation in the form of heat, power sensors for heavy duty machineries are receiving substantial attention in research [9, 10, 14].

Energy management is a process for an efficient system operation. However many of such processes focus upon piece meal projects related to devices or technologies with little or no verification to confirm the effectiveness of the project. Hence it birthed forth a Sustainability Management Program (SMP), which can be carried out in cycle as shown in Fig. 1. It can best be viewed from the systems approach with components having a ripple effect on the system when perturbations occur. An energy management strategy seeks to optimize and obtain a balance between the demand and supply in system operation [15-18]. Market mechanisms could therefore be critically optimized using systems dynamics principles with feedbacks and patterns carefully studied [15, 19, 20-22]. This paper, studied the effectiveness of Sustainability Management Program in improving production efficiency of a manufacturing site located in Malaysia with verified result using the regression analysis.

* Corresponding author: mekbibsams@ump.edu.my

**Fig. 1.** Sustainability management program [23]

The energy matrix is one of the cruxes of an energy management model that levels and organizational action steps across each level. A typical example is as shown in Table 1. Energy policy, systems organization, training, performance management, communication and investment form the primal of energy management matrix. Each level has its standards to meet and best used, as a benchmarking to what it should be as the management system improves. By deploying the energy management matrix, organization is able to set a baseline to where their current status is and how they may improve in building a more resilient if not a sustainable energy management corporation.

**Table 1.** Energy management matrix

| Level | Energy Policy | Organising | Training | Performance Measurement | Communication | Investment |
|---|---|---|---|---|---|---|
| 4 | Energy Policy, Action Plan and regular review have active commitment of Senior Management Team and aligned to SDP | Fully integrated into management structure with clear accountability for energy consumption | Appropriate and comprehensive training tailored to identified needs, with evaluation for all school stakeholders | Comprehensive performance measurement against targets with effective management reporting | Extensive communication of energy issues within and outside of school to wider school stakeholders | Resources routinely commited/sourced to support energy efficiency in school supporting objectives |
| 3 | Formal Policy but no active commitment from Senior Management Team | Clear line management accountability for consumption and responsibility for improvement | Energy training targeted at relevant staff following training needs analysis | Weekly performance measurement for each process, unit, or building | Regular staff briefings, assemblies, performance reporting and energy promotion | Same appraisal criteria used as for other cost reduction projects |
| 2 | Un-adpoted Policy | Some delegation of responsibility but line management and authority unclear | Ad-Hoc training for selected people as required | Monthly monitoring by fuel type of use and cost | Some use of school communication mechanisms to promote energy efficiency | Low or medium cost measures considered if short payback period |
| 1 | An unwritten set of guidelines | Informal, mostly focused on energy supply and billing | Technical staff occasionally attend specialist courses | Invoice checking only | Ad-Hoc informal contacts used to promote energy efficiency e.g. one off assembly/meeting | Only low or no cost measures taken e.g. 'switch off' posters |
| 0 | No explicit Energy Policy | No delegation of responsibility for managing energy | No energy related staff training attended | No measurement of energy costs or consumption | No communication or promotion of energy issues | No investment in improving energy efficiency by the school |

## 2 Materials and Methods

Industrial energy management is factor dependent. Among the factors which governs its success are geographical locations and organizational energy policy. Current pracitce of chosen companies were investigated thorugh an energy auditting process. Achieving a sustainable energy development involves constituting committes accountable for its success. Two committees were constituted: the energy committee and the energy accounting centre committee. Organsisational strucutre to ensure both vertical and horizontal flow of information was developed. Job description of each committee was developed.

### 2.1 Sustainable Energy Savings using AEMAS framework

One of the primal factors to continuous availability of energy systems is the development of energy policy interventions [15]. One of the essential of considerable energy savings is the energy matrix that was devised under the ASEAN Energy Management Scheme (AEMAS).

### 2.2 Energy procurement: electricity and natural gas usage breakdown

In this process, energy costs associated with energy consumption for production were evaluated. At the point of research, two major sources of energy were procured namely electricity and natural gas. Back in 2012 the electricity cost in Malaysia was at RM 0.304 per kWh during peak period, RM 0.187 during off-peak period and RM 27.70 for maximum demand with a blended rate of RM0.32 per kWh . On the other hand average natural gas price for industrial consumption in Malaysia was approximately RM 16.07 for one million British Thermal Unit (MMBTU). Hence selecting the right energy for each application especially for heating ventilation and air conditioning (HVAC) systems are critical towards the company's profitability. Shown in Fig. 2 and 3 are the energy cost movement across four years and how it would have impacted production costs [24].

Electricity costs alone increased by 15% from an average of RM0.32 for each kilowatt-hour to RM0.38 kilowatt-hour. The natural gas cost on the other hand increased by astounding 57% from RM16.07 for one MMBTU to RM26.00 for one MMBTU. The impact was managed through the Sustainability Management Program with results presented at the end of the paper.

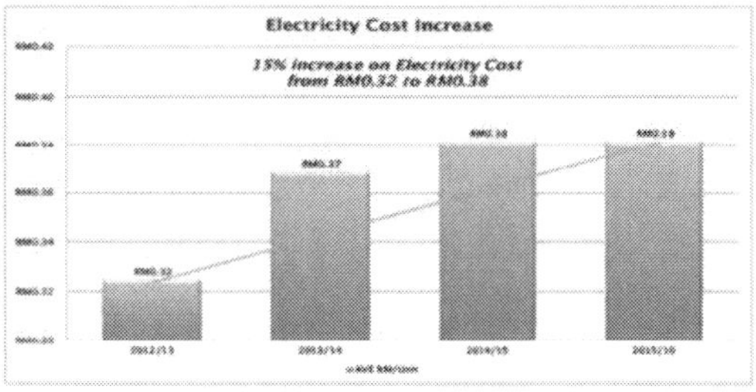


**Fig. 2.** Electricity cost fluctuation for 2012, 2013, 2014 and 2015 [24]

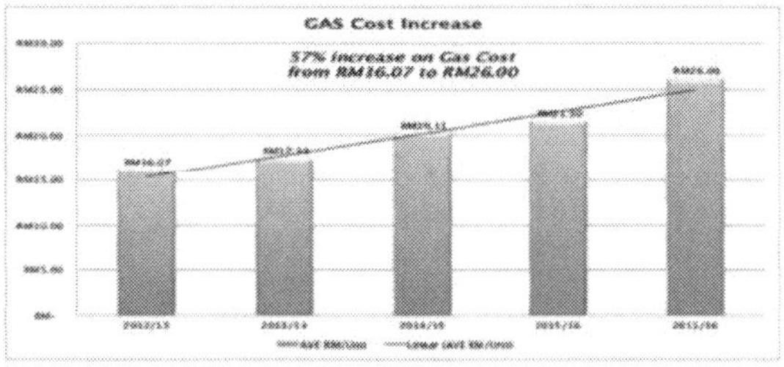


**Fig. 3.** Natural gas cost fluctuation for 2012, 2013, 2014 and 2015 [2]

### 2.3 Energy conservation through empower energy committee

Sustainability in most energy management activity is seen as a challenge. Two main challenges faced were (1) Commitment from top management (2) Measurement and verification on savings achieved.

The paper covers the establishment of a Sustainability Management committee that is headed by the highest management member in the plant. Energy committee was constituted with three management levels with each level having predefined job descriptions as shown in Fig. 4. Thus everyone shall have detailed information of what is required in the role they were appointed governed under the periodic meetings to keep track with progress.

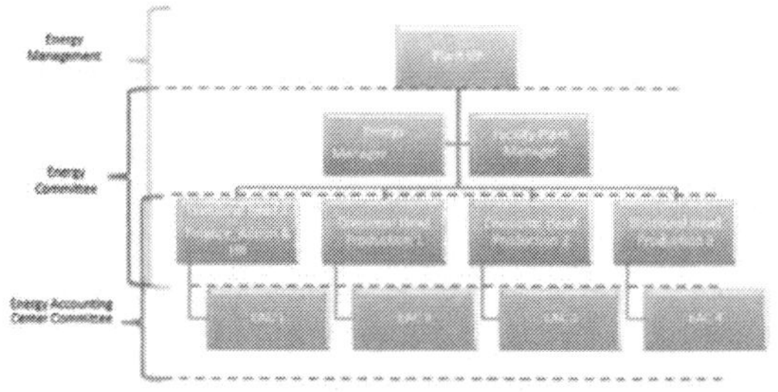

**Fig. 4.** Energy committee management levels

Each committee was saddled with predefined responsibilities as below:

A. Energy Management (EM)
   a. Meet once to twice a year
   b. Decision maker for strategic matters
   c. Setting Target, approve plans, budget
   d. Center for approvals

B. Energy Committee (EC)
   a. Meet once a month (Technical Team)
   b. Carry out Plan to meet overall target
   c. Setting Target for EAC to meet overall target
   d. Update management on Measurement & Verification (M&V)

C. Energy Accounting Center (EAC) Committee
   a. Meet daily or weekly with EAC head
   b. Carry out Plan to meet overall target
   c. Measure & verify savings
   d. Update EC on Measurement & Verification (M&V)
   e. Identify savings opportunity

## 3 Results and Discussion

### 3.1 Energy efficiency by conducting energy audit

The success of implementing new ideas and innovations depends on the outcome of analysis of the current practice. Most organizations are often faced with challenges concomitant to the formulation of energy performance indicators, integrating energy management and technology [16, 20-22]. Organizational energy audit requires selection of a target facility and actions plan developed around it before implementation. Energy generation and outlets of consumptions were selected and an audit were carried out in this paper. Having the Certified Energy Manager reporting directly to the top management helps bring energy management as a priority, working alongside with the grass roots personnel in the EAC

### 3.2 Identification of high energy consuming equipment

The manufacturing plant performs its daily operations on two energy sources; the national grid and natural gas. The EAC Managers will brief their committee on a daily basis to determine the production areas where savings can be found and later raise the proposal to the energy committee on their monthly meeting.

Centralized facilities equipment such as the heating, ventilation and air conditioning (HVAC) equipment with high-energy consumption were identified and energy audit conducted on them. The areas identified are as below;

Heat, Ventilation and Air Conditioning (HVAC) systems

a. Condenser Water System
b. Chilled Water Pump System
c. Cooling Tower Fans Optimization

Unit Operation systems

a. Boiler Optimization
b. Air Compressor System
c. Ammonia Compressor System
d. Cooling Water Plant System
e. Cooling Tower Fans Optimization
f. Chiller Pump System

Value system designs were conducted on the selected Change of Organization Investment criteria by looking at projects with payback below 3 years.

Energy consumption rate to production follows a linear relationship. The plot shown in Fig. 5, 6 and 7 reveal an $R^2$ values of 0.62458 in 2012/2013, 0.82654 in 2013/14, and 0.87833 in 2015/16, respectively. The regression model serves as a good indicator on energy efficiency impact across the three years period and better predictor of energy consumption for the manufacturing plant in the future with the regression Equation 1.

$$Y = 56.886x + 836938 \tag{1}$$

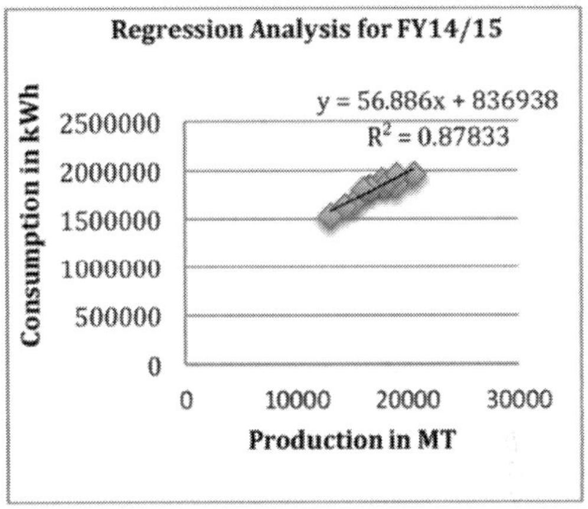


**Fig. 5.** Regression plot of energy consumption on energy production (FY14/15)

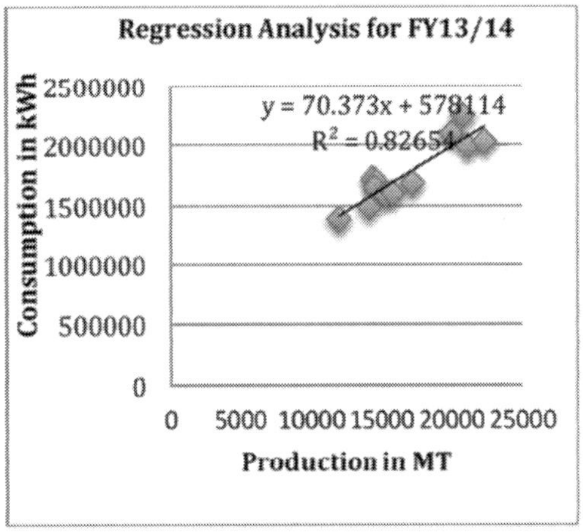


**Fig. 6.** Regression plot of energy consumption on energy production (FY13/14)

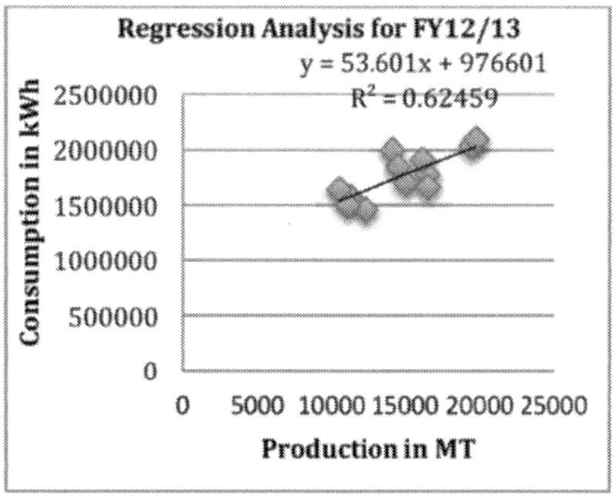


**Fig. 7.** Regression plot of energy consumption on energy production (FY12/13)

## Conclusion

One of the salient ways by which energy is wasted is through its mismanagement. Equipment not needed are often left to run at no load, electric bulbs not needed are left on throughout the day, offices which could be served by one air conditioners are supplied with two and lots more. Identification of energy consuming equipment is therefore needful. This can only be achieved through energy audit of the system. Data collected for this project were subjected to statistical analysis and a regression model developed to this effect.

For the two companies selected, regression models developed has energy production successfully predict the energy consumed. A forecast of energy production from the base period was determined and compared with the achieved production per year. Considerable energy savings was recorded with an energy management framework developed for critical subsystems identified to consume energy the most within the system as a whole.

## Acknowledgement

This paper was supported by the Internal Grant of the University Malaysia Pahang (RDU160393) and Fundamental Research Grant Scheme (FRGS) grant of Malaysian Ministry of Higher Education (RDU140109).

## References

[1] P.A. Adedeji, O.A. Olalere, O.A. Adebimpe, S.O. Olunusi, Int. J. Eng. Sci. **3**, 73-79 (2014)
[2] A. McKane, Ind. Energ. Manage. (2007)
[3] D. Viadmir, The Netherlands, ERSCP-EMSU. (2010)
[4] K. Vikhorev, R. Greenough, N. Brown, J. Cleaner Product. 1-11 (2012)
[5] O.A. Olalere, P.A. Adedeji, O.A. Adebimpe, B. Olayiwola, Int. J. Eng. Sci. **3**, 55-62 (2014)
[6] ECON. Ministry of Industry, Employment and Communications. (2003)
[7] P. Thollander, J. Palm, New York: Spinger. (2013)
[8] F. Douglis, P. Krishnan, Computing Systems. **8**, 381-413 (1995)
[9] F. Jason, M. Satyanarayanan, 17th Symposium on Operating Systems Principles. (2001)
[10] T. Halfhill, Microprocessor. (2000)
[11] T. Heath, 11th International Conference on Parallel Architectures and Compilation Technques. (2002)
[12] R.A. Lebeck, X. Fan, H. Zeng, S.C. Ellis, 9th International Conference on Architectural Support for Programming Languages and Operating Systems (ASPLOS IX). (2000)
[13] R. Bianchini, R. Rajamony, NJ: DARPA. (2003)
[14] A. Oumer, S.M. Atnaw, J.K. Cheng, L. Singh, Mater. Sci. Eng. (2016)
[15] P. Palensky, D. Dietrich, IEEE Transac. Ind. Informatics. **7**, 381-389 (2011)
[16] S.O. Oyedepo, Energ. Sustainab. Society. **2**, 1-17 (2012)
[17] T.Y. Chiu, S.L. Lo, Y.Y. Tsai, Energies. **5**, 5324-5339 (2012)
[18] O. Fuji, Busan Korea, International Electrical Vehicle Symposium. (2002)
[19] J. Moreno, M.E. Ortuzar, J.W. Dixon, IEEE Transac. Ind. Electronics. **53**, 614-624 (2006)
[20] V. Sharma, U. Mrkherji, V. Joseph, S. Gupta, IEEE. 1-23 (2008)
[21] Siemens. Energy mangment and energy optimization in the process industry. Germany: Siemens. (2011)
[22] M. Weiser, B. Welch, A. Demers, S. Shenker, 1st Symposium on Operating System Design and Implementation. (1994)
[23] M.M. Williansen, Professional Safety. (2005)
[24] Tenaga Nasional Pricing & Tariffs (2016), Retrived from http://www.mida.gov.my/env3/uploads/images/Tariff_Rate_Final_01.Jan.2014.pdf

DOI: 10.1051/matecconf/201711102011

# Towards Cooling Tower Efficiency-An Energy Audit Approach

*Alwin* Long Su Weng [1], *Samson* Mekbib Atnaw [2,*], *Roziah* Binti Zailan [2] and *Muhamad* Muhtazam Noor Din [3]

[1]Universiti Kebangsaan Malaysia LESTARI Institute for Environment & Development
[2]Universiti Malaysia Pahang, Faculty of Engineering Technology, Gamtang, Pahang, 26300, Malaysia
[3]Malaysian Green Technology Corporation, Green-Living-Energy Services

**Abstract.** This research studied the power generation trends from national grid and gas for a period of 4 years. Energy audit of critical systems like this is needful for optimal energy utilization. An energy audit was carried out on 6 industrial cooloing towers and their annual operating cost calculated. Variable speed drive suggested was installed and corresponding annual energy savings of 114,900 kWh/year cost saving of RM30,000 was achieved at a case study plant located in Malaysia. Cooling towers with smart systems was recommended for higher energy savings.

## 1 Introduction

Manufacturing industries comprises of many energy-consuming systems and subsystems. To minimize the industry's energy requirements, these systems must operate at optimal overall efficiency. This is practically achievable when such device possess the capability to track load changes in real time [1]. Electricity supply from the mains have been unable to keep pace with the increasing industrial development. The aftermath of this has led to increased concentration of greenhouse gases from anthropogenic actions especially emissions from manufacturing industries [2, 9]. There is therefore an urgent need for reduced energy consumption from cooling systems making them function at optimal level [3, 9, 10-11]. Reducing excessive energy consumption in power drives in turn decreases the lifecycle costs [4].

Cooling towers forms an integral part of the thermodynamic system in petrochemical industries, mining, manufacturing industries not leaving out residential houses [1, 5]. Mechanical forced draft cooling towers with counter-flow systems possess electric motor drives removes heat from cooling water by adding sensible heat from the heated fluid to air up to saturated condition. The cooling rate of mechanical draft cooling towers largely depends on the fan diameter and its speed of operation [7, 8]. The speed of rotation of fans in turn depends on the specification of the motor drive. Cooling in this system is a function of the wet bulb temperature due to psychometric characteristics exhibited by the mixture of fluid and gas.

Shown in Fig. 1 (a) and (b) are two views of a cooling tower equipped with four units of 7.5 kW fans for each cell. A total of 6 fans were running during survey with 4-2 and 3-3 combination on separate occasions. At low production time, only four fans will be in operation. Each Cooling Tower at the point of audit was drawing approximately 43.2 amperes with three cells in operation with all fans at functional conditions.

(a)

(b)

**Fig. 1.** (a) and (b) Vertical draft cooling tower

* Corresponding author: mekbibsams@ump.edu.my

## 2 Materials and Methods

For the success of a new method, it is needful to determine the present operating conditions. To perform this, annual cost of operations at system functionality of 24 hours and 300 days per year was calculated. The trend of energy generation for a period of 4 years (2012-2015) was studied. Energy from the national grid and the gas as a fossil fuel alternative was studied. Electricity cost over the period of study was also considered and related to energy loss per kilowatt.

### 2.1 Energy savings at varying temperature, variable speed drive (VSD)

Energy savings through variable speed drive (VSD) was recommended on the cooling tower fans. Installation of a temperature sensor in the basin to modulate the speed to run efficiently according to process demand. Under all operational procedures, the VSD will be set to provide sufficient cooling to the system. No optimal set-point is provided as the fan speed will be controlled to provide optimal cooling below 32°C as stated in best manufacturing benchmarking data. The system was subjected to varying wet bulb temperatures and future projections of energy consumption was calculated as shown in Table 1.

| Return Temperature caused by production & wet-bulb (°C) | Total Fan Baseline (kW) | Estimated Speed (Hz) | Running (%) | Future Running (kW) |
|---|---|---|---|---|
| 36 | 45 | 47 | 94 | 37.4 |
| 35 | 45 | 42 | 84 | 26.7 |
| 34 | 45 | 40 | 80 | 23.0 |
| 32 | 45 | 37 | 74 | 18.2 |
| 31 | 45 | 35 | 70 | 15.4 |

**Table 1.** Potential savings on temperature fluctuation

### 2.2 Process performance evaluation

On implementing the suggested modifications both to the process and a process performance evaluation based on two models were performed on the system. Energy savings was verified after implementation: using the International Performance Measurement and Verification Protocol (IPMVP) and the Energy Efficiency Index (EEI).

Energy savings were determined after project completion by short-term or continuous measurements taken throughout the term of the contract at the device or system level; performance and operations factors were monitored. Savings were calculated with engineering calculations using metered data. Cost of this option is dependent on number and type of systems measured and the term of analysis metering; typically 3-10% of project construction cost of items subject to M&V. Variable speed drive and controls were applied in the process to a motor to adjust fan speed.

Electric power was measured with a kilowatt meter installed on the electrical supply to the motor, which reads the power every minute. In the baseline period this meter was in place for a week to verify constant loading. The meter was in place throughout the reporting period to track variations in power use.

## 3 Results and Discussion

### 3.1 Costs of operations

Operating the cooling tower for 24 hours a day and 300 days in a year, the following breakdown remains valid;
Average Power Consumption for each fan = 14 ampere
Average Power Consumption for each fan = 7.5 kW
Total units running per time = 6 units
Total Power Consumption for the system = 45 kW
Average Cost of Electricity = 0.30 RM/kWh
Annual Total Energy Consumption = 302,400 kWh
Annual Cost = 114,900 RM
Carbon Emission = 206 Tonnes of $CO_2$

The carbon emission was estimated using a factor based on the total electricity consumption. Shown in Fig. 2 is the electricity tariff within the period of study. Here exist an increase from the base period to the period of the study termination. Significant financial loss is incurred for a kilowatt of energy wasted per year. The plot shown in Fig. 3 reveals an increasing trend of the cost of gas. Production systems running on fossil fuels tend to accrue more operational cost compared with the national grid as the national grid supply is assumed cheaper than the fossil fuel. A kilowatt waste in energy generated therefore amounts to a fortune.

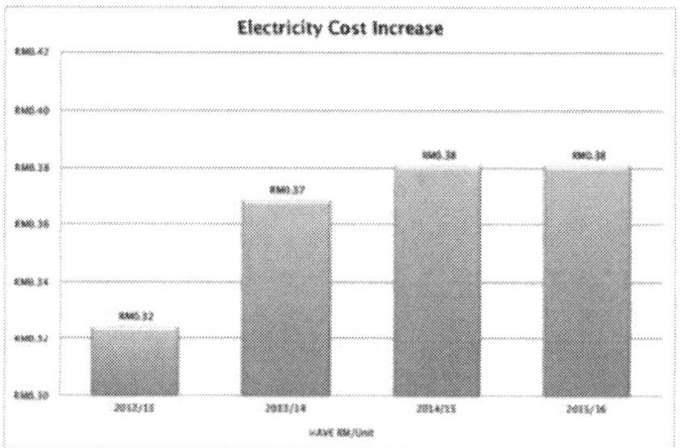


**Fig. 2.** Electricity cost from 2012-2015

DOI: 10.1051/matecconf/201711102011

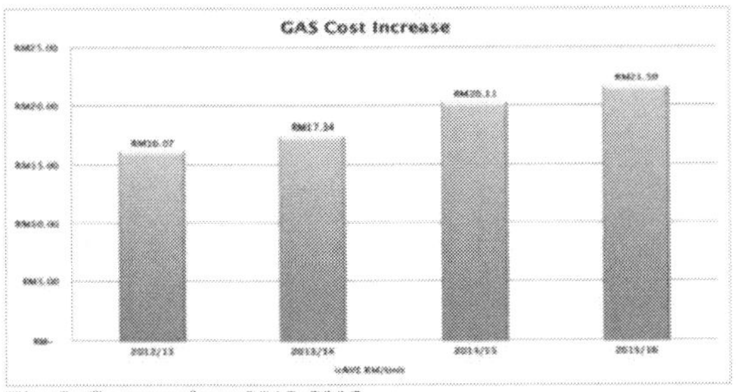


**Fig. 3.** Gas cost from 2012-2015

### 3.2 Monthly monitoring and verification using EEI (Factored data)

Electricity production followed a sinusoidal trend (Fig. 4). A decreasing trend is observed from 2012 to 2015 in power generation. Power generated in 2015 appears to be the least amount the years. Optimal utilization of the power generated is needful for manufacturing industries not only to minimize production cost but also satisfy energy demand.

Energy efficiency index (EEI) reveals energy cost against production increased by 1.8% and 2.2% over the 3 years period and energy Consumption against production improved by 9.3% and 2.9% over the 3 years period. Energy cost increased by an average of 15% for electricity and 34% on gas making cost of production (RM/ton) increase inevitably year-on-year. Production efficiency shows improvement by 11.9%.

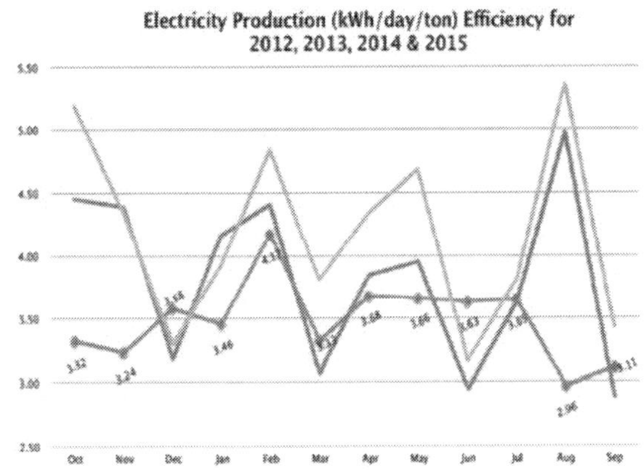


**Fig. 4.** Electricity production efficiency from 2012-2015

In the reverse, gas production increased from 2012 to 2015 (Fig. 5). This accounts for the fall in electricity production from the mains as manufacturing processes need to run on alternative energy source which fossil fuel is more reliable. Increase in the gas production is concomitant to increased environmental pollution among other environmental impact it has. Energy generated should therefore be optimally utilized.

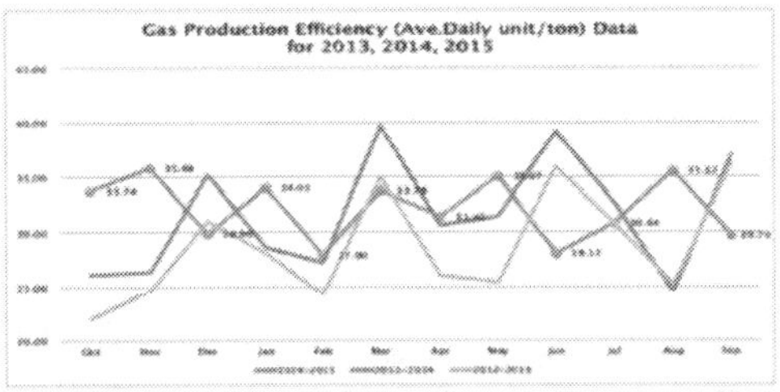


**Fig. 5.** Gas production efficiency from 2013-2015

### 3.3 Cost saving, investment and payback

The use of variable speed drive gave a considerable energy savings which when translated to cost is worth a fortune. The payback period being 3.5 years from installation period and investment cost are as follows;
Energy Saving (estimated 30% Savings 42 Hz):
114,900 kWh/ year

| | |
|---|---|
| Cost Saving: | 34,400 RM/ year |
| Investment Cost (Budget): | 120,000 RM |
| ITA Savings addition: | 30,000 RM |
| Payback: | 3.5 years |
| Annual Savings in $CO_2$ Emission: | 78 Tonnes of $CO_2$ |

### 3.4 Shared savings model

Energy Saving (estimated 30% Savings 42Hz):
121,000 kWh/ year
Cost Saving: 36,000 RM/ year
Investment Cost: 0 RM
Maintenance Cost: 0 RM
Shared Savings: 20% FNDM
7,200 RM/year
80% ATECH
28,800 RM/year
Duration: 5 years
Annual Savings in $CO_2$ Emission: 83 Tonnes of $CO_2$
Evaluations done by two companies reveals 80% cost savings by ATECH and 20% savings by FNDM. Up to 83 tonnes of carbon (iv) oxide was saved on annual basis thereby reducing environmental pollution.

## 4 Conclusion

Increased industrial growth has attracted increased energy demands from industrial systems. Energy production when optimally utilized indirectly affects the unit prices of finished products. Energy audit of energy consuming subsystems in a manufacturing system is of high necessity for increased profitability of the business. Variable speed drives on cooling towers triggered by varying wet bulb temperature readings is recommended in cooling tower systems. Energy supply from national grid decreases over the years with fossil fuels taking the market. Smart systems are recommended for high energy consuming equipment for minimal energy consumption.

## Acknowledgement

This paper was supported by the Internal Grant of the University Malaysia Pahang (RDU: 160393) and Fundamental Research Grant Scheme (FRGS) grant of Malaysian Ministry of Higher Education (RDU140109).

## References

[1] M.S. Bhatt, Appl. Therm. Eng. **20**, 297-307 (2000)

[2] D.Y.L. Chan, J. Energy Policy. **35**, 202-209 (2007)

[3] J. Facao, A.C. Oliveira, Appl. Therm. Eng. **20**, 1225-1236 (2000)

[4] B.B. Ekici, U.T. Aksoy, Adv Eng Softw. **40**, 356-362 (2008)

[5] EERE. U.S. Department of Energy (2011)

[6] P. Rai, I.A. Khan, International Analysis of Cooling Tower. **2**, 96-102 (2016)

[7] BEE. Bureau of Energy Efficiency (2013)

[8] S. Arunkumar, D.B. Sivakumar, T. Senthikumar, Int. J. Eng. Sci. Comput. **6**, 5359-5362 (2016)

[9] Z.K. Morvay, D.D. Gvozdenac, Appl. Ind. Energ. Environ. Manage. (2010)

[10] G.E.D Plessis, L. Leinbenberg, E.H. Matthews, Appl. Energ. 111, 16-27 (2013)

[11] A. Oumer, S. M. Atnaw, J. K. Cheng, L. Singh, Mater. Sci. Eng. (2016)

# *Material Balance And Reaction Kinetics Modeling For Penex Isomerization Process In Daura Refinery*

*Adel Sharif* **Hamadi** [1] **and** *Rawnak Adnan* **Kadhim** [2]

[1] University of Technology, Chemical Engineering Department, Baghdad - Iraq
[2] Midland Refineries Company, Daura Refinery, Baghdad - Iraq

**Abstract.** Penex Deisohexanizer isomerization of light straight run naphtha is a significant process for petroleum refining and proved to be effective technology to produce gasoline components with a high octane number. Modeling of the chemical kinetic reactions is an important tool because it is a better tool for optimization of the experimental data into parameters used for industrial reactors. The present study deals on the isomerization process in Daura refinery. Material balance calculations were done mathematically on the unit for the kinetics prediction purpose. A kinetic mathematical model was derived for the prediction rate constants $K_1$ and $K_2$ and activation energy Ea at operating temperatures range 120-180°C. According to the model, the results show that with increasing of temperature leads to increased $K_1$ directly, where the $K_2$ values proportional inversely. The activation energy results show that $Ea_{1\,(nC6)} < Ea_{1\,(C5)} < Ea_{1\,(CH)}$, and for $Ea_{2\,(iC5)} < Ea_{2\,(2,2\text{-}DMB)} < Ea_{2\,(2,3\text{-}DMB)} < Ea_{2\,(MCP)}$.

## 1 Introduction

The Penex Deisohexanizer (DIH) isomerization is a process operated using the catalytic reaction of n-pentanes, n-hexanes, and mixtures thereof to produce isomerate hydrocarbons. The reactions take place in a hydrocatalytic fixed bed reactor to promote conversion and minimize hydrocracking [1]. The process of light naphtha can produce isomerate hydrocarbons with higher octane number, in addition, that the typical Penex unit product can blend into gasoline pool, and the non-converted low octane components ($nC_6$, CH and MP's) from the deisohexanizer column (DIH) can be recycled to the reactor section for further upgrading [2].

The major elements of Penex (DIH) isomerization processes are reactors operated using chlorinated–alumina ($Pt/Al_2O_3$-Cl) as catalyst [3, 4]. This catalyst is proven to be active at lower temperatures (120-180°C) in which equilibrium favorable to produce iso-paraffins. [5]. The scheme of Penex (DIH) process is illustrated in Fig. 1. [6]

The most important process variables in Penex isomerization unit are the reactor temperatures. The higher temperatures than equilibrium lead to increase the amount of hydrocracking and increase the carbon formation on the catalyst.

A typical UOP Penex unit is provided with two reactors in series. All of the benzene rings in the LSRN feed are hydrogenated in the first reactor and some conversion of cyclohexane (CH) and methyl cyclopentane (MCP) to hexanes also occurs, as does some hydrocracking of $C_7$ components to $C_3$ and $C_4$.

The hydrogenation of benzene rings, naphthene conversion to hexane, and $C_7$ hydrocracking are exothermic reactions and, for a typical feedstock, contribute more to the temperature rise in the reactor than exothermic reaction of paraffin isomerization.

The concentrations and outlet temperature will be influenced by the catalyst and by the mass of $C_6$ cyclic and $C_7$ components in the feed.

## 2 Penex Unit Reaction Mechanism

### 2.1 $nC_5$ Reaction Mechanism [7]

The Isomerization reaction of normal paraffins to isoparaffins is slightly exothermic and thermo-dynamically favorable at a temperature range 120-180°C. The kinetics of the reaction proceed in series through an olefin as an intermediate product formed by dehydrogenation of n-paraffins through an adsorption mechanism on the surface of metal site catalyst.

$$n\,pentane \xrightarrow{pt} n\,pentene$$

Because of low equilibrium conversion of paraffin isomerization, n-olefins can be converted to carbonium ion with the aid of injection of strong chloride acid as initiator ion.

$$n\,pentene + [H^-]\,[A^-] \rightarrow carbonium\ ion + [A^-]$$

This initiator ion allows equilibrium forward to removes n-olefin from the first hydrogenation reaction, and then to rearrange the molecules.

$$carbonium\ ion \rightarrow iso\ pentane\ carbonium\ ion$$

The high catalytic acidity causes a hydrogenation reaction to proceed at a higher reaction rate. Then, the carbonium ion converted iso-paraffin to iso-olefin by dehydrogenation step.

$$iso\ pentane\ carbonium\ ion \rightarrow iso\ pentene$$

In the last reaction, the iso-olefin is hydrogenated again to form iso-paraffin in the presence of surface of the catalyst.

$$iso\ pentene \xrightarrow{pt} iso\ pentane$$

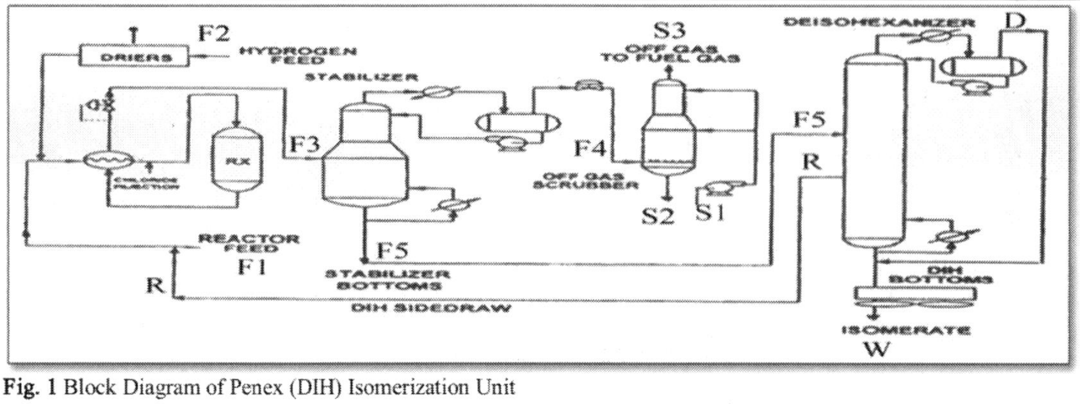


**Fig. 1** Block Diagram of Penex (DIH) Isomerization Unit

Generally, the final steps of (n-$C_6$) isomerization proceed similarly, thus (n-$C_6$) several iso-products like 2-MP, 3-MP, 2,2-DMB, and 2,3-DMB is can be produced.

## 2.2 nC6 Reaction Mechanism [8]

$$n\ pentane_{(RON=24.8)} \overset{pt}{\Leftrightarrow} 2MP_{(RON=73.4)}$$

$$n\ pentane_{(RON=24.8)} \overset{pt}{\Leftrightarrow} 3MP_{(RON=74.5)}$$

$$n\ pentane_{(RON=24.8)} \overset{pt}{\Leftrightarrow} 2{,}2\ DMB_{(RON=91.8)}$$

$$n\ pentane_{(RON=24.8)} \overset{pt}{\Leftrightarrow} 2{,}3\ DMB_{(RON=104.3)}$$

The relationship between equilibrium conversion vs. isomerization reaction temperatures of ($C_5$ and $C_6$) paraffins are shown in Fig. 2 and 3, respectively [5].

## 2.3 Other Reactions [9]

There are several reactions occurring inside and outside of the reactors.

### *2.3.1 Inside Reactors:*

a) Naphthene ring of methyl cyclopentane (MCP) and cyclohexane (CH) present in the Penex feed will hydrogenate to n-paraffins, and then to iso-paraffins. The increasing in reactor temperature leads to increase Naphthene ring opening reactions.

b) As the temperature increased, the naphthenes shift forwards to (MCP) production.

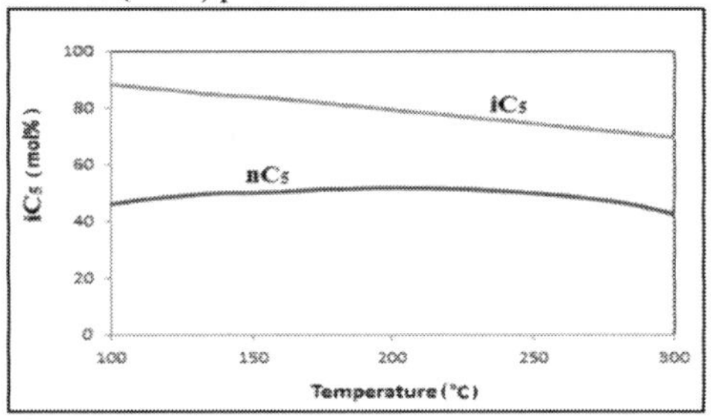


**Fig. 2:** $C_5$ Paraffin Equilibrium Curves [5].

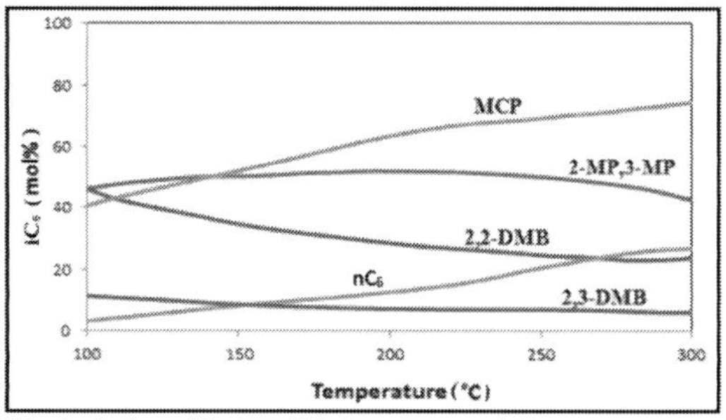


**Fig 3**: $C_6$ Paraffin Equilibrium Curves [5].

$$cyclohexane_{(RON=84)} \overset{pt}{\Leftrightarrow} MCP_{(RON=104.3)}$$

c) Very quickly hydrogenation reaction of the benzene ring in the presence of a catalyst ($Pt/Al_2O_3$-Cl) proceed at very low temperature.

$$benzene_{(RON>100)} + 3H_2 \overset{pt}{\Leftrightarrow} cyclohexane_{(RON=84)}$$

d) Hydrocracking of Penex feed according to types of feed quality. For example, $C_7$ molecules tend to hydrocrack easily than smaller ones, paraffins of $C_5$ and $C_6$ hydrocrack to some extent. The severity of hydrocracking tends to reduce yield and increase product temperature.

### *2.3.2 Outside Reactors :*

a) Hydrogen chloride is injected annually at period of maintenance time, to limit corrosion problems in units, the reaction run until removed any rust present in the unit.

$$6\ HCl + Fe_2O_3 \rightarrow 2FeCl_3 + 3\ H_2O$$

b) Perchloroethylene ($C_2Cl_4$) is injected before reactors at approximately the temperature of (110°C) or higher. The present hydrogen will react with this promoter in the presence of a catalyst ($Pt/Al_2O_3$-Cl) to produce hydrogen chloride.

$$C_2Cl_4 + 5H_2 \xrightarrow{(Pt/Al2O3-Cl)} 4HCl + C_2H_6$$

c) Neutralization of hydrogen chloride formed in above reaction using caustic soda (NaOH) to form salt and water in the scrubber.

$$HCl + NaOH \rightarrow NaCl + H_2O$$

## 3 Material Balance

As shown by Fig. 1, there are three streams entering to the rector of Penex unit, the feed (LSRN), makeup gas ($H_2$) and recycle (from Deisohexanizer) and outlet stream exit as a product from the reactor. The product stream (F3) exits from the reactor, enters to stabilizer column. In this process, overhead light gases ($CH_4$,$C_2H_6$, $C_3H_8$, and i-$C_4H_{10}$) and some of $nC_5$ and $iC_5$ which is removed and exit from the top of column, in addition to $H_2$ and HCl , which then enter to scrubber.

The outlet product from the top of stabilizer column enters to the scrubber column where NaOH solution injected into the scrubber. From the bottom of the stabilizer, the product stream contains ($nC_5$, $iC_5$, $nC_6$, 2-MP, 3-MP, 2,2-DMB, 2,3-DMB ,MCP and CH) which goes to Deisohexanizer (DIH) column. The outlet product from the bottom of stabilizer enters to the DIH column. There are two streams exit from DIH column, top stream (D) and bottom stream (W).

The mass balance calculations on the unit assuming feed rate of Penex unit =10, 000 BPD is presented in Table (1) and (2) respectively.

**Table 1**: Material Balance Calculations on the Reactor and Stabilizer Unit Assuming Feed Rate =10,000 BPD

| | Comp. | Mwt | Density ($Kg/m^3$) | Reactor (kg/hr) | | | | Stabilizer (kg/hr) | | |
|---|---|---|---|---|---|---|---|---|---|---|
| | | | | *F1 in* | *F2 in* | *R in* | *F3 out* | *F3 in* | *F4 out* | *F5 out* |
| 1 | $nC_5$ | 72 | 626 | 11276,16 | 0 | 199.731 | 5278.9 | 5278.9 | 18.47 | 5260.43 |
| 2 | $iC_5$ | 72 | 616 | 5688.17 | 0 | 135.817 | 12020.96 | 12020.96 | 281.52 | 11739.44 |
| 3 | $nC_6$ | 86 | 655 | 14808.4 | 0 | 17440.5 | 17736.79 | 17736.79 | 0 | 17736.79 |
| 4 | 2-MP | 86 | 653 | 4899.84 | 0 | 24902.4 | 29802.27 | 29802.27 | 0 | 29802.27 |
| 5 | 3-MP | 86 | 664 | 3915.52 | 0 | 21467.1 | 25382.58 | 25382.58 | 0 | 25382.58 |
| 6 | 2,2-DMB | 86 | 674 | 139.37 | 0 | 719.031 | 12145.51 | 12145.51 | 0 | 12145.51 |
| 7 | 2,3-DMB | 86 | 695 | 1572.30 | 0 | 6559.16 | 11356.35 | 11356.35 | 0 | 11356.35 |
| 8 | MCP | 84 | 749 | 39.199 | 0 | 23.968 | 501.12 | 501.12 | 0 | 501.12 |
| 9 | BZ | 78 | 876 | 483.451 | 0 | 8444.62 | 0 | 0 | 0 | 0 |
| 100 | CH | 84 | 778 | 731.71 | 0 | 199.731 | 8738.38 | 8738.38 | 0 | 8738.38 |
| 11 | $H_2$ | 2 | 0.0899 | 0 | 208.52 | 0 | 170.55 | 170.55 | 170.55 | 0 |
| 12 | $CH_4$ | 16 | 0.668 | 0 | 227.39 | 0 | 227.39 | 227.39 | 227.39 | 0 |
| 13 | $C_2H_6$ | 30 | 1.264 | 0 | 343.92 | 0 | 343.92 | 343.92 | 343.92 | 0 |
| 14 | $C_3H_8$ | 44 | 1.882 | 0 | 392.32 | 0 | 392.38 | 392.38 | 392.38 | 0 |
| 15 | i-$C_4$ | 58 | 2.5326 | 0 | 230.77 | 0 | 230.77 | 230.77 | 230.77 | 0 |
| 16 | $C_2Cl_4$ | 166 | 1622 | 0 | 0 | 0 | 0 | 0 | 0 | 0 |
| 17 | HCl | 36.5 | 1.63 | 0 | 0 | 0 | 11.43 | 11.43 | 11.43 | 0 |
| 18 | NaOH | 40 | 2130 | 0 | 0 | 0 | 0 | 0 | 0 | 0 |
| 19 | NaCl | 58.5 | 2165 | 0 | 0 | 0 | 0 | 0 | 0 | 0 |
| 20 | $H_2O$ | 18 | 1000 | 0 | 5.15 | 0 | 0 | 0 | 0 | 0 |
| | Total | | | 43554.12 | 1408.07 | 79892.3 | 123838.18 | 123838.18 | 1676.44 | 43554.12 |

**Table 2**: Mass Balance Calculations on the Scrubber and Deisohexanizer Unit Assuming Feed Rate =10,000 BPD

| | Comp. | Mwt | Density ($Kg/m^3$) | Scrubber (kg/hr) | | | Deisohexanizer (kg/hr) | | | |
|---|---|---|---|---|---|---|---|---|---|---|
| | | | | *F4 in* | *S1 out* | *S2 out* | *F5 in* | *R out* | *W out* | *D out* |
| 1 | $nC_5$ | 72 | 626 | 18.47 | 18.47 | 0 | 5260.43 | 199.731 | 0 | 5060.70 |
| 2 | $iC_5$ | 72 | 616 | 281.52 | 281.52 | 0 | 11739.44 | 135.817 | 0 | 11603.62 |
| 3 | $nC_6$ | 86 | 655 | 0 | 0 | 0 | 17736.79 | 17440.5 | 48.14 | 248.15 |
| 4 | 2-MP | 86 | 653 | 0 | 0 | 0 | 29802.27 | 24902.4 | 6.95 | 4892.92 |
| 5 | 3-MP | 86 | 664 | 0 | 0 | 0 | 25382.58 | 21467.1 | 20.00 | 3895.48 |
| 6 | 2,2-DMB | 86 | 674 | 0 | 0 | 0 | 12145.51 | 719.031 | 0 | 11426.48 |
| 7 | 2,3-DMB | 86 | 695 | 0 | 0 | 0 | 11356.35 | 6559.16 | 1.61 | 4795.58 |
| 8 | MCP | 84 | 749 | 0 | 0 | 0 | 501.12 | 23.968 | 823.67 | 0 |
| 9 | BZ | 78 | 876 | 0 | 0 | 0 | 0 | 8444.62 | 0 | 0 |
| 100 | CH | 84 | 778 | 0 | 0 | 0 | 8738.38 | 199.731 | 52.75 | 0 |
| 11 | $H_2$ | 2 | 0.0899 | 170.55 | 170.55 | 0 | 0 | 0 | 0 | 0 |
| 12 | $CH_4$ | 16 | 0.668 | 227.39 | 227.39 | 0 | 0 | 0 | 0 | 0 |
| 13 | $C_2H_6$ | 30 | 1.264 | 343.92 | 343.92 | 0 | 0 | 0 | 0 | 0 |
| 14 | $C_3H_8$ | 44 | 1.882 | 392.38 | 392.38 | 0 | 0 | 0 | 0 | 0 |
| 15 | i-$C_4$ | 58 | 2.5326 | 230.77 | 230.77 | 0 | 0 | 0 | 0 | 0 |
| 16 | $C_2Cl_4$ | 166 | 1622 | 0 | 0 | 0 | 0 | 0 | 0 | 0 |
| 17 | HCl | 36.5 | 1.63 | 11.43 | | 0 | 0 | 0 | 0 | 0 |
| 18 | NaOH | 40 | 2130 | 0 | 0 | 2.5 | 0 | 0 | 0 | 0 |
| 19 | NaCl | 58.5 | 2165 | 0 | 0 | 18.32 | 0 | 0 | 0 | 0 |
| 20 | $H_2O$ | 18 | 1000 | 0 | 0 | 140.91 | 0 | 0 | 0 | 0 |
| | Ttotal | | | 1676.44 | 1665 | 161.73 | 43554.12 | 79892.3 | 122662.9 | 42798.85 |

The concentration of reactants in the isomerization reactions ($nC_5$, $nC_6$ and CH) in the inlet stream to the reactor is calculated from the mass balance, these concentrations are used to calculate $K_1$ (rate constant for the formation of intermediate olefin). The results are shown in Table (3).

Whereas the mole fraction of the produced isomers outlet from reactor ($iC_5$, 2,2-DMB 2,3-DMB and MCP) are calculated from the mass balance ,these mole fractions are used to calculate $K_2$ (rate constant for the formation of isomers).

**Table 3** Concentration of Reacted Components.

| Component | $F_{Ao}$ (Kmole) | V ($m^3$) | $C_{Ao}=F_{Ao}/V$ (Kmole/$m^3$) |
|---|---|---|---|
| $nC_5$ | 156.613 | 18.013 | 8.694 |
| $nC_6$ | 172.191 | 22.608 | 7.161 |
| CH | 104.03 | 11.23 | 9.26 |

## 4 Development Model

In the typical isomerization process, several reactions are taken place: paraffin isomerization; naphthene hydrogenation; naphthene isomerization, benzene saturation, hydrocracking; and naphthene alkylation. Table (4) list the reactions and equilibrium conversion suggest for isomerization of normal paraffinic components in the feed of Penex process. [10]

$$n\,paraffin \overset{K1}{\rightarrow} Olefin \overset{K2}{\rightarrow} i\,paraffin$$

The model developed in this study are taken following assumptions: 1st order reactions at steady state, isothermal conditions, gas phase reaction, constant physical properties, and neglected gradient in pressure drop, temperature and concentration respectively.

Mass balance has been developed over the cross section of the segment of the catalyst bed, as illustrated in Fig. 3.

$$C_{Ao} = C_A + C_N + C_{ISO} \quad (8)$$

$$C_A = C_{Ao}\exp(-K_1 t) \quad (9)$$

$$C_N = \frac{K_1 C_{Ao}}{K_2 - K_1}[\exp(-K_1 t) - \exp(-K_2 t)] \quad (10)$$

$$C_{ISO} = C_{Ao}\left[1 - \exp(-K_1 t) - \frac{K_1}{K_2 - K_1}[\exp(-K_1 t) - \exp(-K_2 t)]\right] \quad (11)$$

**Table 4** A Typical Reactions for LSRN Isomerization

| Reactions | Conv. % | Equ. No. |
|---|---|---|
| $nC_5 \Leftrightarrow iC_5$ | 54 | (1) |
| $nC_6 \Leftrightarrow$ 2,2-DMB | 35 | (2) |
| $nC_6 \Leftrightarrow$ 2,3-DMB | 10 | (3) |
| $BZ+3H_2 \Leftrightarrow$ CH | 100 | (4) |
| CH $\Leftrightarrow$ MCP | 59 | (5) |
| 2-MP $\rightarrow$ 3-MP | | (6) |
| 3-MP $\rightarrow$ 2,2-DMB + 2,3-DMB | | (7) |

The design equation for volume of tubular reactor:

$$V = F_{Ao}\frac{\int dx}{-K_1 C_{Ao}} \quad (12)$$

And reactant concentration of $C_A$ in gas phase:

$$C_A = C_{Ao}\frac{(1-x)T_o}{(1-\varepsilon x)T} \quad (13)$$

Where ε is refer to voidage in reactor bed.

Putting equation (13) into (12) and re-arrange, to get the reaction rate constant K1 as shown in equation (14). From Arrhenius equation (15) ,the activation energy ($Ea_1$) and frequency factor ($K_o$) can be calculated by plot (ln $K_1$) vs. (1/T), as illustrated in Fig. 4.

$$K_1 = \frac{F_{Ao}}{V\,C_{Ao}}\left[ln\,\frac{1}{1-x}(1+\varepsilon) - \varepsilon x\right]\frac{T}{T_o} \quad (14)$$

$$ln\,K_1 = ln\,K_o - \frac{E_a}{RT} \quad (15)$$

## 5 Results And Discussion

The model shown by equation (14) can be used to calculate isomerization reaction rate constant $K_1$, using data given from Penex unit in the Daura Refinery: Volume of catalyst in each reactor =35.66 $m^3$ ,Hold up = 75 %, ε = 0.393, and $T_o$= 50°C.

Table (5) represents the results of $K_1$ at temperatures range of 120-180 °C. It was shown that calculated reaction constants of ($K_1$) values increased with increasing temperature and reach maximum value at a temperature (180°C).

Applying equation (15) and plot ($lnK_1$) vs (1/T) for components listed in Table (6) are shown in Fig. 5.

The activation energy ($Ea_1$) and Frequency factor ($Ko_1$) are calculated from these plots and tabulated in Table (7). The results show that $Ea_{1(nC6*)} < Ea_{1(nC6**)} < Ea_{1(C5)} < Ea_{1(CH)}$.

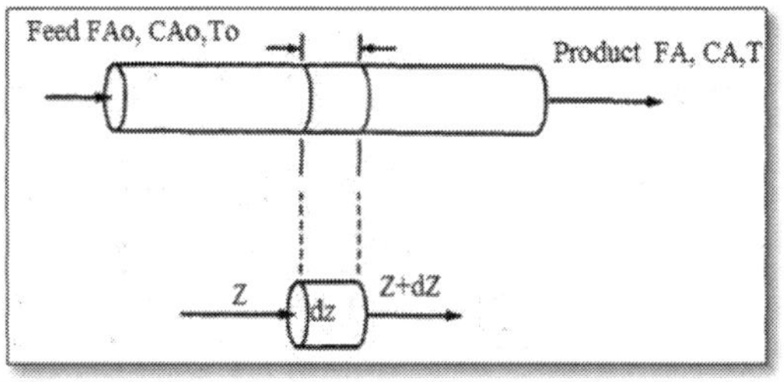


**Fig. 3**: Segment of Tubular Catalyst Bed Reactor.

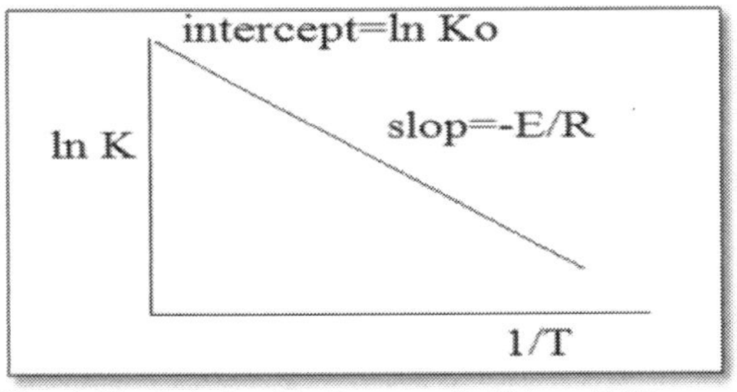


**Fig. 4:** The Relationship Between (ln $K_1$) vs. (1/T)

**Table 5:** Results of $K_1$.

| T(°C) | T(°K) | $K_1$ ($h^{-1}$) | | | |
|---|---|---|---|---|---|
| | | *$nC_5$* | *$nC_6$** | *$nC_6$*** | *CH* |
| 120 | 393 | 0.398 | 0.241 | 0.056 | 0.289 |
| 135 | 408 | 0.413 | 0.250 | 0.058 | 0.300 |
| 150 | 423 | 0.429 | 0.260 | 0.060 | 0.311 |
| 165 | 438 | 0.444 | 0.269 | 0.062 | 0.322 |
| 180 | 453 | 0.459 | 0.278 | 0.065 | 0.333 |

*Where $nC_6$* for $nC_6$ ↔ 2,2-DMB at x=0.35,and $nC_6$** for $nC_6$ ↔ 2,3-DMB at x=0.1*

**Table 6:** ln$K_1$ and 1/T Values.

| 1/T | ln $K_1$ | | | |
|---|---|---|---|---|
| | *$nC_5$* | *$nC_6$** | *$nC_6$*** | *CH* |
| 0.0025 | -0.9202 | -1.4164 | -2.8759 | -1.2429 |
| 0.0025 | -0.8828 | -1.3789 | -2.8385 | -1.2055 |
| 0.0024 | -0.8467 | -1.3428 | -2.8024 | -1.1694 |
| 0.0023 | -0.8118 | -1.3080 | -2.7675 | -1.1345 |

Equation (11) can be used to calculate the rate constant $K_2$ for reaction on olefin to iso-paraffin. The required data were taken from Daura Refinery; as LHSV =1.5, residence time t =1/ LHSV = 0.667 h. The values of $K_2$ are shown in Table (8), which was shown that $K_2$ values decreasing with increasing temperature, which means that product isomerate concentrations were increased as a result of temperature sensitivity of isomerization reactions.

Applying equation (15) and plot (ln$K_2$) vs (1/T) for components listed in Table (9) are shown in Fig. 6.
The values of the activation energy ($Ea_2$) and Frequency factor ($Ko_2$) are calculated from these plots and tabulated in Table (10). The results show $Ea_{2\,(iC5)} < Ea_{2(2,2\text{-DMB})} < Ea_{2(2,3\text{-DMB})} < Ea_{2(MCP)}$.

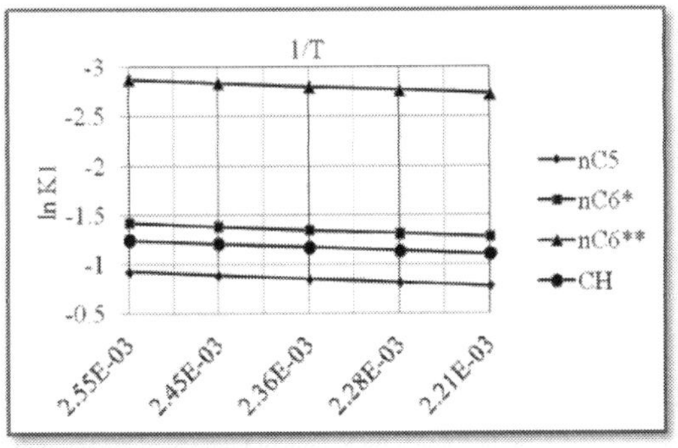


**Fig. 5:** Plot (ln $K_1$) vs. (1/T)

**Table 7:** Activation Energy and $Ko_1$ Results.

| Component | $Ea_1$ (KJ/mole) | $Ko_1$ |
|---|---|---|
| $nC_5$ | 2738.822 | 0.460 |
| $nC_6$* | 1525.743 | 0.263 |
| $nC_6$** | 1835.182 | 0.061 |
| CH | 2880.394 | 0.334 |

**Table 8:** Results of $K_2$.

| T(°C) | T(K) | $K_2$ ($h^{-1}$) | | | |
|---|---|---|---|---|---|
| | | *$iC_5$* | *2,2-DMB* | *2,3-DMB* | *MCP* |
| 120 | 393 | 0.1547 | 0.2566 | 1.2062 | 0.0147 |
| 135 | 408 | 0.1480 | 0.2477 | 1.1512 | 0.0124 |
| 150 | 423 | 0.1430 | 0.2389 | 1.0998 | 0.0113 |
| 165 | 438 | 0.1384 | 0.2308 | 1.0529 | 0.0099 |
| 180 | 453 | 0.1346 | 0.2232 | 1.0099 | 0.0097 |

**Table 9:** ln$K_2$ and 1/T Values

| 1/T | ln $K_2$ | | | |
|---|---|---|---|---|
| | *$iC_5$* | *2,2-DMB* | *2,3-DMB* | *MCP* |
| 0.0025 | -0.8106 | -0.5907 | 0.0814 | -1.8327 |
| 0.0025 | -0.8298 | -0.6062 | 0.0612 | -1.9066 |
| 0.0024 | -0.8446 | -0.6218 | 0.0413 | -1.9469 |
| 0.0023 | -0.8588 | -0.6368 | 0.0224 | -2.0044 |
| 0.0022 | -0.8711 | -0.6513 | 0.0043 | -2.0132 |

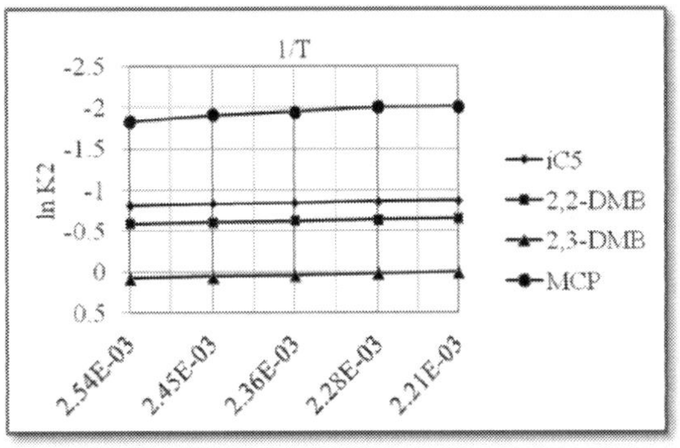


**Fig. 6:** Plot (ln $K_2$) vs. (1/T)

**Table 10:** Activation Energy and $Ko_2$ Results.

| Component | $Ea_2$ (KJ/mole) | $Ko_2$ |
|---|---|---|
| $iC_5$ | 1.503 | 0.281 |
| 2,2-DMB | 1.519 | 0.349 |
| 2,3-DMB | 1.931 | 0.602 |
| MCP | 4.632 | 0.038 |

## 6 Conclusions

1. The increasing temperature leads to increased $K_1$ according to reactor model (proportional directly).
2. The $K_2$ values decreased when the temperature increased for the isomerate (proportional inversely) which indicates the increasing in the isomerate concentration.
3. The activation energy results show that $Ea_{1(nC6)} < Ea_{1(C5)} < Ea_{1(CH)}$, and for $Ea_{2\ (iC5)} < Ea_{2(2,2\text{-}DMB)} < Ea_{2(2,3\text{-}DMB)} < Ea_{2(MCP)}$, which means the reaction of lower activation energy have higher reaction rate.

## Abbreviations

| | |
|---|---|
| dz | Integration Step for the Reactor Length |
| 2,2-DMB | 2,2-Dimethylbutane |
| 2,3-DMB | 2,3-Dimethylbutane |
| 2-MP | 2-Methylpentane |
| 3-MP | 3-Methylpentane |
| BZ | Benzene |
| $nC_6$* | Concentration for $nC_6$ ↔ 2,2-DMB at x=0.35, |
| $nC_6$** | Concentration for $nC_6$ ↔ 2,3-DMB at x=0.1 |
| CA | Concentration of n-paraffin present at time t |
| CAo | Concentration of inlet LSRN stream ($mole/m^3$) |
| Ciso | Concentration of i-paraffin. |
| CN | Concentration of n-olefin |
| CH | Cyclohexane |
| D | Top stream of DIH shown in Fig. 1 |
| DIH | Deisohexanizer (DIH) |
| $Ea_1$ | Activation energy for n paraffin $\xrightarrow{K1}$ Olefin |
| $Ea_2$ | Activation energy for Olefin $\xrightarrow{K2}$ i paraffin |
| F1 | Reactor feed shown in Fig. 1 |
| F2 | Hydrogen feed shown in Fig. 1 |
| F3 | Stabilizer feed shown in Fig. 1 |
| F4 | Scrubber feed shown in Fig. 1 |
| F5 | DIH feed shown in Fig. 1 |
| FAo | Mole rate of inlet component LSRN |
| Fiso | Mole rate of isomerate product |
| HC | Hydrocarbon |
| i-$C_4$ | Iso-butane |
| $iC_5$ | Iso-pentane |
| $iC_6$ | iso-hexane |
| K1 | Rate Constant for Paraffins (hr-1) |
| K2 | Rrate Constant for Olefins (hr-1) |
| Ko | Frequency Factor |
| LHSV | Liquid Hourly Space Velocity |
| LSRN | Light straight run naphtha |
| MCP | Methyl cyclopentane |
| $nC_5$ | Normal pentane |
| $nC_6$ | Normal hexane |
| Pt | Platinum |
| $Pt/ZrO_2$- | Sulfated zirconia |
| $SO_4$ | |
| $Pt/Al_2O_3$-Cl | Chlorinated alumina |
| **R** | DIH sidedraw shown in Fig. 1 |
| R | Gas Constant (8.314 J/K.. mole) |
| RON | Research Octane Number |
| t | Residence time (hr) |
| S1 | NaOH solution feed to scrubber shown in Fig. 1 |
| S2 | Bottom of scrubber shown in Fig. 1 |
| S3 | Off gas outlet from the scrubber shown in Fig. 1 |
| T | Temperature (K) |
| To | Initial temperature (K) |
| V | Volume of reactor ($m^3$) |
| W | Bottom stream of DIH shown in Fig. 1 |
| x | reaction conversion |
| Zt | Length of reactor (Cm) |
| ε | Voidage |

## References

[1]. H. Weyda and E. Koehler, "Modern Refining Concepts - an Update on Naphtha-Isomerization to Modern Gasoline Manufacture". Catalyst Today, **81**, (2003).

[2]. Hallo., J. Hancsok and D. Kallo, "Kinetics of hydroisomerization of C5-C7 alkanes and their mixtures over platinum containing mordenite". Applied Catalyst. A:General, **229**, 93, (2002).

[3]. W. McCabe, J. Smith, and P. Harriott, "Unit Operations of Chemical Engineering",7th ed.. McGraw Hill., (2004).

[4]. G. Valavarasua and B. Sairamb "Light Naphtha Isomerization Process: A Review". Petroleum Science and Technology , **31**, 6, (2013).

[5]. Z. Rui, M. Xianghai, L. Zhichang, M. Jiaying and X. Chunming ,"Isomerization of n-Pentane Catalyzed by Acidic Chloroaluminate Ionic Liquids". Industrial and Engineering Chemistry Research, **47**, 21, pp 8205–8210, (2008).

[6]. T. Bui., V. Nguyen and A. Akhmetov, "A Comparative Analysis of the Various Schemes Isomerization of Pentane-Hexane Fraction". Refining and Petrochemicals, **2**, pp.22-25, (2008).

[7]. Cusher N., UOP Penex Process in: R. A. Mayers, Handbook of petroleum refining processes, McGraw-Hill, 9.16-9.27, (2004).

[8]. Elena, E. Alexander, S. Shilov,V. Dmitrievich and Z. Gennady, "Chemical And Biological Kinetics New Horizons: Vol. 1, Chemical Kinetics 1st Edition", 1991.

[9]. G. James, "The Chemistry and Technology of Petroleum", 4th ed. CRC Press, 2006.

[10]. M. Marios, G. Denis,G. Pierre, and S. Daniel," Single-Event Microkinetic Model for Long-Chain Paraffin Hydrocracking and Hydroisomerization on an Amorphous Pt/SiOR2R AlR2ROR3R Catalyst", India Engineering Chemical Research, **48**, pp. 3284–3292, (2009).

# Statistical Optimization for Biobutanol Production by *Clostridium acetobutylicum* ATCC 824 from Oil Palm Frond (OPF) Juice Using Response Surface Methodology

*Nur Syazana* Muhamad Nasrah[1], *Mior Ahmad Khushairi* Mohd Zahari[1,*], *Nasratun* Masngut[1] and *Hidayah* Ariffin[2]

[1]Faculty of Chemical and Natural Resources Engineering, Universiti Malaysia Pahang, 26300 Gambang, Pahang
[2] Faculty of Biotechnology and Biomelucular Sciences, Universiti Putra Malaysia, 43400 Serdang, Selangor.

**Abstract.** The interaction between incubation temperature, yeast extract concentration and inoculum size was investigated to optimize critical environmental parameters for production of biobutanol from oil palm frond (OPF) juice by *Clostridium acetobutylicum* ATCC 824 using response surface methodology (RSM). A central composite design (CCD) was applied as the experimental design and a polynomial regression model with quadratic term was used to analyse the experimental data using analysis of variance (ANOVA). ANOVA analysis showed that the model was very significant ($p < 0.0001$) for the biobutanol production. The incubation temperature, yeast extract concentration and inoculum size showed significant value at $p < 0.005$. The results of optimization process showed that a maximum biobutanol production was obtained under the condition of temperature 37 °C, yeast extract concentration 5.5 g/L and inoculum size 10%. Under these optimized conditions, the highest biobutanol yield was 0.3054 g/g after 144 hours of incubation period. The model was validated by applying the optimized conditions and 0.2992 g/g biobutanol yield was obtained. These experimental findings were in close agreement with the model prediction, with a difference of only 9.76%.

**Keywords**: Biobutanol, Response Surface Methodology, Central Composite Design, Oil Palm Frond Juice, *Clostridium acetobutylicum* (ATCC 824)

## 1 Introduction

Biofuels production has gained interest as an alternative supplies due to the expected rising demand of oil because of the depletion of fossil fuels and increasing oil prices. Biobutanol has been growing more popularity to be one of the alternative biomass fuels since it has many advantageous properties. It has a high energy content, low vapor pressure, less corrosive, can be used directly or blended with gasoline, can be shipped through existing pipelines and can be used in existing engines without any modifications [1, 2].

At the moment, agricultural products and wastes such as sugar canes, cassavas, rice straws, corncobs, corn, wheat straw, sweet sorghum, sweet cheese whey, rice bran, date fruit, oil palm trunk sap, and palm oil mill effluent are used as raw materials for the production of butanol [3-14]. Therefore, lignocelluloses, substances commonly found in agricultural wastes, appear to be a feasible resource compare to agricultural products which are used as foods. Since oil palm is one of industrial crops that have been planted extensively throughout the Malaysia, it leads to the increasing amount of palm oil wastes. MPOC (2010) reported that oil palm frond (OPF) is the most generated biomass wastes with 83 million tonnes (wet weight) annually [15]. OPF is available daily because it is obtained during pruning for harvesting fresh fruit bunch (FFB). As for the management of the OPF waste, the pruned fronds are just left in the plantation to be used for nutrient recycling and soil conservation. However, OPF does not contain high metal contents, but it contains high carbohydrates in the form of simple sugar which can be utilized into biobutanol [16]. The renewable sugars from OPF can simply obtained by pressing the fresh OPF to get the juice. The sugars content reported up to 76.09 g/L [17].

Therefore, in this study, OPF juice was used as fermentation feedstock by *Clostridium acetobutylicum ATCC 824* for production biobutanol. The different strains of *Clostridium* were used for the production of biobutanol from different lignocellulosic biomass such as *C. saccharoperbutylicum,* and *C. beijerinckii.* However, *C. acetobutylicum* has been chosen in this study to optimize the environmental parameters of the fermentation process for the biobutanol production.

Response surface methodology (RSM) is one of the most useful statistical optimization tools in biological and chemical process. RSM has been increasingly used for various phases of an optimization process in fermentation. RSM is a collection of statistical and mathematical techniques useful for designing experiments, developing models and evaluating the effects of variables in which a response of interest is influenced by several variables and the objective is to optimize this response. Basically, RSM consists of central composite design (CCD), Box-Behnken design, one factor design, D-optimal design, user-defined design and the historical data design. The most used statistical methods are CCD and Box-Behnken design. For one numeric variable, CCD has 5 levels ($-\alpha$, $-1$, 0, +1, $+\alpha$) whereas the Box- Behnken design only has 3 levels ($-1$, 0, +1). RSM also provides an experimental model that predicts the correlation and interaction between a set of experimental variables and observed results, and subsequently provides optimized conditions [18, 19].

The current research aimed to optimize biobutanol production using RSM with CCD by investigating the

* Corresponding author: ahmadkhushairi@ump.edu.my

effect of inoculum size, incubation temperature, and yeast extract concentration by *C. acetobutylicum* ATCC 824 from OPF juice.

## 2 Material and methods

### 2.1 Microorganism and culture media

*Clostridium acetobutylicum* ATCC 824 was obtained from American Type Culture Collection (ATCC) in freeze-dried form and maintained in 10% glycerol stock at -80°C. Spore suspension was prepared by transferring 1 mL of glycerol stock into 90 mL RCM for 3 days at 37°C under anaerobic conditions. Spores were heat shocked for 90 s in 90°C water bath and incubated at 37°C under static condition for 18-20 h [7]. The inoculum was deemed to be ready for fermentation when the optical density was between 1.5 and 2.0.

### 2.2 OPF juice preparation

The oil palm frond (OPF) was obtained from oil palm plantation at Felda Lepar Hilir, Kuantan, Pahang. Malaysia. The OPF was cut into small size and the juice was prepared by pressing the fresh OPF (without leaves) using sugarcane pressing machine by following the previous method described earlier by Zahari et al [16]. 72 mL of OPF juice was then distributed into 125 ml serum bottles, sparged with oxygen-free nitrogen gas, sealed and sterilized at 121°C for 15 minutes. It was then mixed with synthetic P2 medium in known proportion to prepare the final fermentation medium. Synthetic P2 medium had the following composition (in g/L): yeast extract, 5; $KH_2PO_4$, 0.5; $K_2HPO_4$, 0.5; para aminobenzoic acid, 0.001; thiamin, 0.001; biotin, $1x10^{-5}$; $MgSO_4.7H_2O$, 0.2; $MnSO_4.7H_2O$, 0.01; $FeSO_4.7H_2O$, 0.01; NaCl, 0.01; and ammonium acetate, 2.2 [7].

### 2.3 Fermentation

The fermentation was conducted in 125 ml serum bottle containing 100 mL final volume of the OPF juice, P2 medium and inoculum with different properties depending on the runs based on Table 2, flushed with $N_2$ gas and sterilized at 121°C for 15 min. The total sugars concentration and initial pH conditions were fix with 50g/L and pH 6. Initial pH of the fermentation broth was adjusted using 2 M NaOH or HCl. 15 ml of sample were withdrawn at 0 and 144 hr for analysis of acetone, butanol, ethanol (ABE), organic acids and sugar. The samples were centrifuged at 8,500 rpm, 4°C for 20 min before analysis.

### 2.4 Experimental design and statistical analysis

A total of twenty (20) experiments were performed according to the range and levels of independent variables, inoculum size, incubation temperature and yeast extract concentration studied for biobutanol production as stated in Table 1. Table 3 shows the experimental design and the results of the response variable studied. The experimental design and analysis of data were done using Design Expert version 7.1 (State-Ease, Inc., Minneapolis, MN).

**Table 1.** Experimental range and levels of the independent variables.

| Independent variables | Range and levels | | | | |
|---|---|---|---|---|---|
| | -α | -1 | 0 | +1 | +α |
| Temperature (°C) | 31 | 34 | 37 | 40 | 43 |
| Yeast extract (g/L) | 1.5 | 3.5 | 5.5 | 7.5 | 9.5 |
| Inoculum size (%) | 6 | 8 | 10 | 12 | 14 |

All experiments were performed in triplicates and the results were recorded as mean values of the biobutanol yield. The biobutanol yield can be expressed as a function of independent variables by a second order polynomial equation:

$$Y = \beta_0 + \sum \beta_0 x_j + \sum \beta_{jj} x_j^2 + \sum \beta_{jk} x_j x_k \quad (1)$$

where, Y is the response (biobutanol yield), $\beta_0$ $\beta_j$, $\beta_{jj}$ and $\beta_{jk}$ represent the regression coefficient for intercept, linear, quadratic and interaction terms, respectively. Adequacy of the developed model was evaluated by using analysis of variance (ANOVA) technique. The significant terms in the model equation were determined with the help of p-value and F-value and adequacy of model fitting was evaluated by the coefficient of determination $R^2$ and the lack of fit test. Response surfaces and optimum conditions for biobutanol production were obtained through this model.

### 2.5 Analytical procedures

The samples of fermentation were centrifuged at 10,000 rpm and filtered through 0.02μm nylon syringe filter before used for analysis. The products including solvents (acetone, butanol and ethanol) and acids (acetic acid and butyric acid) were analysed by gas chromatography (Agilent Technologies, 6890N network GC system). A microliter of the sample was injected into a 30 m x 0.32 mm x 0.5 μm HP-INNOWAX capillary column at 250°C. The column carrier was helium, 40 cm/sec, 11.7 psi (60°C) with 2.5 ml/min constant flow. A flame ionization detector (FID) at 275°C was used for signal detection. Glucose, sucrose, and fructose were determined by HPLC(Agilent; 1200) with Rezex ROA – organic acid H+ (8%) column (Phenomenex) (300 x 7.80 mm) with a flow rate of 0.5 ml/min and RI detector at 30°C. The mobile phase consists of 0.005 M $H_2SO_4$, recommended for the column used.

## 3 Results and discussion

### 3.1 Sugars composition of oil palm frond (OPF) juice

The total sugars concentration of OPF juice was determined by HPLC analysis as shown in Table 2. The total sugars concentration was 68.58 g/L. Glucose was found to be dominant sugar (48.19 g/L) followed by sucrose (11.91 g/L) and fructose (8.48 g/L). This result was almost similar with the findings reported by Zahari et al. [16]. They also found that glucose was the dominant sugar in OPF juice followed by sucrose and fructose.

**Table 2.** Amount of sugar contained in OPF juice

| OPF juice | Sugar (g/L) | | | Total sugar (g/L) |
|---|---|---|---|---|
| | Glucose | Fructose | Sucrose | |
| Sample 1 | 47.78 | 8.37 | 11.78 | 67.93 |
| Sample 2 | 48.60 | 8.58 | 12.04 | 69.22 |
| Average | 48.19 | 8.48 | 11.91 | 68.58 |

However, Zahari et al. [16] showed slightly higher total sugars concentration (76.09 g/L) compared to this study where sugars composition was reported as; 68.58 g/L. This might be due to the different location of collecting the OPF. The OPF in this work was obtained from the oil palm plantation in Gambang, Pahang, meanwhile Zahari et al. [16] obtained from Serdang, Selangor.

### 3.2 Model equation for biobutanol production

Inoculum size, temperature, and yeast extract concentration were optimized by central composite design (CCD) of RSM around the ranges as shown in Table 1. The mathematical relationship for the biobutanol production from OPF juice by *C. acetobutylicum* ATCC 824 was developed by considering three independent variables: temperature, yeast extract concentration and inoculum size; and one dependent variable: biobutanol yield obtained from per gram biobutanol concentration per sugars concentration consumption. The respective models are shown below:

$$Y = +0.27 +0.017 * A - 0.030 * B + 0.025 * C - 0.014 * A * B + 4.388E\text{-}003 * A * C - 0.012 * B * C - 0.046 * A^2 - 0.049 * B^2 - 0.037 * C^2 \quad (2)$$

where Y is biobutanol yield, A is inoculum size, B is temperature and C is yeast extract concentration. The quadratic model was selected to provide the best fit with the experimental results. Eq. (2) was formulated using the complete experimental design and response matrix through Table 3 and Eq. (1). Analysis of variance for response surface quadratic model and parametric conditions for validating the model of biobutanol production process were represented in Tables 4 and 5 respectively.

### 3.3 Analysis of variance (ANOVA) of biobutanol production model

The results obtained by CCD were analysed by standard analysis of variance (ANOVA) as in Table 4. Regression analysis indicated a coefficient of determination ($R^2$) of 0.9497 explaining that approximately 94.97% of the variability in the response, as well as a high value of the adjusted determination coefficient (adjusted $R^2$ = 0.9044). It indicated a good agreement between experimental and predicted values. From Table 4, it was clearly shows that the model was statistically significant regarding highest F-value of 20.97, very low probability value ($p < 0.0001$) and sum of square of 0.14. It was also observed that all the linear (A, B and C) and quadratic ($A^2$, $B^2$ and $C^2$) coefficients were significant on biobutanol production as the p-values calculated for this factor was less than 0.05. Therefore, changes in this parameter could significantly impact the biobutanol production from OPF juice fermentation. Meanwhile, all the interaction coefficients were insignificant; indicating that these terms had little impact on butanol production. The most significant effect of the linear coefficients is temperature (B), followed by yeast extract concentration (C) and inoculum size (A). Generally, the lack of fit p-value of 0.7229 implied that the lack of fit is not significant relative to the pure error. The non-significant lack of fit is positive because it demonstrates a good fit of the model to the data. A good fit means that the generated models adequately explained the variation of data. The actual vs. predicted and residuals vs. predicted responses are shown respectively in Fig. 1 and 2.

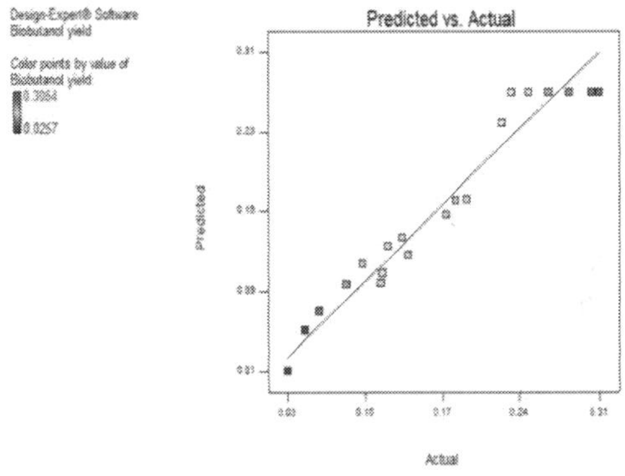


**Fig. 1.** Actual vs. predicted response of biobutanol production.

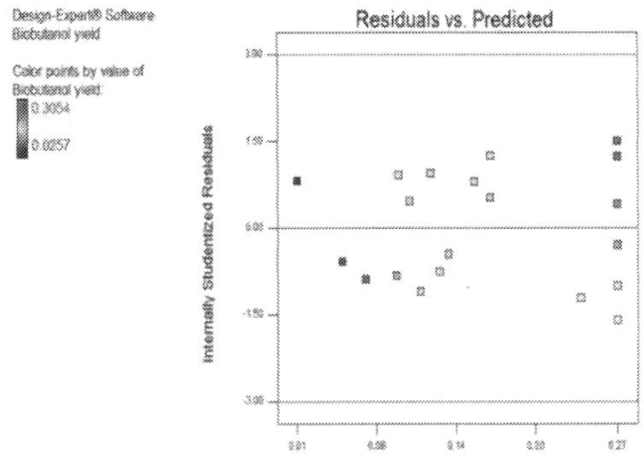


**Fig. 2.** Residual plot of biobutanol production.

**Table 3**. Experimental design matrix using RSM with CCD and response for biobutanol production.

| Run No. | Variables | | | Response | |
|---|---|---|---|---|---|
| | Inoculum size | Temperature | Yeast extract | Biobutanol yield (g/g) | |
| | (%) | (°C) | (g/L) | Experimental value | Predicted value |
| 1 | 8 | 34 | 3.5 | 0.1107 | 0.1000 |
| 2 | 12 | 34 | 3.5 | 0.1694 | 0.1600 |
| 3 | 8 | 40 | 3.5 | 0.1091 | 0.0940 |
| 4 | 12 | 40 | 3.5 | 0.0789 | 0.0930 |
| 5 | 8 | 34 | 7.5 | 0.1781 | 0.1700 |
| 6 | 12 | 34 | 7.5 | 0.2199 | 0.2400 |
| 7 | 8 | 40 | 7.5 | 0.0932 | 0.1100 |
| 8 | 12 | 40 | 7.5 | 0.1150 | 0.1300 |
| 9 | 6 | 37 | 5.5 | 0.0419 | 0.0510 |
| 10 | 14 | 37 | 5.5 | 0.1339 | 0.1200 |
| 11 | 10 | 31 | 5.5 | 0.1282 | 0.1400 |
| 12 | 10 | 43 | 5.5 | 0.0257 | 0.0130 |
| 13 | 10 | 37 | 1.5 | 0.0551 | 0.0690 |
| 14 | 10 | 37 | 9.5 | 0.1887 | 0.1700 |
| 15 | 10 | 37 | 5.5 | 0.2289 | 0.2700 |
| 16 | 10 | 37 | 5.5 | 0.3054 | 0.2700 |
| 17 | 10 | 37 | 5.5 | 0.2786 | 0.2700 |
| 18 | 10 | 37 | 5.5 | 0.2437 | 0.2700 |
| 19 | 10 | 37 | 5.5 | 0.2986 | 0.2700 |
| 20 | 10 | 37 | 5.5 | 0.2611 | 0.2700 |

**Table 4**. Analysis of variance (ANOVA) for the experimental results of CCD quadratic model.

| Source | Sum of square | DF | Mean square | F Value | Prob > F | |
|---|---|---|---|---|---|---|
| Model | 0.14 | 9 | 0.015 | 20.97 | < 0.0001 | Significant |
| A : Inoculum size | 4.76E-003 | 1 | 4.76E-003 | 6.59 | 0.0281 | |
| B : Temperature | 0.015 | 1 | 0.015 | 20.48 | 0.0011 | |
| C : Yeast extract | 0.010 | 1 | 0.010 | 14.19 | 0.0037 | |
| AB | 1.482E-003 | 1 | 1.482E-003 | 2.05 | 0.1828 | |
| AC | 1.540E-004 | 1 | 1.540E-004 | 0.21 | 0.6544 | |
| BC | 1.193E-003 | 1 | 1.193E-003 | 1.65 | 0.2280 | |
| $A^2$ | 0.053 | 1 | 0.053 | 72.92 | < 0.0001 | |
| $B^2$ | 0.059 | 1 | 0.059 | 81.89 | < 0.0001 | |
| $C^2$ | 0.035 | 1 | 0.035 | 48.36 | <0.0001 | |
| Residual | 7.234E-003 | 10 | 7.234E-004 | | | |
| Lack of fit | 2.631E-003 | 5 | 5.262E-004 | 0.57 | 0.7229 | Not significant |
| Pure error | 4.603E-003 | 5 | 9.206E-004 | | | |
| Cor total | 0.14 | 19 | | | | |

| | | |
|---|---|---|
| Standard deviation = 0.027 | PRESS = 0.028 | Pred $R^2$ = 0.8077 |
| Mean = 0.16 | $R^2$ = 0.9497 | Adequate precision = 13.409 |
| C. V. % = 16.48 | Adjusted $R^2$ = 0.9044 | |

The actual vs. predicted responses exhibit almost a linear relationship which predicts the reasonable precision of fitted empirical model. Residuals vs. predicted responses in Fig. 2 represent unusual structure and equally scattered points above and below the x-axis but all these points are between ±3.0 which imply adequacy and reliability of proposed models. Hence, it can be resolved that developed models are adequate in predicting biobutanol production from OPF juice by *C. acetobutylicum* ATCC 824.

* Corresponding author: ahmadkhushairi@ump.edu.my

DOI: 10.1051/matecconf/201711103001

### 3.4 Parametric interaction effect

Evaluations of the interactions between various factors using RSM quantifies in terms of three dimensional response surfaces and contour lines. Figure 3, 4 and 5 were plotted to demonstrate the interactions among the three factors and to estimate biobutanol yield over the independent variables. These plots demonstrate the effects of two factors on the response at a time and assist in arbitration of degree of parametric interactions on the desired responses. Three response surfaces were generated depending on three variables involved in the process. Fig. 3 shows three dimensional response surface relationships between inoculum size and temperature on the biobutanol production at the center level of the fermentation temperature. It can be observed that biobutanol yield increases with increasing of inoculum size and the temperature has no steeper effect on biobutanol production.

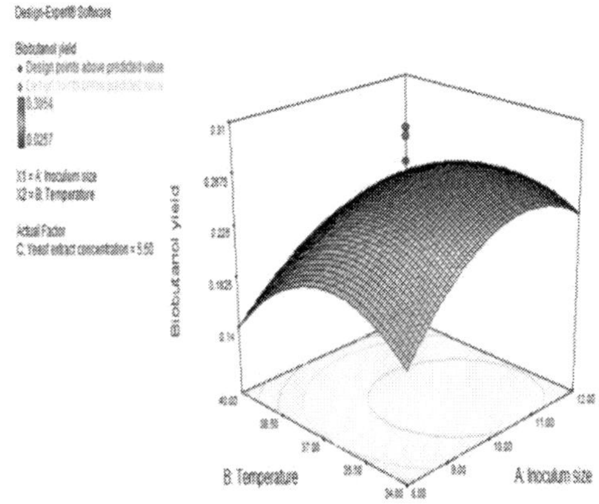


**Fig. 3.** Response surface and contour plot showing the effect of inoculum size and temperature on biobutanol yield.

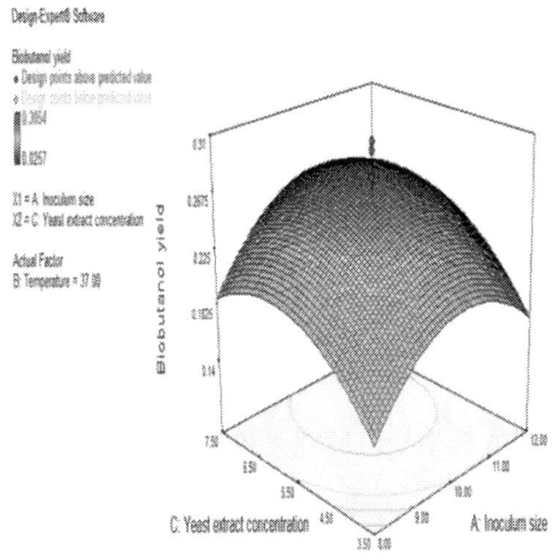


**Fig. 4.** Response surface and contour plot showing the effect of inoculum size and yeast extract concentration on biobutanol yield.

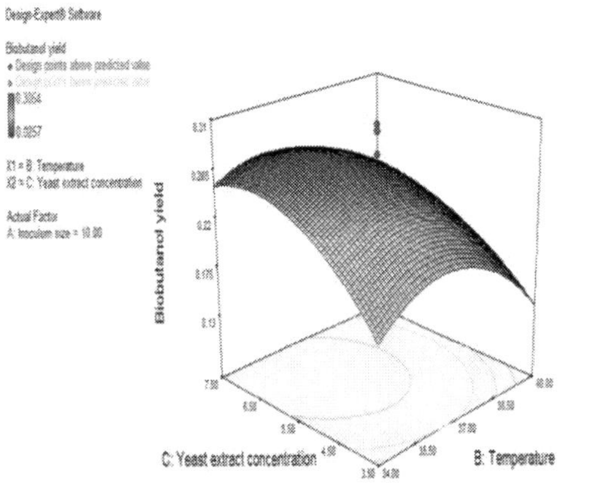


**Fig. 5.** Response surface and contour plot showing the effect of temperature and yeast extract concentration on biobutanol yield.

The three dimensional response surface relationships between inoculum size and yeast extract concentration at the center level of reaction time is illustrated in Fig. 4. It can be seen that biobutanol yield increases with increasing of inoculum size and yeast extract concentration. Biobutanol yield then reaches to a maximum point and again starts decrease with increasing the yeast extract concentration. Fig. 5 displays three dimensional response surface interactions between temperature and yeast extract concentration at the center level of biobutanol yield. It is evident that biobutanol production increasing with increasing of yeast extract concentration but it reduces with rising the temperature. The effects of temperature and yeast extract concentration on biobutanol yield are more significant than inoculum size. From Fig. 3 and 5, it can be observed that there were very low biobutanol productions at high temperature. This phenomenon happened could be attributed to the fact that high temperature inhibits enzyme catalysed reactions that are essential for cellular metabolism of acid and solvent production in ABE fermentation [21].

### 3.5 Optimization of biobutanol production

The model generated from experimental results (Table 3) represented maximum biobutanol yield is at the center point conditions with temperature 37°C, yeast extract concentration 5.5 g/L and inoculum size 10%. At these conditions, the biobutanol yield is up to 0.3054 g/g.

The optimum inoculums size is varied depending upon the microorganism and substrate used. A study was found by Al-Shorgani et al. (2015), who observed that the maximum butanol production by *C. saccharoperbutylacetonicum* N1-4 (ATCC 13564) from palm oil mill effluent in ABE fermentation is at an optimum inoculum size of 15% [20]. On the other hand, the production of butanol by *C. acetobutylicum* MTCC 481 from rice straw hydrolysate was investigated by Ranjan et al. (2013) and they found that an inoculum size of 5% was the optimum inoculum size [21]. Besides that, studies performed by Razak et al. (2013) revealed that the maximum butanol production by *C. acetobutylicum* from oil palm decanter cake hydrolysate in ABE fermentation was achieved at an optimum inoculum size of 16.2% (v/v) [22]. The improving butanol production with the increasing inoculum size due to the fact that is because of an increase in the inoculation of Clostridial cells decreases the lag phase of microbial cell growth. This condition improves Clostridial growth resulting in the enhancement of solventogenesis phase and also the butanol production [23]. As can be seen in Fig. 3 and 4, increasing the inoculum size higher than the critical level of 10% did not show an improvement in butanol production. This phenomenon can be related to the fact that an increase in the inoculum size more than the critical level (optimum value) had no considerable effects on lag phase of microbial growth and cell activity, which resulted in the production of biobutanol [21].

The optimum incubation temperature in this study is observed at 37°C. A similar pattern was observed by Ni et al. (2013) who cultivated *Clostridium saccharobutylicum* DSM 13864 using cane molasses found the optimum temperature for maximum butanol synthesis was 37°C [24]. Studies performed by Sheng et al. (2011) and Khamaiseh et al. (2013) showed that the highest butanol production by *C. acetobutylicum* from corn straw hydrolysate and date fruit medium was detected at incubation temperature of 35°C [25, 26]. The optimum incubation temperature is varied depending upon the microorganism and substrate used.

Yeast extract, enriched with nitrogen sources as proteins and amino acids, is one of the common used nitrogen sources for cell culture and fermentation processes which can promote phase shift from acidegonesis to solventogenesis and indirectly enhance biobutanol production. It was reported that the addition of yeast extract could promote phase shift occurrence and improve fermentation by *C. acetobutylicum* ATCC 824 performances comprehensively when using cassava substrate as high as 80% of productivity [27]. In this study, the maximum biobutanol yield by *C. acetobutylicum* ATCC 824 from OPF juice was produced at 5.5 g/L yeast extract concentration (Table 3). This result similar to the study reported by Madihah et al. (2001) who observed 0.34 g/g biobutanol yield by *C.acetobutylicum* from gelatinized sago starch with additional 5 g/L of yeast extract in the substrate [28]. Ibrahim et al. (2012) also reported that the ABE production by *C. butyricum* EB6 using oil palm empty fruit bunch as alternative substrate with supplemented basal medium which consisted 6 g/L of yeast extract produce as high as 0.24 g/g of ABE yield [29]. Therefore, yeast extract can stimulate the growth of the cells, resulting in increasing the cells to produce the solvents. However, as Fig. 3 and 5, at too high yeast extract concentration, more than 5.5 g/L, the production seems to be decreasing. This is because of the extensive sugar consumption by the cells in the growth phase that make the reduction in cells to produce the solvents.

### 3.6 Validation of biobutanol optimization

To further validate the accuracy of RSM prediction, an experiment was performed under the optimal conditions obtained from Table 3. This validation was also used to verify the accuracy of the model. Validation was carried out under conditions as follows: temperature 37°C, the yeast extract concentration 5.5 g/L and the inoculum size 10%. Under the above optimized condition, the maximum production of biobutanol was estimated as 0.27 g/g. The results were verified by triplicate experiments and the maximum butanol yield was 0.3012 g/g was obtained from the validation test (Table 5). These experimental findings were in close agreement with the model prediction, with a difference of only 9.76%. This validated that the RSM approach was effective for optimizing the operational conditions for the ABE fermentation process.

**Table 5.** Model validation of the biobutanol production.

| Parameters | | Value |
|---|---|---|
| Inoculum size (%) | | 10 |
| Temperature (°C) | | 37 |
| Yeast extract (g/L) | | 5.5 |
| Biobutanol yield (g/g) | Predicted | 0.2700 |
| | Actual | 0.2992 |

## 4 Conclusions

Using RSM based on CCD model, optimum conditions for biobutanol production from OPF juice by *C. acetobutylicum* ATCC 824 were; inoculum size 10%, incubation temperature 37°C and yeast extract concentration 5.5 g/L. The interactions between the factors can be observed that biobutanol yield increases with increasing of temperature, inoculum size and yeast extract concentration and reach to the maximum point which are the optimum conditions for the fermentation process. The linear and quadratic effects of inoculum size, temperature and yeast extract concentration are significant. The maximum biobutanol production observed was 0.2992 g/g of biobutanol yield.

The authors are grateful to Universiti Malaysia Pahang for financial support (grant numbers RDU 140139).

## References

1. P. Durre, Fermentative production of butanol – the academic perspective, Current Opinion in Biotechnology **22**, 331-336 (2011)
2. S. Y. Lee, J. H. Park, S. H. Jang, L. K. Nielsen, J. Kim, K. S. Jung, Fermentative butanol production by clostridia, Biotechnology and Bioengineering **101**, 2 (2008)
3. W. Jiang, J. Zhao, Z. Wang, S. T. Yang, Stable high-titer n-butanol production from sucrose

and sugarcane by *Clostridium acetobutylicum* JB200 in repeated batch fermentation, Bioresource Technology **163**, 172-179 (2014)
4. Y. Gu, S. Hu, J. Chen, L. Shao, H. He, Y. Yang, S. Yang, W. Jiang, Ammonium acetate enhances solvent production by *Clostridium acetobutylicum* EA 2018 using cassava as a fermentation medium, J Ind Microbial Biotechnol **36**, 1225-1232 (2009)
5. L. D. Gottumukkala, B. Parameswaran, S. K. Valappil, K. Mathiyazhakan, A. Pandey, R. K. Sukumaran, Biobutanol production from rice straw by a non acetone producing *Clostridium sporogenes* BE01, Bioresource Technology **145**, 182-187 (2013)
6. M. Kumar, D. Kumar, B. Singh, Utilization of agro residue corncob for production of acetone-butanol using *Clostridium acetobutylicum* and process optimization through RSM, J Microbial Biochem Technol **S8**, 005 (2014)
7. N. Qureshi, H. P. Blaschek, ABE production from corn: a recent economic evaluation, Journal of Industrial Microbiology & Biotechnology **27**, 292-297 (2001)
8. C. Bellido, M. L. Pinto, M. Coca, G. G. Benito, M. T. G. Cubero, Acetone-butanol-ethanol (ABE) production by *Clostridium beijerinckii* from wheat straw hydrolysates: Efficient use of penta and hexa carbohydrates **167**, 198-205 (2014)
9. B. Ndaba, I. Chiyanzu, S. Marx, Direct fermentation of sweet sorghum juice by *Clostridium acetobutylicum* and *Clostridium tetanomorphum* to produce bio-butanol and organic acids, Biofuel Research Journal **6**, 248-252 (2015)
10. M. Becerra, M. E. Cerdan, M. I. G. Siso, Biobutanol from cheese whey, Microbial Cell Factories **14**, 27 (2015)
11. N. K. N. Al-Shorgani, M. S. Khalil, W. M. W. Yusoff, Biobutanol production from rice bran and de-oiled rice bran by *Clostridium saccharoperbutylacetonicum* N1-4, Bioprocess Biosyst. Eng. **35**, 817-826 (2012)
12. E. I. Khamaiseh, M. S. Kalil, E. Ali, A. A. Hamid, W. M. W. Yusoff, Biobutanol production by *Clostridium acetobutylicum* NCIMB 13357 in modified medium using date fruit as a carbon source, Jordon International Energy Conference (2011)
13. I. Komonkiat, B. Cheirsilp, Felled oil palm trunk as a renewable source for biobutanol production by *Clostridium spp.*, Bioresource Technology **146**, 200-207 (2013)
14. M. S. Kalil, P. W. Kit, W. M. W. Yusoff, Y. Sadazo, R. A. Rahman, Direct fermentation of palm oil mill effluent to acetone-butanol-ethanol by solvent producing Clostridia, Pakistan Journal of Biological Sciences **6**, 14 (2003)
15. MPOC, Palm Oil: A Success Story in Green Technology Innovations. http://www.akademisains.gov.my/download/asmic/asmic2010/Plenary12.pdf (2011)
16. M. A. K. M. Zahari, S. S. S. Abdullah, A. M. Roslan, H. Ariffin, Y. Shirai, M. A. Hassan, Efficient utilization of oil palm frond for bio-based products and biorefinery, Journal of Cleaner Production **65**, 252-260 (2014)
17. M. A. K. M. Zahari, M. R. Zakaria, H. Ariffin, M. N. Mokhtar, J. Salihon, Y. Shirai, M. A. Hassan, Renewable sugars from oil palm frond juice as an alternative novel fermentation feedstock for value-added products, Bioresource Technology **110**, 566-571 (2012)
18. M. A. Bezerra, R. E. Santelli, E. P. Oliveira, L. S. Villar, L. A. Escaleira, Response surface methodolody (RSM) as a tool for optimization in analytical chemistry, Talanta **76**, 965-977 (2008)
19. D. Bas, I. H. Boyac, Modeling and optimization I: usability of response surface methodology, J. Food Eng. **78**, 836–845 (2007)
20. N. K. N. Al-Shorgani, H. Shukor, P. Abdeshahian, M. Y. M. Nazir, M. S. Kalil, A. A. Hamid, W. M. W. Yusoff, Process optimization of butanol production by *Clostridium saccharoperbutylacetonicum* N1-4 (ATCC 13564) using palm oil mill effluent in acetone-butanol-ethanol fermentation, Biocatalysis and Agricultural Biotechnology **4**, 244-249 (2015)
21. A. Ranjan, S. Khanna, V. S. Moholkar, Feasibility of rice straw as alternate substrate for biobutanol production, Applied Energy **103**, 32-38 (2013)
22. M. N. A. Razak, M. F. Ibrahim, P. L. Yee, M. A. Hassan, S. A. Aziz, Statistical optimization of biobutanol production from oil palm decanter cake hydrolysate by *Clostridium acetobutylicum* ATCC 824, BioResource **8**, 2 (2013)
23. H. Shukor, N. K. N. Al-Shorgani, P. Abdeshahian, A. A. Hamid, N. Anuar, N. A. Rahman, M. S. Kalil, Production of butanol by *Clostridium saccharoperbutylacetonicum* N1-4 from palm kernel cake in acetone-butanol-

ethanol fermentation using empirical model, Bioresource Technology **170**, 565-573 (2014)
24. Y. Ni, Z. Xia, Y. Wang, Z. Sun, Continuous butanol fermentation from inexpensive sugar-based feedstocks by *Clostridium saccharobutylicum* DSM 13864, Bioresource Technology **129**, 680-685 (2013)
25. L. Y. Sheng, W. Jing, W. X. Ming, S. X. Hong, Optimization of butanol production from corn straw hydrolysate by Clostridium acetobutylicum using response surface method, Chinese Sci. Bull. **56**, 1422-1428 (2011)
26. E. I. S. Khamaiseh, A. A. Hamid, W. M. W. Yusoff, M. S. Kalil, Effect of some environmental parameters on biobutanol production by Clostridium acetobutylicum NCIMB 13357 in date fruit medium, Pakistan Journal of Biotechnology Sciences **16**, 20 (2013)
27. X. Li, Z. Li, J. Zheng, Z. Shi, L. Li, Yeast extract promotes phase shift of bio-butanol fermentation by Clostridium acetobutylicum ATCC 824 using cassava as substrate, Bioresource Technology **125**, 43-51 (2012)

# Synthesis and application of dual functionalized task specific ionic liquid for bamboo dissolution

*Nor Shahroon* Hameed Sultan[1,*], *Lethesh* Kallidanthiyil Chellappan[2], *Wan Suzaini* Wan Hamzah[1], *Leveque* Jean Marc[3] and *Mohamed Ibrahim* Abdul Mutalib[1].
[1] *Centre of Research of Ionic liquid, University Teknologi Petronas, 32610 Seri Iskandar, Perak, Malaysia*
[2] *Centre for Biofuel & Biochemical, University Teknologi Petronas, 32610 Seri Iskandar, Perak, Malaysia*
[3] *Université de Savoie Mont-Blanc, 73376 Le Bourget du Lac cede*

**Abstract.** A new class of dual functionalized imidazolium based ionic liquid (IL) namely 3-(2-cyano-ethyl)-1-(2-ethoxy-ethyl)-3-imidazolium bromide [CNEIM][Br], was synthesized and characterized to study their potential in bamboo dissolution. The chemical structure for the IL was characterized using NMR ($^{1}H$ and $^{13}C$). Thermal properties, surface morphology and functional group of the native bamboo and IL treated bamboo were analyzed by Thermal Gravimetric Anaylysis (TGA), Scanning Electron Microscopy (SEM) and Fourier Transform Infrared Spectroscopy (FTIR) respectively. The new IL was able to dissolve up to 5wt% of bamboo biomass within 48 hours and 100°C.

## 1 Introduction

Presently, the world facing of deficiency of non-renewable resources such as coal, petroleum and natural gas. Hence, it is necessary to utilize renewable resources to fulfill the energy needs of our society [1]. Cellulosic biomass is one of the possible alternatives to petroleum-based feedstocks for fuels and chemicals because it is renewable, abundant, and inexpensive [2]. Cellulose comprise of recalcitrant polymer of 1–4 linked β-d-glucose repetitive units with inter- and intramolecular hydrogen bonding [3]. It is very hard to convert cellulose into biofuels and biochemical because its recalcitrant nature due to the presence of inter and intramoleculer H-bonding network. In recent years, significant efforts have been made to develop new techniques for the conversion of cellulose into biofuels and platform chemicals [4, 5]. Numerous pre-treatment procedures such as alkaline and acid pretreatment, ammonia fiber explosion and steam treatment have been established for the utilization of biomass [6]. However, an effective, inexpensive and sustainable process still needs to be developed. Recently, pretreatment of biomass using ionic liquids (ILs) has appeared as an alternative to the conventional methods due to its high capacity and capability.

Ionic liquids (ILs) are molten salts that are entirely ionic in nature, containing both cationic and anionic species and mostly having a melting point below 100 °C [7, 8]. The choice of cations and anions has a large influence on their properties. Hence, by changing the cation or anion of the ILs, their physical properties can be tailored per the requirements of a process. These properties include melting point, density, viscosity, solubility and hydrophobicity. ILs is unique due to their certain remarkable properties which differentiate them from conventional organic solvents. They have a very low vapor pressure, wide liquids range, flammability and recyclability property [9, 10]. ILs has been recently investigated for a wide range of potential applications such as catalytic synthesis [11], nanotechnology [12] and electrolytes for dye-sensitized solar cell [13]. Functionalized ILs may be defined as ionic liquids in which a functional group is covalently attached either to the cation or to the anion or even to both. The advantage of introducing a functional group into ILs is the fine-tuning of their properties for a particular application.

The types of anions and cations and functional groups attached to the cations/anions can affects the biomass dissolution potential of ILs. Imidazolium-based ILs has been used widely for dissolution of hardwoods and softwoods [14]. Some of the ILs cation increase the

* Corresponding author: norshahroon23@gmail.com

cellulose dissolution by forming hydrogen bonds with the hydroxyl and ether oxygen of cellulose [3]. Besides, cyano (nitrile) group shows better dissolution ability compared to allyl group because of its smaller size and electron withdrawing nature [10].

The attachment of ether / PEG functional group to the side chain of imidazole cation would results in lowering the viscosity of IL compared to simple alkyl side chain [15]. Researcher has shown that hydrogen bond accepting ability of the IL anions related to its ability to dissolve cellulose. ILs containing anions with H-bond acepting ability such as chloride, formate, acetate or alkylphosphonate showed high biomass dissolution potential [16].

The present work focus on designing a new functionalized ILs containing nitrile and ether group for the dissolution of bamboo biomass. Bamboo was selected for this research due to the widely available in Malaysia. Besides, bamboo is very cheap, renewable and sustainability properties [17]. In this work, 3-(2-cyano-ethyl)-1-(2-ethoxy-ethyl)-3-imidazolium bromide was successfully synthesized and characterized. The synthesized ionic liquid was finally tested for the biomass dissolution process.

## 2 Experimental

### 2.1 Materials and Chemicals

Acetone, 99%, 2-bromoethyl ether, 90%, and imidazole, 99%, were acquired from Sigma Aldrich. Chloroform, 99.8% acyrolinitrile, 99%, and ethyl acetate, 99.8% was purchased from Fisher Scientific. Bamboo (Gigantochloa scortechini) was gained from local market in Perak, Malaysia. The bamboo size was standardise using Restesh Test sieve less than 500μm. All chemicals (analytical reagent grade) were commercially obtained and used without further purification unless otherwise stated.

### 2.1 Synthesis of Propionitrile imidazole

Imidazole (0.74mol, 50gm) was dissolved in chloroform (50 mL) in a round bottom flask and acrylonitrile (0.74mol, 38 gm) was added dropwise and stirred for 48 hours at 65°C. The viscous liquid was cooled and dried using rotary evaporator. Yield about (68g) 77%.

### 2.2 Synthesis of 3-(2-cyano-ethyl)-1-(2-ethoxy-ethyl)-3-imidazolium bromide [CNEIM][Br]

The propionitrile imidazole (0.41mol, 50gm) was added to 2-bromoethyl ether (0.41mol, 63gm) and stirred at 85°C for 72 hours. The viscous ionic liquid was washed with ethyl acetate (4×20mL). The residue is dried using rotary evaporator. Yield about (90g) 79%.

### 2.3 Characterization for IL

$^{1}H$ and $^{13}C$ NMR spectrum of the IL are recorded on a Bruker Avance 500 spectrometer. $^{1}H$-NMR (500Mhz, DMSO-d6): δ= 9.297(s,1H), 7.887(d,2H), 4.566(t,2H), 4.415(t,2H), 3.740(m,2H), 3.473(t,2H), 3.268(t,2H), 1.093(t,3H). $^{13}C$-NMR (125.77Mhz, DMSO-d6) : δ=137.28, 123.65, 122.76, 118.16, 67.86, 66.05, 49.61, 44.93, 19.13, 15.30.

Thermogravimetric analysis is using Perkin Elmer STA600 analyser is carried out on an average of 5-10 mg of samples. The experiment is carry out at a constant heating rate of 10 °C/min from 50 to 800 °C in inert atmosphere (pure nitrogen).

### 2.4 Bamboo dissolution

The bamboo (2wt %, 3wt %, 4wt %, 5wt %) was mixed with 3 gram of IL in 12mL vial. The mixture was then stirred at 100°C for 48 hours using hotplate. The temperature of bamboo dissolution experiment was controlled by using oil bath. The solubility of bamboo in IL was checked visually after 48 hours by using a handheld microscope, Dino-lite Pro AM413T. After complete dissolution of bamboo, acetone (5×10 mL) was added to regenerated cellulose. The regenerated cellulose was filtered and washed with water to remove any trace of ILs. The cellulose was dried in an oven at 60°C for 48 hours.

### 2.5 Characterization for regenerated bamboo

The spectra of the samples was recorded on a FTIR spectrometer model Nicolet iS5 which is equipped with an attenuated total reflectance (ATR) unit (Thermo Scientific). The spectra was conducted over the range of 4000 to 500 $cm^{-1}$.

The morphology of untreated and treated bamboo was obtained using a Hitachi TM 3030 model scanning electron microscope (SEM) operated at 15 kV accelerating voltage. The free and fracture surfaces of the film are sputtered with gold, and then observed and photographed. The samples are sputter coated with gold before analysis to avoid electrical charging of the samples during micrographs by using Quorum SC 7620 sputter coater. The TGA of untreated and treated bamboo was carry out at a constant heating rate of 10 °C/min from 50 to 800 °C in inert atmosphere (pure nitrogen).

## 3 Result and discussion

### 3.1 Thermal properties of [CNEIM][Br]

DOI: 10.1051/matecconf/201711103002

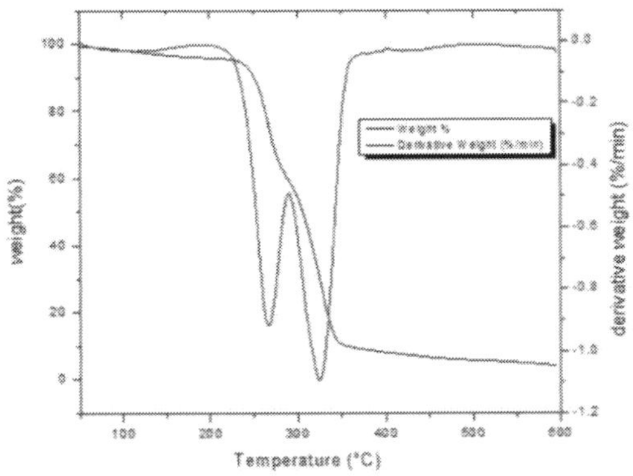


**Fig. 1.** TGA for [CNEIM][Br]

The TGA profile displaying the thermal decomposition of [CNEIM][Br] is shown in Fig. 1. As shown in Fig. 1, there are two thermal decomposition temperature noticed. The decomposition temperature of [CNEIM][Br] IL are at 248.89 $^{O}$C and 306.50 $^{O}$C. According to, Liang et al [18] the two-thermal decomposition is due to the anion. This indicates that the IL is highly stable and suitable for reaction that require high or extreme temperature.

### 3.1 Bamboo Dissolution

The dissolution of bamboo in [CNEIM][Br] IL was studied at different condition. Each mixture of bamboo and IL was heated at 100°C for 48 hour with different loading of bamboo from 1 wt% to 5wt%. After heating for 48 hour at 100°C and under vigorous stirring. The microscopic visual changes of bamboo before and after the dissolution process was observed and showed in **Fig 2.** From the microscopic visual of bamboo dissolution, it can see that there was brownish fiber. Meanwhile after the dissolution it can see that a clear liquid. Thus, it shows that [CNEIM][Br] IL could dissolve bamboo at 100 $^{O}$C within 48 hr. Swatloski et al. [20] conduct a study on cellulose dissolution using 1-butyl-3-methylimidazolium ([Bmim]+) cation with different combination of anion which are $Cl^-$, $PF_6^-$, $Br^-$, $SCN^-$, and $BF_4^-$. It shows that only $Cl^-$, $SCN^-$ and $Br^-$ anion containing ionic liquids could dissolve cellulose at 100 to 110 $^{O}$C. The dissolved biomass was precipitated by adding acetone and the biomass was filtered using filter paper. The biomass was washed with water till the complete removal of ILs. The ILs was regenerated by evaporating the solvent under vacuum. The regenerated ionic liquid was successfully characterized using $^1$HNMR and no changes in the structure is observed **Fig. 3**.

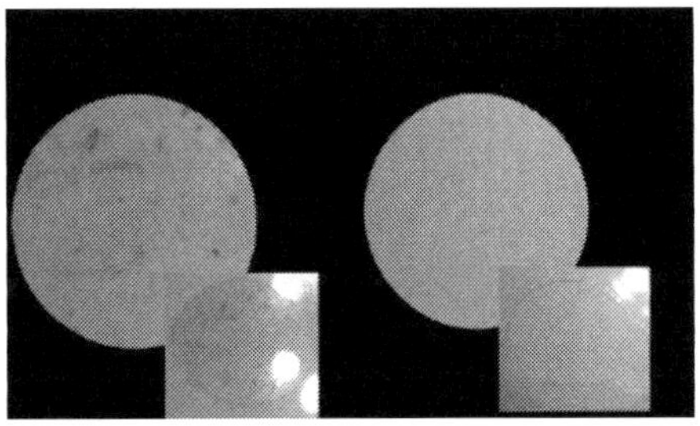

**Fig. 2.** Microscopic visual changes of bamboo before and after dissolution process.

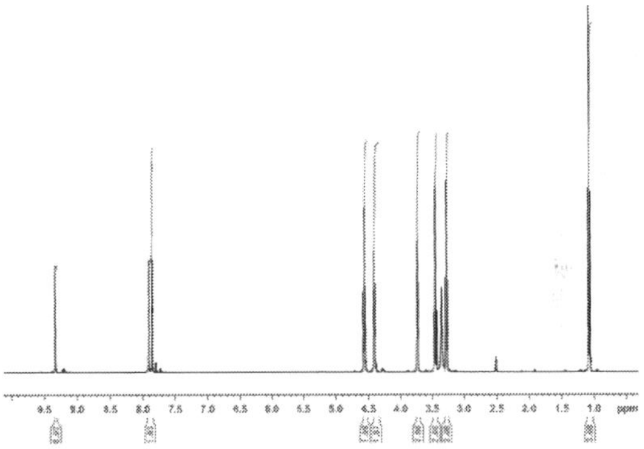

**Fig. 3.** $^1$H-NMR for regenerated [CNEIM][Br]

The capability of the IL to dissolve bamboo is confirmed by studying the morphology of bamboo after the dissolution (**Fig. 4**). The IL treated bamboo showed a significant amendment in morphology compared to the untreated bamboo. The untreated bamboo seemed to have a circular [21], rough surface and compact. Similar observation was reported by using oil palm biomass treated with 1-ethyl-3-methylimidazolium-diethyl phosphate IL. According to Financie et al. [22] it might be due to the lignin covered on cellulose and hemicellulose. The treated bamboo showed modification in the fiber where there is a crack and curly structure [23, 24]. This shows that the [CNEIM][Br] IL was completely disrupting the cell wall of bamboo.

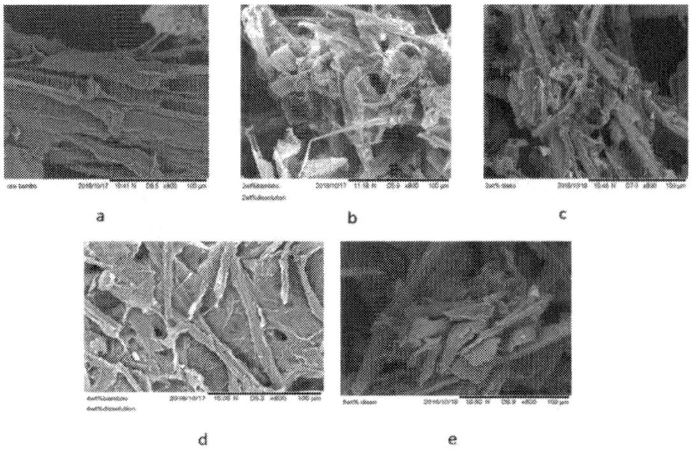


**Fig. 4.** Morphology of (a) untreated bamboo (b) treated bamboo at 2wt% (c) treated bamboo at 3wt% (d) treated bamboo at 4wt% (e) treated bamboo at 5wt%.

FTIR analysis was conducted to analyze any structural changes on bamboo during IL treatment. The FTIR spectra of the untreated and treated bamboo are shown in **Fig. 5**. The FTIR spectrum of treated and untreated bamboo is found to be similar. The presence of hydroxyl, carbonyl, methoxyl and carboxyl functional groups have been reported for identification of lignin [24]. A broad band of hydroxyl and methoxyl groups are observed around 3336 $cm^{-1}$ and 2895 $cm^{-1}$ respectively. The strong intensity bands at 1602 to 1504 $cm^{-1}$ was attribute to the aromatic skeleton from lignin component [ 17, 25, 26, 27]. The peak value at 1240 $cm^{-1}$ is due to phenol ether bonds of lignin while 1158 $cm^{-1}$ and 1031$cm^{-1}$ are due to structural contribution of cellulose and hemicellulose [26]. A band of range (896 - 561) $cm^{-1}$ represent the alkane group (CH-CH) and the deformation of cellulose [26].

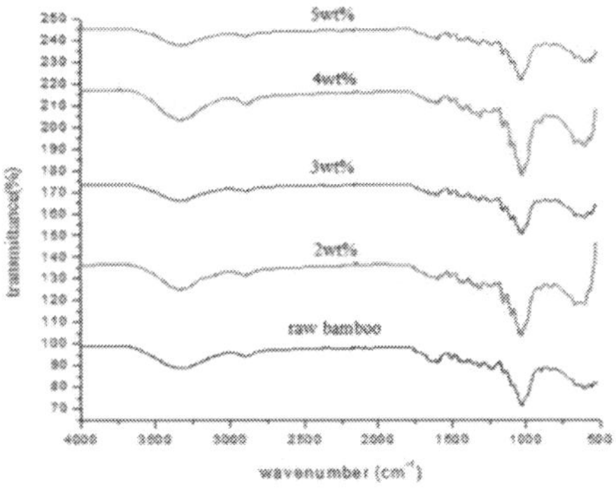


**Fig. 5.** FTIR spectra for untreated and treated bamboo

The thermal stability of treated and untreated bamboo was determined using TGA analysis. The thermogram of treated and untreated bamboo was shown in **Fig. 6**. The results showed the thermal stability of the bamboo reduce after the IL pretreatment. The decomposition of the treated and untreated bamboo could be divided into three stage. The first stage involved moisture evaporation at around 150°C, measured at 4wt% loading. This is followed by the decomposition of the hemicellulose at around 300°C [28, 29]. Hemicellulose can be categorized as the most reactive and easily degrade constituent compared to cellulose and lignin. In the third stage, which covers the temperature range from 350 to 400°C, represents the thermal degradation of the cellulose. This is due to higher amount of hydrogen bonds existed between cellulose chains which leads to more ordered and packed cellulose regions [30]. The temperature range between 300 and 600°C is due to the weight loss stage of lignin composition in bamboo [31].

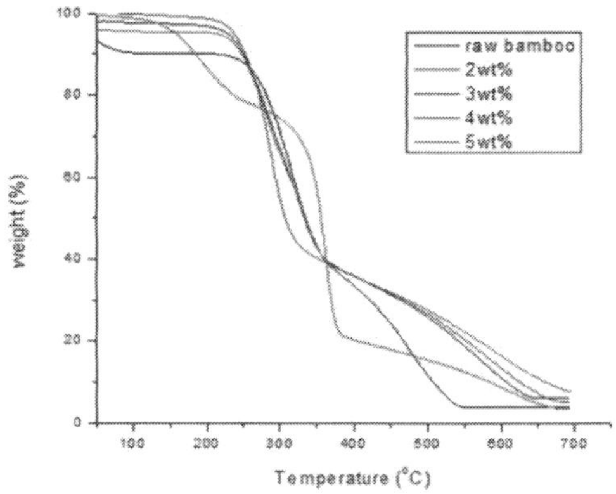


**Fig. 6.** TGA curve untreated and treated bamboo.

* Corresponding author: norshahroon23@gmail.com

**Table 1.** Thermal decomposition of untreated and treated bamboo

| Biomass loading | Thermal decomposition (°C) | |
|---|---|---|
| Raw bamboo | 269.25 | 469.73 |
| 2wt% | 244.18 | 470.50 |
| 3wt% | 238.93 | 491.63 |
| 4wt% | 319.72 | 493.94 |
| 5wt% | 220.25 | 469.73 |

**Fig.** 7 showed the differential thermal gravimetric (DTG) curve for untreated and treated bamboo. DTG shows that at 4wt% of bamboo dissolution shows a high thermal stability compared to untreated bamboo, 2wt%, 3wt% and 5wt%. The present of two distinct peaks in DTG is due to the degradation of hemicellulose followed by cellulose [31]. The thermal degradation between 400 and 600°C may be due to degradation of lignin [32].

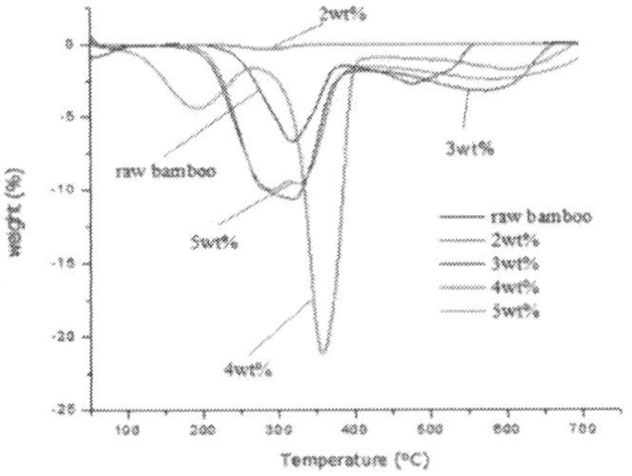


**Fig. 7.** DTG curve for untreated and treated bamboo.

## 4 Conclusion

Synthesizing and characterizing of [CNEIM][Br] has been successfully achieved. Based on the analyzed data, this type of IL with nitrile and ether functionalized group, is proven to have the potential and capability of dissolving the bamboo and the dissolution capacity can reach up to 5wt%. There is no degradation of ILs was observed after the dissolution process. Moreover, significant changes was occurred after pretreatment with IL.

The authors would like to thank to center of research in ionic liquid and also FRGS grant for financial support.

## References

[1] Beltramino, F., Roncero, M. B., Vidal, T., Torres, A. L., & Valls, C. (2015). Increasing yield of nanocrystalline cellulose preparation process by a cellulase pretreatment. Bioresour. Technol., **192**, 574-581.

[2] Minnick, D. L., & Scurto, A. M. (2015). Reversible and non-reactive cellulose separations from ionic liquid mixtures with compressed carbon dioxide. *Chem. Commun.*, **51(63)**, 12649-12652.

[3] Ibrahim, F., Moniruzzaman, M., Yusup, S., & Uemura, Y. (2015). Dissolution of cellulose with ionic liquid in pressurized cell. J. Mol. Liq., **211**, 370-372.

[4] Christensen, C. H., Rass-Hansen, J., Marsden, C. C., Taarning, E., & Egeblad, K. (2008). The Renewable Chemicals Industry. ChemSusChem, **1(4)**, 283-289.

[5] Gallezot, P. (2008). Catalytic Conversion of Biomass: Challenges and Issues. ChemSusChem, **1(8-9)**, 734-737.

[6] Xu, J., Xiong, P., & He, B. (2016). Advances in improving the performance of cellulase in ionic liquids for lignocellulose biorefinery. Bioresour. Technol, **200**, 961-970.

[7] An, Y., Zong, M., Wu, H., & Li, N. (2015). Pretreatment of lignocellulosic biomass with renewable cholinium ionic liquids: Biomass fractionation, enzymatic digestion and ionic liquid reuse. Bioresour.Technol, **192**, 165-171.

[8] Lee, H. V., Hamid, S. B., & Zain, S. K. (2014). Conversion of Lignocellulosic Biomass to Nanocellulose: Structure and Chemical Process. The Scientific World Journal, 2014, 1-20.

[9] Socha, A. M., Parthasarathi, R., Shi, J., Pattathil, S., Whyte, D., Bergeron, M., Singh, S. (2014). Efficient biomass pretreatment using ionic liquids derived from lignin and hemicellulose. PNAS, **111(35)**.

[10] Badgujar, K. C., & Bhanage, B. M. (2015). Factors governing dissolution process of lignocellulosic biomass in ionic liquid: Current status, overview and challenges. Bioresour. Technol., **178**, 2-18.

[11] Cocalia, V., Gutowski, K., & Rogers, R. (2006). The coordination chemistry of actinides in ionic liquids: A review of experiment and simulation. Coor. Chem. Rev., **250(7- 8)**, 755-764.

[12] Wang, P., Wenger, B., Humphry-Baker, R., Moser, J., Teuscher, J., Kantlehner, W,. Grätzel, M. (2005). Charge Separation and Efficient Light Energy Conversion in Sensitized Mesoscopic Solar Cells Based on Binary Ionic Liquids. J. Am. Chem. Soc., **127(18)**, 6850-6856.

[13] Wang, P., Zakeeruddin, S. M., Moser, J., Humphry-Baker, R., & Grätzel, M. (2004). A Solvent-Free, SeCN - /(SeCN) 3 - Based Ionic Liquid Electrolyte for High-Efficiency Dye-Sensitized Nanocrystalline Solar Cells. J. Am. Chem. Soc., **126(23)**, 7164-7165.

[14] Mora-Pale, M., Meli, L., Doherty, T. V., Linhardt, R. J., & Dordick, J. S. (2011). Room temperature ionic liquids as emerging solvents for the pretreatment of lignocellulosic biomass. Biotechnol. Bioeng., **108(6)**, 1229-1245.

[15] Kuhlmann, E., Himmler, S., Giebelhaus, H., & Wasserscheid, P. (2007). Imidazolium Dialkylphosphates — A Class of Versatile, Halogen-Free and Hydrolytically Stable Ionic Liquids. ChemInform, **38(30)**.

[16] Mood, S. H., Golfeshan, A. H., Tabatabaei, M., Jouzani, G. S., Najafi, G. H., Gholami, M., & Ardjmand, M. (2013). Lignocellulosic biomass to bioethanol, a

comprehensive review with a focus on pretreatment. Renewable Sustainable Energy Rev., **27**, 77-93.

[17] Bhat,I., Mustafa, M.T., Mohmod, A. L., Khalil., (2011) Spectroscopic, Thermal, And Anatomical Characterization Of Cultivated Bamboo (Gigantochloa Spp.) Bioresour. Technol., **6(2)**, 1752-1763.

[18] Liang, H.F, Li, G.J, Gou, G.J., & Wang, A.Q. (2013). Study of Preparation and Thermal Stability of Cyano-Functionalized Imidazolium Type Ionic Liquids., Asian J. Chem., **25(9)**, 4779-4782.

[19] Sun, N., Rahman, M., Qin, Y., Maxim, M. L., Rodríguez, H., Rogers, R. (2009) Green Chem.,**11**, 646–655.

[20] R. P. Swatloski, S. K. Spear, J. D. Holbrey, R. D. Rogers, (2002). J. Am. Chem. Soc., **124**, 4974–4975.

[21] Yang, Z., Xu, S., Ma, X., & Wang, S. (2008). Characterization and acetylation behavior of bamboo pulp. *Wood Sci And Technol*, ***42*(8)**, 621-632.

[22] Financie, R., Moniruzzaman, M., & Uemura, Y. (2016). Enhanced enzymatic delignification of oil palm biomass with ionic liquid pretreatment. Biochem. Eng. J., **110**, 1-7.

[23] Hernández-Mena L., Pecora A., Beraldo A., 2014, Slow pyrolysis of bamboo biomass: analysis of biochar properties, Chem. Eng. Trans, **37**, 115-120.

[24] Jiang, Z., Miao, J., Yu, Y., & Zhang, L. (2016). Effective Preparation of Bamboo Cellulose Fibers in Quaternary Ammonium/DMSO Solvent. Bioresources, **11(2).**

[25] Muhammad, N., Man, Z., Bustam, M. A., Mutalib, M. A., & Rafiq, S. (2013). Investigations of novel nitrile-based ionic liquids as pre-treatment solvent for extraction of lignin from bamboo biomass. J. Ind. Eng. Chem., **19(1)**, 207-214.

[26] Xu, G., Wang, L., Liu, J., & Wu, J. (2013). FTIR and XPS analysis of the changes in bamboo chemical structure decayed by white-rot and brown-rot fungi. Appl. Surf. Sci., **280**, 799-805.

[27] Liew, F. K., Hamdan, S., Rahman, M. R., Rusop, M., Lai, J. C., Hossen, M. F., & Rahman, M. M. (2015). Synthesis and Characterization of Cellulose from Green Bamboo by Chemical Treatment with Mechanical Process. J. Chem., 1-6.

[28] Zhang, X., Wang, F., & Keer, L. (2015). Influence of Surface Modification on the Microstructure and Thermo-Mechanical Properties of Bamboo Fibers. Materials, **8(10)**, 6597-6608.

[29] Wang, Q., Chen, Q., & Endo, T. (2015). Thermal Decomposition of Bamboo Phyllostachys Edulis Pretreated with Ionic Liquids-Water Mixtures. Green and Sustainable Chemistry GSC, **05(02)**, 55-62.

[30] Meenatchi, B., Renuga, V., & Manikandan, A. (2016). Cellulose dissolution and regeneration using various imidazolium based protic ionic liquids. J. Mol. Liq.

[31] Gordobil, O., Moriana, R., Zhang, L., Labidi, J., & Sevastyanova, O. (2016). Assesment of technical lignins for uses in biofuels and biomaterials: Structure-related properties, proximate analysis and chemical modification. Industrial Crops and Products, **83**, 155-165.

[32] Strydom, C., Sehume, T., Bunt, J., & van Dyk, J. (2015). The influence of selected biomass additions on the co-pyrolysis with an inertinite-rich medium rank C grade South African coal. J. South. Afr Inst. Min. Metallu., **115(8)**, 707-716.

# Synthesis, characterization and application of 1-(2-cyanoethyl)-3-(3-methoxypropane)imidazolium bromide for $CO_2$ capture

*Shantini* Ravichandar[1,*], *Cecilia Devi* Wilfred[1], Chong *Fai Kait*[1], and Ng *Zhung Gia*[1]

[1]Centre of Research in Ionic Liquids (CORIL), Universiti Teknologi Petronas (UTP), Seri Iskandar, 32610, Perak, Malaysia.

**Abstract.** Amine scrubbing is dominating in carbon dioxide ($CO_2$) capturing technology because of its high affinity towards $CO_2$. However, the drawbacks of amine solvents are its high corrosivity and volatility. Ionic liquids (ILs) have gained a lot of attention in recent years for $CO_2$ capturing and have been proposed to be one of the promising alternative to the conventional solvents. The objective of this research is to design a new imidazolium based ether-nitrile functionalized ionic liquid of low viscosity to improve $CO_2$ capturing. The molecular structure of the ionic liquid were confirmed by $^1$H NMR, $^{13}$C NMR and FTIR. The thermal properties; glass transition temperature, thermal decomposition temperature, and their physical properties; water content and density were determined. The solubility of $CO_2$ in the synthesized ionic liquid was measured using pressure drop method. They showed high thermal stability above 200°C and the glass transition temperature was -49.80°C. The $CO_2$ sorption in the newly synthesized IL was 0.08, 0.12, 0.29, 1.01, 2.30 mol of $CO_2$/mol of IL at pressures 1, 5, 10, 15 and 20 bar respectively.

## 1 Introduction

Carbon dioxide has become abundant in the atmosphere due to the extensive burning of fossil fuels. The $CO_2$ capturing technology is now being the main focus among researchers battling against the dramatic climate change. At present, amine-based solvents are widely used in power plants to capture $CO_2$ [1]. However, the limitation of amine solvents is that they are highly corrosive and volatile, which is unfriendly to the environment [2]. Besides, they acquire large enthalpy of reaction and amine solvents are also sensitive to degradation in the presence of oxygen [3].

Ionic liquids (ILs) are organic salts which melts below 100°C and are liquid form at room temperature [4]. In light of these factors, ILs have been studied intensively as a potential replacement for amine solvents. This is due to the distinctive characteristics of ILs, i.e. low volatility, high chemical and thermal stability. Furthermore, the added characteristic of ILs is the ability of designing the IL according to the specified task of application.

Imidazolium-based ILs are mostly used in $CO_2$ solubility due to the presence of acidic hydrogen in its ring structure. $CO_2$ can act as both Lewis acid and Lewis base. The $CO_2$ solvation is initiated by some carboxylation reaction at the C2 position of the imidazolium ring [5]. Besides, the ether functionality helps in $CO_2$ capture whereby the lone pair of electrons present on the oxygen atom takes part in nucleophilic attack with the carbon atom on $CO_2$ [4]. In a recent article, ILs with ether functionality has the ability to decrease the viscosity of ILs whereby the alkoxy chain is related to the high rotational flexibility and increases the total free volume [6, 7]. Anions play a vital role in $CO_2$ dissolution [8]. The common anions of interest are bromide [$Br^-$], tetrafluoroborate [$BF_4^-$], bis(trifluoromethylsulfonyl)imide [$Tf_2N^-$], hexafluoroposphate [$PF_6^-$] and methide. Carbon dioxide solubility is increased as the quantity of fluoroalkyl groups present in anions increased [9].

In the present research, ether-nitrile functionalized imidazolium-based 1-(2-cyanoethyl)-3-(3-methoxypropane)imidazolium bromide, [MOP-im$C_2$CN][Br] was synthesized, characterized and tested for the possible application on $CO_2$ capturing.

## 2 Methodology

### 2.1 Materials

Chemicals such as imidazole and 1-bromo-3-methoxypropane and solvents, i.e. methanol and acetonitrile, were purchased from Merck, Malaysia. Acrylonitrile was of Sigma Aldrich, Malaysia. All chemicals were used without any further purification.

* Corresponding author: ravishantini@yahoo.com

 DOI: 10.1051/matecconf/201711103003

## 2.2 Synthesis

### *2.2.1 Synthesis of 1-(2-cyanoethyl)imidazole*

The starting material; 1-(2-cyanoethyl)imidazole was synthesized by reacting imidazole with acrylonitrile and methanol as a solvent [10]. The reaction conditions were 50-55°C, under reflux and $N_2$ atmosphere for 10 h and cooled to room temperature. The impurities were removed by heating at 70°C under vacuum.

### *2.2.2 Synthesis of 1-(2-cyanoethyl)-3-(3-methoxypropane] imidazolium bromide, [MOP-imC2CN][Br]*

[MOP-im$C_2$CN][Br] was prepared by the reaction of 1-(2-cyanoethyl)imidazole with 1-bromo-3-methoxy propane at 70-80°, under reflux and $N_2$ atmosphere for 48 h. Then, the product was cooled to room temperature resulting [MOP-im$C_2$CN][Br].

### *2.2.3 Characterization of ILs*

The synthesized IL was characterized using $^1$H NMR, $^{13}$C NMR (Bruker Avance 500 MHz) and FTIR (Perkin Elmer Spectrum One). Water content in the ionic liquid was measured using Karl Fischer titrator model Mettler Toledo DL39. Density was measured using density meter (Anton-Paar, DMA 5000 M). The thermal properties and decomposition temperature ($T_d$) were measured using thermogravimetric analysis (TGA) and glass transition temperature ($T_g$) was measured with differential scanning calorimetry (DSC) (Mettler Toledo model DSC 1).

### *2.2.4 CO2 sorption study*

The possible application of the IL for $CO_2$ capture was investigated. A custom-made instrument was used for the $CO_2$ sorption studies (**Fig. 1**).

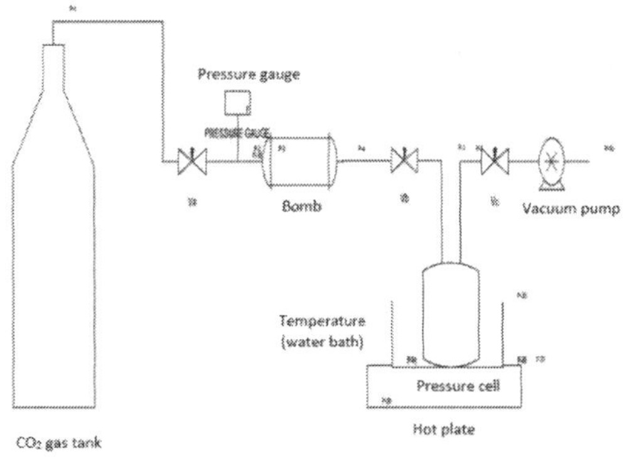


**Fig. 1.** Schematic diagram for $CO_2$ gas sorption cell

The measurement technique used was the pressure drop method. Using this method, temperature was held constant and the pressure difference was recorded during gas sorption into the sample. The initial pressure was set at a range of 1 bar to 20 bar. The amount of gas sorbed was calculated from the initial measurement of temperature, volume and pressure and the final measurement of the variables at equilibrium. The calculation was done by converting all three variables into moles of gas using Peng-Robinson equation of state [11].

The mole of $CO_2$ captured was calculated using equation (1).

$$n = \frac{P_{ini} \cdot V_{tot}}{Z_{ini} \cdot R \cdot T_{ini}} - \frac{P_{eq}(V_{tot} - V_{sample})}{Z_{eq} \cdot R \cdot T_{eq}} \quad (1)$$

where n = mol of $CO_2$ captured, $P_{ini}$ = initial pressure obtained in the absence of sample after expansion of the gas sample from the bomb into the whole system, $V_{tot}$ = total volume of system, $Z_{ini}$ = compressibility factor ($P_{ini} \times T_{ini}$), $Z_{eq}$ = compressibility factor ($P_{eq} \times T_{eq}$), $P_{eq}$ = pressure at equilibrium, $V_{sample}$ = volume of sample, R is the ideal gas constant, $T_{ini}$ = initial temperature and $T_{eq}$ = temperature at equilibrium. Compressibility factor for $CO_2$ modifies the ideal gas to account for real gas behaviour [11].

# 3 Results and discussion

## 3.1 NMR analysis

The $^1$H-NMR data for the synthesized IL is as shown in **Fig. 2** and $^{13}$C-NMR data shown in **Fig. 3** respectively. The solvent used during NMR analysis was deuterated acetonitrile.

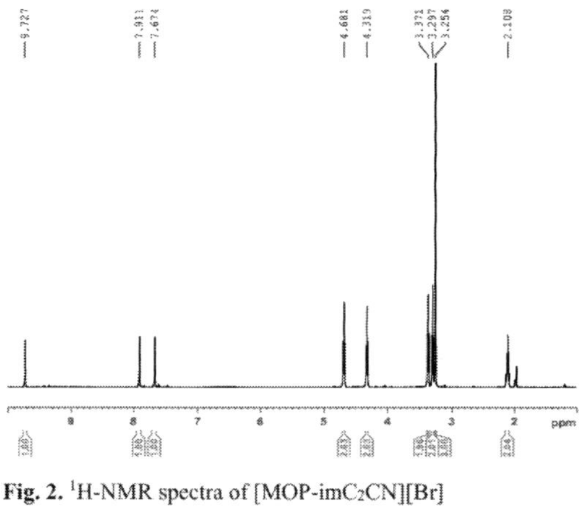


**Fig. 2.** $^1$H-NMR spectra of [MOP-im$C_2$CN][Br]

 DOI: 10.1051/matecconf/201711103003

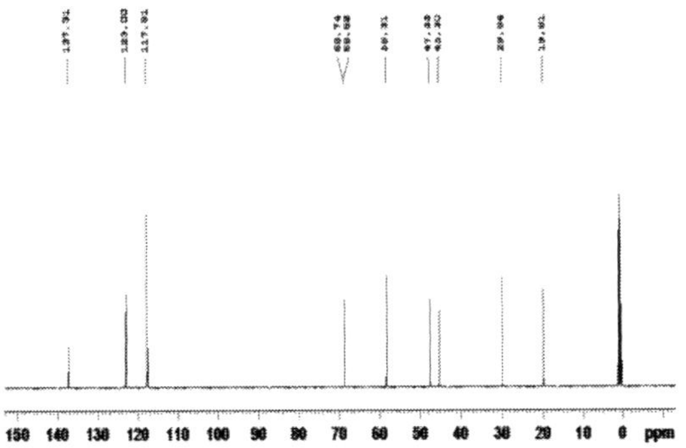

**Fig. 3.** $^{13}$C-NMR spectra of [MOP-im$C_2$CN][Br]

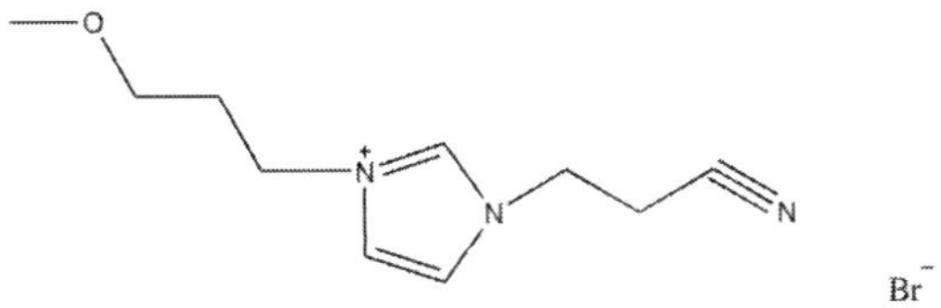


**Fig. 4.** Structure of [MOP-im$C_2$CN][Br]

$^1$H-NMR (500MHz, $CD_3CN$): δ (ppm) 9.73 (s, 1H, NCHN), 7.91 (d, 1H, CHN), 7.67 (d, 1H, CHN), 4.68 (t, 2H, $CH_2CH_2N$), 4.31 (t, 2H, $CH_2CH_2N$), 3.32 (t, 2H, $CH_2CH_2CN$), 3.30 (t, 2H, $CH_3OCH_2$), 3.25 (s, 3H, $CH_3O$), 2.11 (m, 2H, $CH_2CH_2CH_2$); $^{13}$C-NMR (125.77MHz, $CD_3CN$): δ (ppm) 137.31, 123.00, 117.81, 68.74, 58.31, 47.55, 45.30, 29.84, 19.81.

Spectroscopic data for this ionic liquid were consistent with the literature data [12].

### 3.2 FTIR analysis

The structure of [MOP-im$C_2$CN][Br] was analysed using Fourier transform infrared (FTIR) and the FTIR spectra obtained is shown in **Fig. 5**. The peaks of wavenumbers 660, 750 and 825 cm$^{-1}$ are the C-H stretching. A broad peak at about 3400 cm$^{-1}$ is the O-H stretch which could be due to the presence of water. Peak at wavenumber 2250 cm$^{-1}$ is due to C≡N stretching. Wave numbers 3050-3100 cm$^{-1}$ and 1400-1600 cm$^{-1}$ represents =C-H stretch and C=C of the aromatic heterocyclic ring respectively. The ether functionality (C-O-C) shows a strong stretching at about wavenumber 1160 cm$^{-1}$. Based on the FTIR analysis, it was observed that the IL synthesized is [MOP-im$C_2$CN][Br].

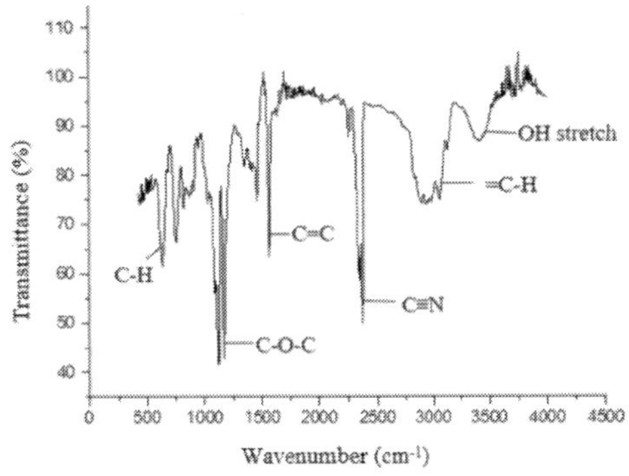


**Fig. 5.** FTIR spectra of [MOP-im$C_2$CN][Br]

### 3.3 Physical properties of [MOP-im$C_2$CN][Br]

The physical properties of [MOP-im$C_2$CN][Br] such as the water content, physical appearance and colour at room temperature are shown in **Table 1** below.

**Table 1.** Physical properties of [MOP-im$C_2$CN][Br]

| IL | Water content (%) | Colour | Form |
|---|---|---|---|
| [MOP-im$C_2$CN][Br] | 0.04 | Yellowish | Liquid |

#### *3.3.1 Density of [MOP-im$C_2$CN][Br]*

The density of [MOP-im$C_2$CN][Br] at different temperatures is shown in **Fig. 6** below. The plot shows that the density of the synthesized IL decreased linearly with increasing temperature. As shown in Fig. 4, the density of the IL is below 1.36 g/cm$^3$ at 293.15 K. Generally, ILs are much denser as compared to water [13].

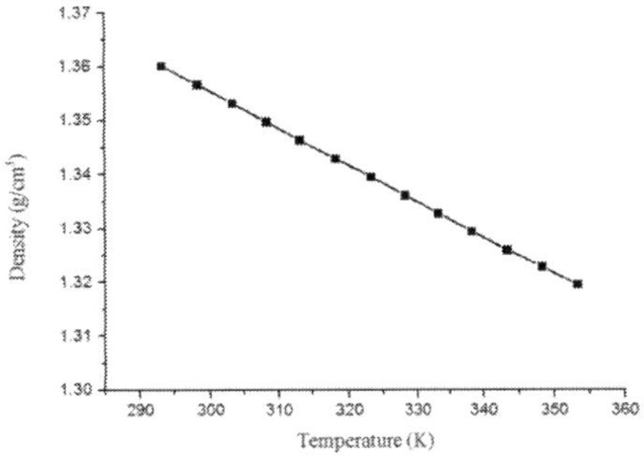


**Fig. 6.** Density of [MOP-im$C_2$CN][Br] as a function of temperature (K)

The density data could be correlated with equation (2).

$$\rho = A + BT \quad (2)$$

where ρ represents the density of [MOP-im$C_2$CN][Br] in g/cm$^3$, T is the temperature in Kelvin (K) whereas A and B are the model parameters.

Fitting results are shown in Fig. 4 and the related model parameters are listed in **Table 2** below.

**Table 2.** Fitting result of the experimental density of [MOP-im$C_2$CN][Br]

| Parameters | [MOP-im$C_2$CN][Br] |
|---|---|
| $A$ | 1.5586 |
| $B \times 10^4$ | -6.774 |
| $R^2$ | 0.9999 |

### 3.4 Derived thermodynamic property

#### *3.4.1 Thermal expansion coefficient*

The thermal expansion coefficient was calculated for the synthesized IL using equation 3 below since the temperature-density relationship was linear.

$$\alpha_P = -\left(\frac{1}{\rho}\right)\left(\frac{\partial \rho}{\partial T}\right)_P = -(B)/(A+BT) \quad (3)$$

where $\alpha_P$ represents the thermal expansion coefficient in K$^{-1}$, $\rho$ is the density, $A$ and $B$ are the fitting parameters of equation (2).

The calculated thermal expansion coefficient is presented in **Table 3**. The thermal expansion coefficient of [MOP-im$C_2$CN][Br] obtained in the present work do not change significantly in the temperature range from 293.15 to 353.15 K. Besides, dual functionalized IL has low thermal expansion coefficient [10].

**Table 3.** Density and thermal expansion coefficients, $\alpha_P$ for [MOP-im$C_2$CN][Br]

| Temperature (K) | Density (g/cm$^3$) | $\alpha_P \times 10^{-4}$ (K$^{-1}$) |
|---|---|---|
| 293.15 | 1.360 | 4.98 |
| 298.15 | 1.357 | 4.99 |
| 303.15 | 1.353 | 5.01 |
| 308.15 | 1.350 | 5.02 |
| 313.15 | 1.346 | 5.03 |
| 318.15 | 1.343 | 5.04 |
| 323.15 | 1.340 | 5.06 |
| 328.15 | 1.336 | 5.07 |
| 333.15 | 1.333 | 5.08 |
| 338.15 | 1.330 | 5.10 |
| 343.15 | 1.326 | 5.11 |
| 348.15 | 1.323 | 5.12 |
| 353.15 | 1.320 | 5.13 |

### 3.5 Thermal analysis

The IL was heated over a temperature range of 45-600°C at a heating rate of 10°C/min. The synthesized ionic liquid, [MOP-im$C_2$CN][Br], has a good thermal stability whereby the onset thermal decomposition started at 246°C with a weight loss of 7.65% and the decompositions ends around 313°C with a weight loss of 97.39%. The $T_g$ of [MOP-im$C_2$CN][Br] was -49.80°C. The TGA profile of [MOP-im$C_2$CN][Br] is as illustrated in **Fig. 7**.

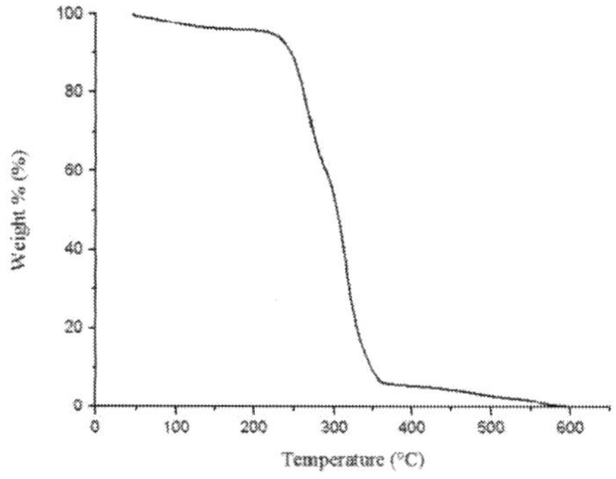


**Fig. 7.** TGA profile of [MOP-im$C_2$CN][Br]

A high thermal decomposition temperature of ILs gives a long-term stability of the ILs. Mostly ionic liquids are highly viscous, however, they exhibit a high thermal stability. Therefore, this property enables ILs to be used for a specified application under high temperature which significantly reduces their viscosity [10].

### 3.6 $CO_2$ solubility

The $CO_2$ sorption using [MOP-im$C_2$CN][Br] was investigated at 25°C using the custom-made gas sorption cell under different pressure condition 1, 5, 10, 15 and 20 bar.

**Fig. 8** shows that $CO_2$ solubility in [MOP-im$C_2$CN][Br] increases with increasing pressure. At equilibrium pressure, $P_{eq}$= 0.73 bar (at 1 bar), the $CO_2$ absorbed was 0.08 mol of $CO_2$/mol of IL whereas at $P_{eq}$=15.82 bar (at 20 bar) the $CO_2$ shows an absorption of 2.30 mol of $CO_2$/mol of IL at 25°C. Ziyada, (2011) has reported that 1-butyl-3-propanenitrile imidazolium dioctylsulfosuccinate [$C_2$CN Bim][DOSS] has $CO_2$ absorption capacity about 2.47 mol of $CO_2$/mol of IL at 20 bar [10].

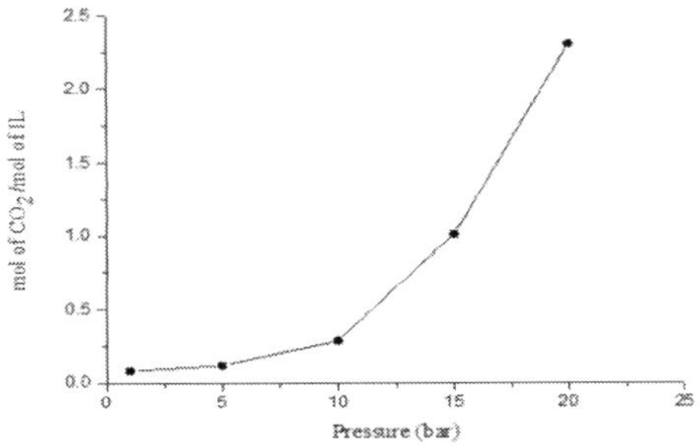


**Fig. 8.** Effect of pressure on $CO_2$ solubility in [MOP-im$C_2$CN][Br] at 25°C

**Table 4** below shows the comparison of experimental solubility data for $CO_2$ in [$C_2$CN Bim][DOSS] and [MOP-im$C_2$CN][Br] at 25°C respectively.

**Table 4.** Experimental solubility data for $CO_2$ in [$C_2$CN Bim][DOSS] and [MOP-im$C_2$CN][Br] at 25°C

| Pressure (bar) | $CO_2$ solubility (mol of $CO_2$/mol of IL) | |
|---|---|---|
| | [$C_2$CN Bim] [DOSS] | [MOP-im$C_2$CN] [Br] |
| **1** | 0.06 | 0.08 |
| **5** | 0.30 | 0.12 |
| **10** | 0.70 | 0.29 |
| **15** | 1.28 | 1.01 |
| **20** | 2.47 | 2.30 |

At 5, 10, 15 and 20 bar, [$C_2$CN Bim][DOSS] has a higher $CO_2$ sorption than [MOP-im$C_2$CN][Br]. This might be due to the significant role of the DOSS anion present which has a higher affinity towards $CO_2$ [10]. However, [MOP-im$C_2$CN][Br] showed equally good $CO_2$ sorption whereby the imidazolium ring of the cation is tethered to the ether and nitrile functionality. In addition, at 1 bar, [MOP-im$C_2$CN][Br] has slightly higher $CO_2$ sorption compared to [$C_2$CN Bim][DOSS]. The presence of ether functionality enables dipole-quadrupole interactions between the oxygen atoms present and $CO_2$ [14]. Besides, high $CO_2$ solubility can also be correlated with the nitrile group that weakens the cation-anion interaction [7].

Literature shows that $CO_2$ sorption capacity increases with pressure because at high pressure, physical sorption becomes more significant [15]. From the trend shown in Fig. 8, it is expected that when the pressure is increased to >20 bar, the $CO_2$ sorption will gradually increase.

## 4 Conclusion

The imidazolium based ether-nitrile functionalized ionic liquid, [MOP-im$C_2$CN][Br] has been successfully synthesized, characterized and evaluated for $CO_2$ sorption. The NMR and FTIR confirmed the structure of the synthesized ionic liquid. The technique used for $CO_2$ sorption was the pressure-drop method. The results obtained from this study implies that [MOP-im$C_2$CN][Br] has a good sorption of $CO_2$ which is 0.08 mol of $CO_2$/mol of IL at $P_{eq}$= 0.73 bar (at 1 bar), and 2.30 mol of $CO_2$/mol of IL at $P_{eq}$=15.82 bar (at 20 bar), at 25°C. Besides, the $CO_2$ sorption can be further enhanced by anion exchange using selective anions which is $CO_2$-philic.

The authors acknowledge support from Centre of Research in Ionic Liquids (CORIL) and Fundamental & Applied Sciences Department (FASD), Universiti Teknologi Petronas (UTP). The authors also would like to appreciate and thank E-Science, MOSTI for funding this research.

## References

1. A.B. Rao, E.S. Rubin, Environ. Sci. Technol. **44**, 20 (2002).
2. M. Wang, A. Lawal, P. Stephenson, J. Sidders, C. Ramshaw, Chem. Eng. Res. Des. **89**, 9 (2011).
3. H. Gao, W. Rongwong, C. Peng, Z. Liang, K. Fu, R. Idem, P. Tontiwachwuthikul, Energy Procedia **63**, 1911-1918 (2014).
4. P. Sharma, S.D. Park, K.T. Park, S.K. Jeong, S.C. Nam, I.H. Baek, Bull. Korean Chem. Soc. **33**, 7 (2012).
5. C. Cadena, J.L. Anthony, J.K. Shah, T.I. Morrow, J.F. Brennecke, E.J. Maginn, J. Am. Chem. Soc. **126**, 16 (2004).
6. P. Sharma, S.H. Choi, S.D. Park, I.H. Baek, G.S. Lee, Chem. Eng. J. **181**, 834-841 (2012).
7. Y. Park, K.A. Lin, A.A. Park, C. Petit, Front. Energy Res. **3**, 42 (2015).
8. M. Kanakubo, T. Umecky, Y. Hiejima, T. Aizawa, H. Nanjo, Y. Kameda, J. Phys. Chem. B **109**, 29 (2005).
9. M.J. Muldoon, S.N. Aki, J.L. Anderson, J.K. Dixon, J.F. Brennecke, J. Phys. Chem. B, **111**, 30 (2007).
10. A.B. Ziyada, *Thermophysical properties and carbon dioxide solubility of novel room temperature ionic liquids* (2011).

11. E. Torralba-Calleja, J. Skinner, D. Gutierrez-Tauste J. Chem. (2013).
12. S. Ravichandar, C.D. Wilfred, F.K. Chong, M.A. Abdul Halim, Z.G. Ng, *Contemporary of Sciences: Physics, Chemistry and Mathematics*, **2**, (2016).
13. J.S. Wilkes, J. Mol. Catal. A: Chem. **214**, 1 (2015).
14. Y.J. Heintz, L. Sehabiague, B.I. Morsi, K.L. Jones, D.R. Luebke, H.W. Pennline, Energy Fuels **23**, 10 (2009).
15. Y. Cheng-Hsiu, H. Chih-Hung, T. Chung-Sung, Aerosol Air Qual Res, **12** (2011).

# Effect of Sacrificial Anodes on Hydrodynamic Coefficients of Jacket Members Subjected to Regular Waves

*Ahmad* Raziqin, *Kurian* Velluruzhathil John*, and *Ng* Cheng Yee

Universiti Teknologi PETRONAS, Civil Engineering Department, 32610 Seri Iskandar, Perak Darul Ridzuan, Malaysia

**Abstract.** Designing a fixed offshore jacket platform requires the approximation of wave forces acting on jacket members which is influenced by multiple factors such as wave height, wave period, shape and size of the jacket member. An important part of determining wave forces using Morrison's equation is the accurate determination of hydrodynamic coefficients. The presence of sacrificial anodes increases the hydrodynamic coefficient of jacket member and therefore this research focuses on improving the design of fixed offshore platforms by conducting experimental investigation on rigid cylinders in order to obtaining better estimates for hydrodynamic coefficients. Steel members of size 27 and 34mm diameters were tested in the wave basin 20 m x 10 m x 1 m. Different set of regular waves with varying height and frequency were generated. Clean members were fitted with small welded steel bars of fixed length and size along the member simulating sacrificial anodes. The results show that the coefficients obtained were close to the values proposed by PTS and API standards but are generally lower than the suggested value of 2.0 for both $C_d$ and $C_m$.

## INTRODUCTION

Compared to land structures, offshore structures face various environmental loads, interactions and dynamic responses on top of their static loads which cause the calculation of hydrodynamic forces more complex. Due to the highly indeterministic and un-predictable characteristic of waves, the wave force estimation needs to be done with caution and precision. The wave and current forces acting on these platform members are evaluated using the hydrodynamic drag and inertia coefficients, assumed as constant at all locations on these members for all the wave periods. All these members are covered with increasing thickness of marine growth during operation years and the braces are fitted with sacrificial anodes for corrosion protection. However, due to lack of experimental supporting data, the hydrodynamic coefficients used by the codes are comparatively higher and more conservative. The prediction of accurate values of these coefficients at various depths of water for different wave heights and wave periods will facilitate optimum design for integrity analysis of these platforms, leading to great economy. Hydrodynamic coefficients are found to be influenced by the Reynolds number, the Keulegan-Carpenter (KC) number, and other parameters such as surface roughness, surface area exposed to the waves and current speed [1]. There is no single parameter with which the drag and inertia coefficients may be correlated without the need for other parameters, thus showing the complexity of the determination of hydrodynamic force coefficients [2].

Generally, Morison equation is used to determine the external hydrodynamic force on small structures subjected to wave loadings [3]. This equation assumes that the total in-line hydrodynamic force on a structure consists of semi-empirical values of inertia and drag forces added linearly. The inertia force is developed as the water moving in a wave carry a momentum with it, while the drag force component is caused due to the presence of wake region on the streamside of the cylinder. The total hydrodynamic force on the jacket member can be estimated from Equation 1.

$$F = \int_0^d f ds = \int_0^d [\frac{C_m \rho \Pi D^2}{4}\frac{du}{dt} + C_d \frac{\rho D}{2} | u + u_c | (u + u_c)] ds \quad (1)$$

Where, d = water depth, $C_m$ = hydrodynamic coefficient of mass $C_d$ = hydrodynamic coefficient of drag, ds = integration is done over the full wetted length of the cylinder, ρ = density of water, D = pipe diameter, u= horizontal water particle velocity, du/dt = water particle acceleration, and $U_c$ = current velocity. Airy wave theory is used in determining the wave kinematics and the most appropriate combinations of $C_m$ and $C_d$ corresponding to phase angles that satisfy Equation 1 was selected as the best drag and inertia coefficients for each loading case since both are a function of phase angle Θ.

According to [4] several methods were used in determining the drag and inertia coefficient but it was determined that the Least Square Method produced an overall best. Results by [5] also concluded that Least Square Method produced the best results. It is found that on an average, the wave by wave Weighted Least Squares method gives both the lowest bias and root mean square error.

* kurian_john@utp.edu.my

## EXPERIMENTAL DETAILS

The experiments were carried out in the wave basin of Universiti Teknologi PETRONAS (UTP) of size 20 m by 10 m with a water depth of 1m. Various regular waves were ran using 2 different types of cylindrical members. Details for the regular waves used can be seen in Table 1. The specimens used consist of pipe with outer diameters of 27 and 34mm, both having one clean pipe member and another fitted with anodes. Three wave probes were used to measure the wave elevations during experimentation. The first probe was placed 1.5m in front of the model and the other two probes were placed in line at 1.5m to the left and right of the model. The anode fittings were 1cm x 3.5cm and spaced 12cm from each other centre to centre.

**Table 1.** Wave combination table

| Wave Characteristics | |
|---|---|
| Wave Height, (m) | Wave Period, (s) |
| 0.1 | 1 |
| 0.1 | 1.5 |
| 0.1 | 2 |
| 0.1 | 2.5 |
| 0.1 | 3 |
| 0.15 | 1 |
| 0.15 | 1.5 |
| 0.15 | 2 |
| 0.15 | 2.5 |
| 0.15 | 3 |
| 0.2 | 1 |
| 0.2 | 1.5 |
| 0.2 | 2 |
| 0.2 | 2.5 |
| 0.2 | 3 |

Figure 1 shows the set up used during this experiment. First the clean member was fitted into the coupler which is set up at the bottom of the bridge on the wave tank. The cylinder was mounted vertically using the coupler while the free end was immersed in 1m water depth. The model cylinder pipes were made of steel with thickness of 2.5mm. The total wave force on the cylinder was measured by a wave force sensor located between the coupler and the bridge. The same set up is used for the anode fitted members. Figure 2 shows the anode fitted members used in this experiment.

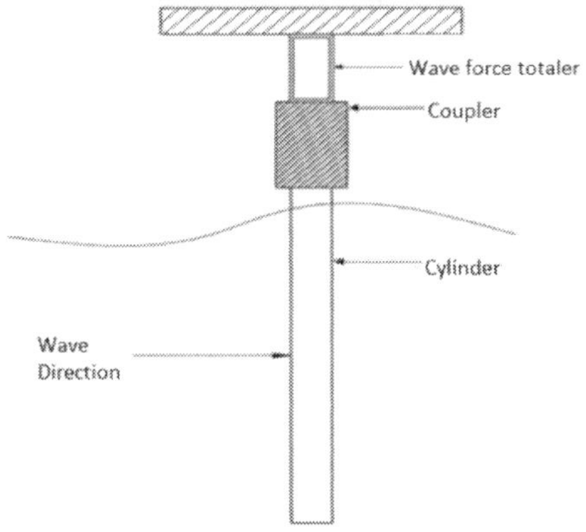


**Fig. 1.** Model set up in wave basin

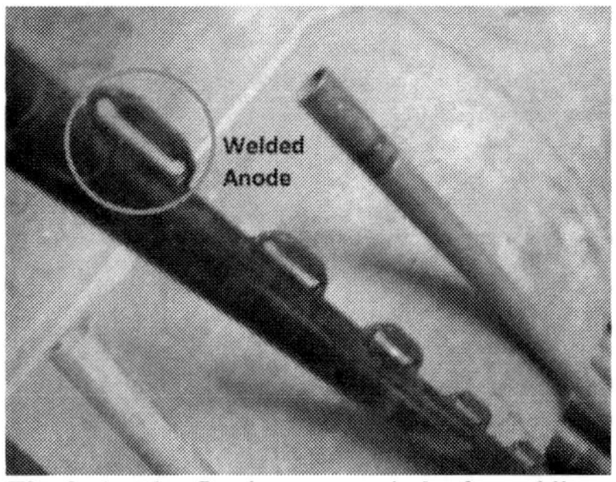


**Fig. 2.** Anodes fixed onto a steel pipe by welding

Determination of the hydrodynamic coefficients are done by estimation of wave kinematics using Airy Wave Theory and using Least Square Method to provide a good estimation for the hydrodynamic coefficients. Load cells were used to measure the force acting on the cylinder members. The configuration of the load cell is made in such a way that total load acting on the cylinder is read by the load cell in terms of volts, which is further converted in terms of Newton, N by using appropriate calibration constant. The equation proposed by Morisson gives the Force per unit length but in order to determine the total load acting on cylinder, the Morison equation is integrated from the water surface to the depth up to which structure extends from surface.

The total wave force on the member can be estimated by combining the drag force expressed by Equation 2 with the inertia force in Equation 3.

$$F_D(t) = \frac{1}{2} C_D d \frac{h^2\omega^2}{4} \frac{\{\frac{\sinh(2kH)}{4} + \frac{kH}{2}\}}{k \sinh^2(kH)} \cos(\omega t)\,|\cos(\omega t)| \quad (2)$$

$$F_I(t) = -\rho C_M \pi \frac{\pi}{4} d^2 \frac{h}{2k} \omega^2 \sin(\omega t) \quad (3)$$

## ANALYSIS

The methods that can be used for determination of values can be divided into two groups, frequency domain techniques and time domain techniques. Frequency domain techniques use the entire record. The spectra and cross spectra of measured force and particle velocity is determined to estimate the force coefficients. Time domain techniques may also be used for a wave by wave analysis and can be further divided into two-point method and least square approximation. The two points at which the acceleration or, respectively, the velocity is zero, are evaluated. At points of zero acceleration, the corresponding force is purely drag and therefore the drag coefficient can be calculated directly. The same applies for the point in time of zero velocity at which the inertia coefficient can be determined. Due to its simplicity, the coefficients can be obtained with low computational effort. When using the least square approach, an iteration is carried out until values are found for which the force determined with the Morisson equation makes the best fit with the measured force. Hydrodynamic coefficients are varied until the sum of the squares of the difference between measured force ($F_M$) and the estimated Morisson force ($F_P$) is minimized

$$\varepsilon^2 = \frac{1}{N}\sum_{t=1}^{N}(F_P(t) - F_M(t))^2 \qquad (4)$$

with ε being the squared error and N the number of time steps. The evaluation can be carried out for one wave or a record of waves. In the least square approach, every data point has an equal influence on the result.

The following figures shows the effect of anodes on the hydrodynamic coefficients with respect to different wave conditions. Figure 3 and Figure 4 illustrates the mass and drag coefficient of member size 27mm. It can be seen for the most part both mass and drag coefficients were higher in anode fitted member when compared to clean members but the changes were more significant in drag coefficients. The lowest inertia coefficient of anodes fitted members was 1.5 with a maximum of 2.4 while clean members were 1.77 and 2.5 respectively. The drag coefficients for member diameter 27mm obtained are depicted in figure 4 show the highest value of drag obtained was 1 with both types of members with a minimum value of 0.1056 and 0.13 for anode and clean member respectively.

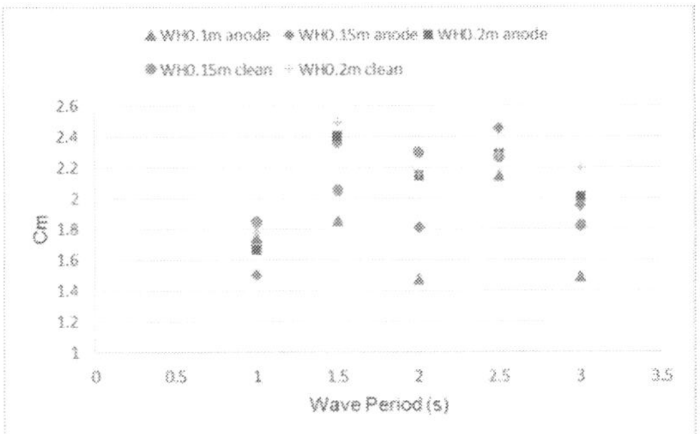


**Fig. 3.** Mass coefficients for 27mm anode fitted cylinder and clean members plotted against different wave periods

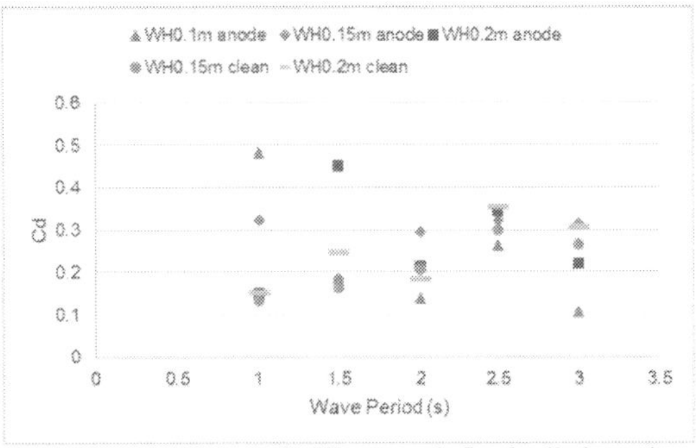


**Fig. 4.** Drag coefficients for 27mm anode fitted cylinder and clean members plotted against different wave periods

Figure 5 and Figure 6 that shows the mass and drag coefficients of anode fitted members and clean members of size 34mm. It can be seen that the presence of anodes increases the mass coefficient values for most wave periods used. The lowest coefficient obtained was 1.4076 for anode and 1.56 for clean member with a peak of 2.46 and 2.4 respectively. Results presented in Figure 6 show similar trends in which the presence of anodes increased the drag coefficient. The highest drag coefficient obtained was 0.43 for anodes and 0.35 for clean members wiles the minimum values obtained are 0.113 and 0.035 respectively.

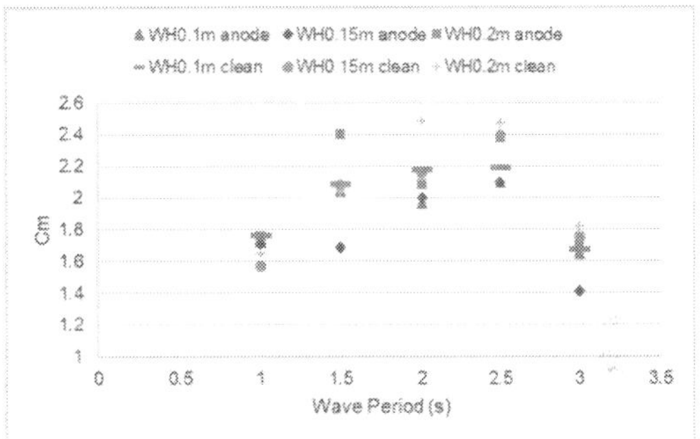


**Fig. 5.** Mass coefficients for 34mm anode fitted cylinder and clean members plotted against different wave periods

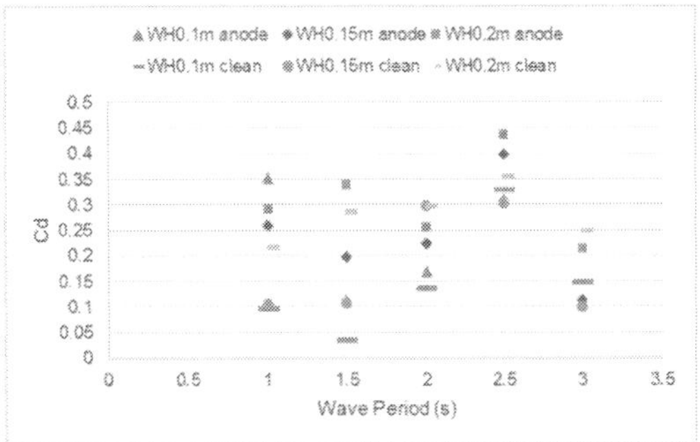


**Fig. 6.** Drag coefficients for 34mm anode fitted cylinder and clean members plotted against different wave periods

Figure 7 and 8 compare the mass and drag coefficient for clean member of size 27mm and 34m. The two different pipe diameter show a general trend of increase in both mass and drag coefficient as pipe diameter increases.

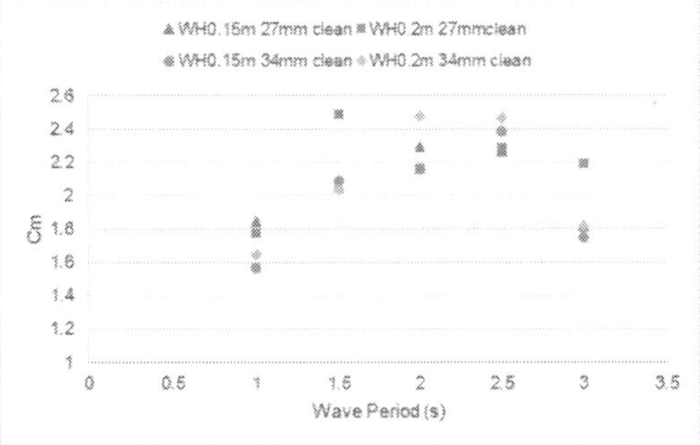


**Fig. 7.** Mass coefficients for 27mm and 34mm clean members plotted against different wave periods

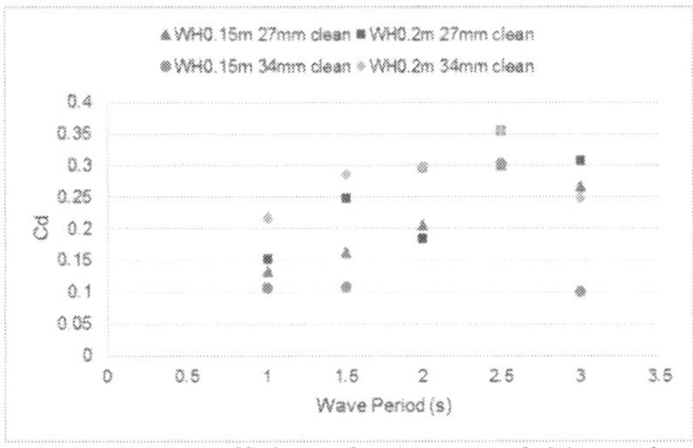


**Fig. 8.** Drag coefficients for 27mm and 34mm clean members plotted against different wave periods.

From the results obtained, it can be seen that anode fittings have increased the overall of both drag coefficients and inertia coefficient in both 27mm and 34mm pipe member but has a higher effect on drag compared to inertia coefficient. Although the mass coefficient obtained in this experiment was higher than recommended values in PTS, the total force produced by the combination of drag and inertia were the same. Results obtained from this experiment are in agreement with values recommended by [6] but a large part of the findings suggest that the standard hydrodynamic coefficients for anode fitted members used in designing were still conservative.

## CONCLUSION

The results of the experiment show that the fitting of anodes onto a clean member generally increases the hydrodynamic coefficient of the cylinder. The overall effect of anode fittings on inertia coefficients were smaller when compared to drag coefficient. The drag and inertia coefficients obtained are in agreement with values suggested by PTS. However, a large portion of the results suggests that the standard values for anode fitted members are conservative because the force measured was less than the combination of Cm=2.0 and Cd=2.0.

## Acknowledgement

This research was funded by the Fundamental Research Grant Scheme (FRGS 2014-2) under the Ministry of Higher Education (MOHE).

## REFERENCES

[1] Chakrabati, S.K., (1991). Wave force on offshore structures. *Handbook of Coastal and Ocean engineering*, 2, pp.1-54.

[2] Shafieefar, M. and Massie, W.W., 2001. In-Line Force from Combined Wave and Current Flow on Oscillating Cylinders. *International Journal of Offshore and Polar Engineering*, 11(02).

[3] Morison, J.R., Johnson, J.W. and Schaaf, S.A., (1950). The force exerted by surface waves on piles. *Journal of Petroleum Technology*, 2(5), pp.149-154.

[4] Isaacson, M., Subbiah, K. and Baldwin, J., 1991, January. Force coefficient estimation from random wave data. *The First International Offshore and Polar Engineering Conference*. International Society of Offshore and Polar Engineers.

[5] Naghipour, M., (2001). Estimation of hydrodynamic force on rough circular cylinders in random waves and currents. *International Journal of Engineering*, ***14(1)***, pp.25-34.

[6] PTS, Petronas Technical Standards, *Design of Fixed Offshore Structures*. (2012).

[7] Li, Y. C., & Ming, H. (1995). The In-Line Wave Forces on Small Square Cylinder. *Proc. Of the Fifth International Offshore and Polar Engineering Conf.* City: The Hague, The Netherlands, pp. 289-296.

[8] Borgman, L.E., (1958). Computation of the ocean-wave forces on inclined cylinders. *Eos, Transactions American Geo-physical Union*, **39(5)**, pp.885-888.

# AUTHOR INDEX